INTERNATIONAL SERIES OF MONOGRAPHS IN
NATURAL PHILOSOPHY
General Editor: D. ter Haar

Volume 17

TYPE II SUPERCONDUCTIVITY

*OTHER TITLES IN THE SERIES
IN NATURAL PHILOSOPHY*

Vol. 1. DAVYDOV — Quantum Mechanics

Vol. 2. FOKKER — Time and Space, Weight and Inertia

Vol. 3. KAPLAN — Interstellar Gas Dynamics

Vol. 4. ABRIKOSOV, GOR'KOV and DZYALOSHINSKII — Quantum Field Theoretical Methods in Statistical Physics

Vol. 5. OKUN' — Weak Interaction of Elementary Particles

Vol. 6. SHKLOVSKII — Physics of the Solar Corona

Vol. 7. AKHIEZER et al — Collective Oscillations in a Plasma

Vol. 8. KIRZHNITS — Field Theoretical Methods in Many-body Systems

Vol. 9. KLIMONTOVICH — The Statistical Theory of Non-equilibrium Processes in a Plasma

Vol. 10. KURTH — Introduction to Stellar Statistics

Vol. 11. CHALMERS — Atmospheric Electricity (2nd. edition)

Vol. 12. RENNER — Current Algebras and their Applications

Vol. 13. FAIN and KHANIN — Quantum Electronics, Vol. 1: Basic Theory

Vol. 14. FAIN and KHANIN — Quantum Electronics, Vol 2: Maser Amplifiers and Oscillators

Vol. 15. MARCH — Liquid Metals

Vol. 16. HORI — Spectral Properties of Disordered Chains and Lattices

TYPE II SUPERCONDUCTIVITY

D. SAINT-JAMES AND
G. SARMA
Centre d'Etudes Nucléaires de Saclay

E. J. THOMAS
*Physical Laboratories
Manchester University*

PERGAMON PRESS
OXFORD · NEW YORK · TORONTO
SYDNEY · BRAUNSCHWEIG

Pergamon Press Ltd., Headington Hill Hall, Oxford

Pergamon Press Inc., Maxwell House, Fairview Park, Elmsford, N.Y. 10523

Pergamon Press of Canada Ltd., 207 Queen's Quay West, Toronto 1

Pergamon Press (Aust.) Pty. Ltd., 19a Boundary Street, Rushcutters Bay, N.S.W. 2011, Australia

Vieweg & Sohn GmbH, Burgplatz 1, Braunschweig

Copyright © 1969

Pergamon Press Ltd.

All Rights Reserved. No part of this publication may be reproduced, stored in a retrieval system, or transmitted, in any form or by any means, electronic, mechanical, photocopying, recording or otherwise, without the prior permission of Pergamon Press Ltd.

First edition 1969

Reprinted 1970

Library of Congress Catalog Card No. 67–27491

Reprinted lithographically by A. Wheaton & Co., Exeter

08 012392 9

CONTENTS

PREFACE

PART I REVERSIBLE PROPERTIES

By D. SAINT-JAMES and G. SARMA

1. INTRODUCTION   3

1.1 Introduction to superconductivity   3
1.2 Meissner effect — diamagnetism   3
1.3 Some considerations about the normal-superconducting transition   4
1.4 The London equation   7
1.5 The coherence length and the energy gap   10
1.6 Classification of superconductors   12
1.7 Magnetic and thermodynamic properties of type II superconductors   13
1.8 Ideal type II superconductors and hard superconductors   18
    Bibliography   21

2. THE PHENOMENOLOGICAL THEORY OF GINZBURG AND LANDAU

2.1 The free energy and the Ginzburg–Landau equations   22
2.2 The two characteristic lengths $\xi(T)$ and $\lambda(T)$ and the validity of the Ginzburg–Landau scheme   26
2.3 The Ginzburg–Landau parameter $\kappa(T)$ 29
2.4 The surface energy   31
2.5 Laminar model and filamentary structure   37
    Bibliography   40

3. MAGNETIC PROPERTIES OF TYPE II SUPERCONDUCTORS
   THE VORTEX LINE STRUCTURE

3.1 Introduction   41
3.2 Onset of superconductivity. The critical field H_{c2} 41
3.3 Structure of the mixed state   43
3.4 General properties of the vortex line structure. The magnetization curve   62
3.5 Experimental observation of the vortex line structure   70
    Appendix   76
    Bibliography   80

4. SURFACE SUPERCONDUCTIVITY

4.1 Nucleation in a semi-infinite medium. The field H_{c3} 81
4.2 Nucleation in a slab   91
4.3 Angular dependence of the nucleation field   100
4.4 Persistence of the surface sheath below $H_{//}$ 109
4.5 Superconductor coated by a normal metal   117
    Bibliography   121

Contents

5. MICROSCOPIC THEORY

5.1 Introduction	122
5.2 General features of the microscopic theory. The Cooper pairs	122
5.3 Formulation of the microscopic theory. The self-consistent method	126
5.4 Derivation of the Ginzburg–Landau equations. The distinction between clean and dirty superconductors	134
5.5 Conclusion and discussion	152
Bibliography	156

6. MISCELLANEOUS PROPERTIES OF TYPE II SUPERCONDUCTORS IN HIGH FIELDS

I *The Paramagnetic Effect*

6.I 1. Introduction	157
2. Calculation of the various kernels	161
3. The Fulde and Ferrell state	163
4. The transition field H_{c2} in the dirty case	169
5. Experimental situation. The spin-orbit scattering effect	174

II *Gapless Superconductivity*

6.II 1. Introduction	177
2. Order parameter and excitation spectrum	178
3. Conclusions	190
Bibliography	192

7. THERMAL PROPERTIES

7.1 Specific heat	193
7.2 Thermal conductivity	196
Bibliography	205

CONTENTS

PART II IRREVERSIBLE PROPERTIES
By E. J. Thomas

8. Flux Trapping and the Irreversible Magnetization Characteristic

8.1 Observation of trapped flux	209
8.2 Trapping in a filamentary system	212
8.3 Flux trapping in a singly connected system	214
8.4 Pinning points	218
8.5 Variation of trapped flux with temperature	223
Bibliography	226

9. Flux Movement within Type II Superconductors

9.1 Steady movement of flux lines	227
9.2 Flux jumping	238
9.3 Magneto-thermal effects	244
Bibliography	247

10. Critical Current Characteristics

10.1 Variation of the critical current with field	249
10.2 Training	264
10.3 Degradation	266
Bibliography	271

11. Applications

11.1 High field magnets	272
11.2 Other magnetic field applications	277
11.3 D. C. power transmission	280
11.4 A. C. applications	281
Appendix	284
Bibliography	285

Author Index	287
Subject Index	291

Preface

For the past few years one good paper on the subject of superconductivity has been published almost every day. By far the largest proportion of these have been concerned with the properties of the superconductors known as type II. Our object is to present an up to date description of type II superconductivity based as far as possible on the phenomenological approach first proposed by Ginzburg and Landau. In the first part of the book, which deals with reversible properties, the Ginzburg–Landau theory leads to a good theoretical understanding but with the irreversible properties, described in part II, often no satisfactory theoretical description of any kind exists and we have had to resort to a qualitative explanation of the effects.

The book is written with experimentalists in mind and we hope that it will be of particular use to physicists and engineers who are working, or about to work, in the subject. The microscopic theory is used only in Chapters 5 and 6 and there mainly to justify the Ginzburg–Landau equations from first principles and to deduce the Ginzburg–Landau parameters in terms of microscopic quantities.

Two of us have attended Professor de Gennes' lectures at Orsay and we expect that the reader will often recognize his influence in the method of presentation. We are much indebted to him for his valuable advice on many parts of this work and we wish to express our grateful thanks to him.

Special thanks are also due to our colleagues and friends Drs. Caroli, Burger, Corsan, Cribier, Deutscher, Dew-Hughes, Goldsmid, Guyon Hillel, Hurault, Jacrot, Matricon, Rose-Innes and Roubeau who have given us much help in the various stages of the preparation of the book. The two French writers want also to express their thanks to the English co-writer for his advice on the language of the English edition. We wish to acknowledge the co-operation of Mrs. Guggenberg and her staff and of Miss Sandra Cato and Miss Ann Nicholls in preparing the typescript. It is a pleasure also to thank the many workers in the subject who have given us permission to reproduce diagrams from their papers. Each is acknowl-

Preface

edged in the text. Two of the diagrams in Chapter 11 first appeared in an article on the Applications of Superconductivity in the journal *British Communications and Electronics* in April, 1965: we wish to thank the Editor for allowing us to reproduce them here. Finally, one of us (E.J.T.) wishes to thank his wife for her encouragement and help with the proof reading and preparation of many of the diagrams.

E.J.T.
D.S.J.
G.S.

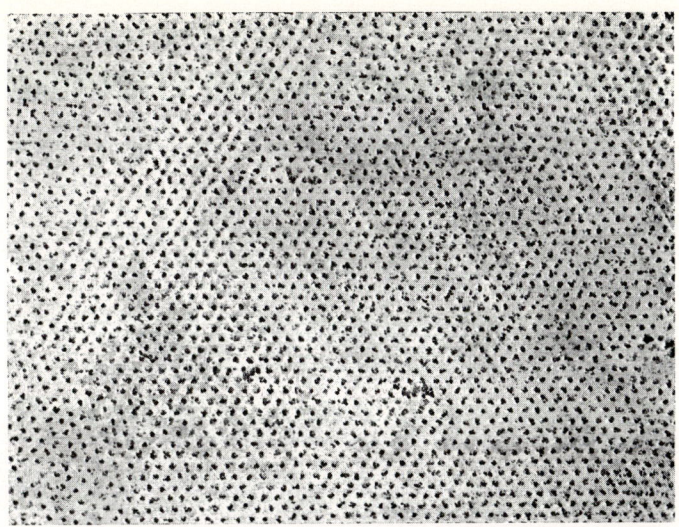

Electron microscope photograph of the Abrikosov triangular vortex line structure. This was observed at the surface of a lead-4 at. % indium crystal at 1.1 °K in the remnant state using a replica technique. Photograph by courtesy of Dr. U. Essmann and Dr. H. Trauble.

Magnetic field penetration into a type II superconductor containing pinning points. The black regions are normal (see section 8.1). Photograph by courtesy of Dr. W. DeSorbo. FIG. 8.1 b.

PART I

REVERSIBLE PROPERTIES

By
D. Saint-James
and
G. Sarma

A la S. A. B. R. M.

CHAPTER 1

Introduction

1.1. Introduction to superconductivity

In 1908 Kamerlingh-Onnes, in Leiden, succeeded in liquefying helium. This achievement provided Onnes and his school with a method of studying a number of phenomena in the temperature interval 1°K to 14°K. In order to check the Drude–Lorentz theory which predicted that the electrical resistivity of all materials should vary as the half power of the temperature, Onnes studied the variation of metallic resistance.

Mercury was chosen as the first metal to be investigated since it could be obtained in a very pure state and, in 1911, Onnes discovered that at about 4°K the resistance of his sample fell abruptly to zero. This special behaviour was naturally called superconductivity. Since then it has been found that a large number of metallic elements and alloys, and even some heavily doped semiconductors are superconductors.

The striking property of all superconductors is that below a well-defined critical temperature T_0 the electrical resistance is zero. [In some recent work (Quinn and Ittner, 1962), it has been reported that the resistivity, of superconducting lead for example, if not zero, must at least be less than 10^{-23} ohm-cm.] In metals and alloys T_0 ranges from less than 1°K to about 18°K. An obvious consequence of the zero resistance is the existence of permanent currents: closed currents set up in a ring of superconducting material have flowed for up to 2 years without showing any trace of decay.

1.2. Meissner effect—diamagnetism

In itself, the vanishing of the resistance has not lead to a good understanding of the phenomenon of superconductivity. The important discovery, in this respect, is the so-called Meissner effect, discovered by Meissner and Ochsenfeld in 1933. This effect shows the difference of behaviour between a perfect conductor and a superconductor, in the presence of a magnetic field.

Figure 1.1 shows the variation of the magnetic induction in the interior of a long solid cylinder of a superconductor when the applied field H_e is parallel to the axis of the cylinder. When the field is increased from zero to a certain field H_c surface currents oppose the penetration of the field and the induction B is zero in the interior of the sample. Up to now the superconductor behaves exactly like a perfect conductor. At $H = H_c$ the superconductor becomes normal and therefore the field penetrates and B becomes equal to H_c. Let the applied field now be lowered below H_c. If the superconductor were a perfect conductor, the induction would be maintained to the value $B = H_c$ by the surface currents. In practice, however, it is found that the superconductor expels the field and that $B = 0$ for $0 < H_e < H_c$.

Thus, at a given temperature $T < T_0$, an ideal superconductor expels any field $H < H_c$. This does not depend on the previous history of the specimen, i.e. whether it is first cooled and the field then switched on, or

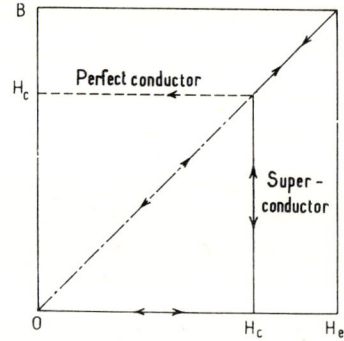

FIG. 1.1. Variation of the induction versus field for a superconductor and a perfect conductor

the field switched on and then the temperature reduced. The superconductor behaves as a perfect diamagnet.

The Meissner effect proves that the "superconducting state" is a reversible equilibrium state, a stable thermodynamic one.

1.3. Some considerations about the normal-superconducting transition

The reversibility of the expulsion of a magnetic field from an ideal superconductor implies that the transition between normal and superconducting state is reversible in T, H and p, where p is the pressure. However, superconductors undergo only very small volume changes and the pressure dependence can be neglected. Thus the two phases are separated by a threshold curve $H = H_c(T)$. $H_c(T)$ has approximately a parabolic variation with T (Fig. 1.2), i.e.

$$H_c(T) = H_c(0)\left[1 - \frac{T^2}{T_0^2}\right]. \tag{1.1}$$

For $T > T_0$ the material is normal even in zero field.

Introduction

As a consequence of the existence of $H_c(T)$, there is a critical current density $J_c(T)$ flowing through a superconductor, which will drive it into the normal state. This current density is simply that which will produce the critical field $H_c(T)$ at the surface (Silsbee, 1916).

Let $F_s(T)$ be the free energy in zero field for the superconductor and $F_n(T)$ the corresponding quantity in the normal phase. The difference $F_s(T) - F_n(T)$ can be calculated for any temperature (Gorter and Casimir, 1934). The thermodynamic potential per unit volume may be written as

$$G(H,T) = F(T) - \frac{1}{4\pi} \int_0^H B(H)\, dH. \tag{1.2}$$

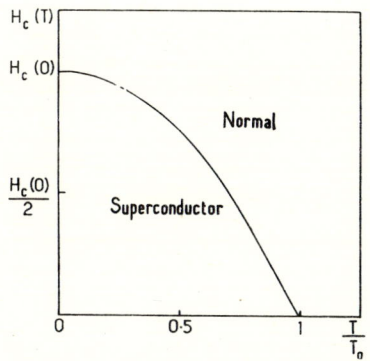

FIG. 1.2. Variation of the critical field H_c versus temperature (arbitrary units)

In the superconductor for $H < H_c$ (neglecting surface effects) $B = 0$ and

$$G_s(H,T) = F_s(T). \tag{1.3}$$

In the normal state we have $B = H$ (neglecting susceptibility) and

$$G_n(H,T) = F_n(T) - \frac{H^2}{8\pi}. \tag{1.4}$$

For $H = H_c(T)$ the two potentials are equal:

$$F_n(T) - F_s(T) = \frac{H_c^2(T)}{8\pi}. \tag{1.5}$$

This equation defines the thermodynamic critical field $H_c(T)$ as a function of the difference of free energy between the normal and superconducting phases.

Reversible Properties

The entropy density is $S = -(\partial G/\partial T)_H$, and the latent heat per unit volume is given by

$$Q = T(S_s - S_n) = \frac{T}{4\pi} H_c \frac{dH_c}{dT}, \quad (1.6)$$

while the variation of specific heat is

$$C_n - C_s = T\left[\left(\frac{\partial^2 G}{\partial T^2}\right)_s - \left(\frac{\partial^2 G}{\partial T^2}\right)_n\right]_H = -\frac{T}{8\pi}\frac{d^2(H_c^2)}{dT^2}. \quad (1.7)$$

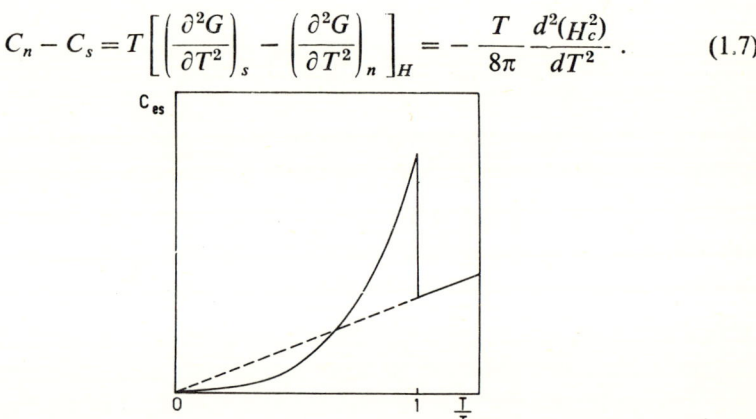

FIG. 1.3. Schematic variation of the electronic specific heat C_{es} versus temperature

As $dH_c/dT < 0$, we get $Q < 0$, i.e. the body releases heat in the transition from the normal to the superconducting state. dH_c/dT is finite and $H_c(T) = 0$ for $T = T_0$; the superconducting transition in *zero field* has no latent heat. The specific heat has the discontinuity (1.7) and the transition is classified as a "transition of the second order" (see Fig. 1.3).

For $T = 0$, the difference

$$F_n(0) - F_s(0) = \frac{H_c^2(0)}{8\pi}, \quad (1.8)$$

is the so-called condensation energy. Indeed, it has been shown that the superconducting transition corresponds to a kind of condensation of the electrons near the Fermi surface.† It is interesting to note that the condensation energy is of order $(k_B T_0)^2/E_F$ (where E_F is the Fermi energy of the conduction electrons in the normal state) so that only a small fraction $k_B T_0/E_F (\sim 10^{-3})$ of the electrons has its energy modified in the condensation.

Equations (1.6) and (1.7) agree with experiment, confirming the reversibility of the Meissner effect and that thermodynamics may be applied to superconductivity.

† This point is discussed in Chapter 5.

1.4. The London equation

In this section the effect of supercurrents $\boldsymbol{j}_s(\boldsymbol{r})$ and magnetic field $\boldsymbol{h}(\boldsymbol{r})$ together in a sample is considered.

The Maxwell equation can be written:

$$\operatorname{curl} \boldsymbol{h} = \frac{4\pi}{c} \boldsymbol{j}_s, \qquad (1.9)$$

where c is the velocity of light.

The free energy of the system has the form:

$$F = F_s + E_{\text{kin}} + E_{\text{mag}}. \qquad (1.10)$$

F_s is the energy of the electrons in the condensed state at rest. E_{kin} is the kinetic energy associated with the permanent currents. For electrons in a parabolic band of drift velocity $\boldsymbol{v}(\boldsymbol{r})$ at the point $\boldsymbol{r}$:

$$n_s e \boldsymbol{v}(\boldsymbol{r}) = \boldsymbol{j}_s(\boldsymbol{r}), \qquad (1.11)$$

where e is the electron charge, and n_s is the number of superconducting electrons per unit volume.

$$E_{\text{kin}} = \frac{1}{2} \int_{\text{vol}} d\boldsymbol{r} \, m v^2 n_s, \qquad (1.12)$$

where m is the effective mass of the electrons. Equation (1.12) is only strictly valid for a uniform flow ($\boldsymbol{v}$ is constant). It can be used, however, if $\boldsymbol{v}(\boldsymbol{r})$ is a slowly varying function of $\boldsymbol{r}$. In (1.10) E_{mag} is the magnetic energy:

$$E_{\text{mag}} = \int \frac{h^2}{8\pi} d\boldsymbol{r}. \qquad (1.13)$$

The free energy can be written in the following form:

$$F = F_s + \frac{1}{8\pi} \int [h^2 + \lambda_L^2 |\operatorname{curl} \boldsymbol{h}|^2] d\boldsymbol{r}, \qquad (1.14)$$

where λ_L is a length defined by

$$\lambda_L = \left[\frac{mc^2}{4\pi n_s e^2} \right]^{1/2}. \qquad (1.15)$$

Minimizing the free energy with respect to the variations of the field distribution $\delta h(r)$ to find the equilibrium state we have

$$\delta F = \frac{1}{4\pi} \int [h \cdot \delta h + \lambda_L^2 \operatorname{curl} h \cdot \operatorname{curl} \delta h] \, dr$$

$$= \frac{1}{4\pi} \int [h + \lambda_L^2 \operatorname{curl} \operatorname{curl} h] \cdot \delta h \, dr \quad (1.16)$$

and the minimum condition yields

$$h + \lambda_L^2 \operatorname{curl} \operatorname{curl} h = 0. \quad (1.17)$$

(1.17) is the London equation (F. and H. London, 1935).

Let us now investigate the properties of a semi-infinite specimen, the surface being the x–y plane, the region $z < 0$ being the empty space (Fig. 1.4).

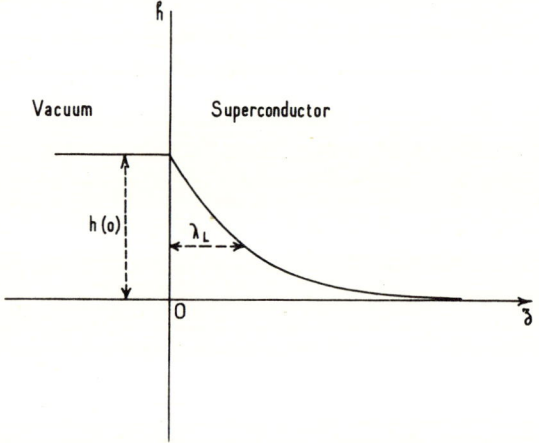

FIG. 1.4. Penetration of the field inside a superconductor according to the London equation

To (1.17) the Maxwell equation

$$\operatorname{div} h = 0 \quad (1.18)$$

can be added. Owing to the symmetry of the problem, h is a function of z only. Equation (1.18) yields $dh_z/dz = 0$ and the z component of (1.17) gives $h_z = 0$.

Thus h has no normal component to the surface of the specimen. Let us suppose then that h is directed along the x axis. Equation (1.18) is satisfied. The current j_s is directed along the y axis and (1.17) is written as

$$\frac{d^2h}{dz^2} = \frac{-h}{\lambda_L^2},$$

i.e.

$$h(z) = h(0) \, e^{-z/\lambda_L}$$

Introduction

Hence the field falls off exponentially inside the superconductor as shown in Fig. 1.4. λ_L is called the penetration depth.

At $T=0$, $n_s = n$ the number of conduction electrons per unit volume. We can then compute λ_L. Values for λ_L range from several hundred of angströms in pure metals where m is small to several thousands in transition metals with narrow band and in intermetallic compounds.

Since the number n of conduction electrons is related to the Fermi energy and the density of states $N(0)$ by

$$n = \frac{1}{3\pi^2}\left[\frac{2m}{\hbar^2}E_F\right]^{3/2} = \frac{2mv_F^2}{3}N(0),$$

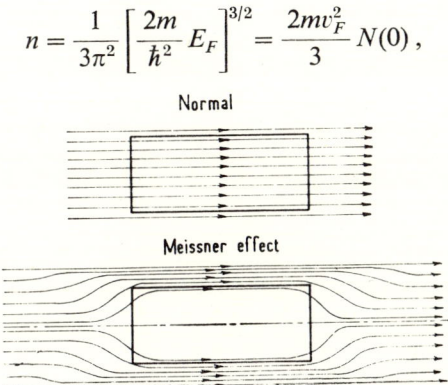

FIG. 1.5. Exclusion of the field from a superconductor (Meissner effect)

where v_F is the Fermi velocity, the London penetration depth at $T=0$ is given by

$$\lambda_L(0) = \left[\frac{3c^2}{8\pi N(0) v_F^2 e^2}\right]^{1/2} \tag{1.19}$$

This expression will often be used in the following chapters.

Thus a weak magnetic field only slightly penetrates the bulk of the superconductor. For all practical purposes, one can consider that the lines of force are excluded (Fig. 1.5). So the superconductor achieves an equilibrium where the sum of the kinetic and magnetic energy is a minimum.

It may be noted that for a perfect conductor, using the Maxwell equation:

$$\frac{\partial h}{\partial t} = \frac{\partial h(0)}{\partial t} e^{-x/\lambda_L}$$

showing that at $x \gg \lambda_L$ the magnetic field cannot change with time. This is in agreement with the discussion of section 1.2.

We can also note that in order to account for the perfect conductivity one should add the equation of motion of the electron in the presence of the electric field $\boldsymbol{E}$, i.e.

$$\frac{m\partial \boldsymbol{v}}{\partial t} = e\boldsymbol{E} = \frac{4\pi\lambda_L^2}{c^2}\frac{\partial \boldsymbol{j}_s}{\partial t}.$$

1.5. The coherence length and the energy gap

In establishing the London equation $v(r)$ is required to be a slowly varying function of r. The same condition holds for $h(r)$ and $j_s(r)$. The London equation (1.17) shows that $h(r)$ varies in a range λ_L.

In the condensed state, it is expected that there are strong correlations between the superconducting electrons. Let the range of these correlations be denoted by ξ_0. Then the condition of validity of the London equation can be written as $\lambda_L \gg \xi_0$.

The concept of the coherence length ξ_0 was introduced by Pippard (1950), in order to account for the failure of the London equation for certain superconductors. In 1955 Faber and Pippard estimated ξ_0 from the uncertainty principle. The correlation distance of the superconducting electrons ξ_0 is related to the range of momentum δp by

$$\xi_0 \delta p \sim \hbar.$$

In the condensation process the electrons involved are those within a distance $k_B T_0$ of the Fermi surface, i.e.

$$k_B T_0 \sim v_F \delta p,$$

where $v_F = p_F/m$ is the Fermi velocity. The coherence length ξ_0 is such that:

$$\xi_0 \sim \frac{\hbar v_F}{k_B T_0} \tag{1.20}$$

Another important feature of superconductivity is the existence of a gap in the low energy excitations. The existence of this gap has been related by Bardeen, Cooper and Schrieffer [B.C.S.] (1957) to the formation of condensed pairs of electrons. In most superconductors it is necessary in order to create an electron-hole pair close to the Fermi surface, to furnish an energy $\varepsilon_{pp'}$ such that:

$$\varepsilon_{pp'} \geq 2\Delta \tag{1.21}$$

where Δ is the so-called energy gap.

The existence of a gap in the low energy excitations is implied by a number of experiments which yield a measurement of Δ:

The specific heat at low temperatures is proportional to $e^{-\Delta/k_B T}$.

The electromagnetic absorption in the far infrared ($\lambda \sim 1$ mm) occurs only for photons of energy $\hbar\omega \geq 2\Delta$ (the photon creates an electron-hole pair).

The spin-lattice relaxation of nuclear spins proceeds predominantly in an interaction with the conduction electrons and has a frequency proportional to $e^{-\Delta/k_B T}$ at low temperatures.

The ultrasonic attenuation proceeds in a similar way.

Introduction

TABLE 1.1

Element	T_0 (°K)	Ratio $2\Delta(0)/k_B T_0$				
		a	b	c	d	e
Aluminium	1·19			3·16	2·9	~3·4
Cadmium	0·56					
Gallium	1·09					
Indium	3·41	4·1	3·9		3·9	3·45–3·63
Lanthanum β	5·95					
Lead	7·18	4·14	4·0			4·26–4·33
Mercury α	4·15	4·6				
Mercury β	3·95					
Niobium	9·46	2·8			4·4	
Rhenium	1·70					
Ruthenium	0·49					
Tantalum	4·48	≤3·0			3·6	
Thallium	1·37				3·2	
Tin	3·72	3·6	3·3	3·5	3·6	3·46
Vanadium	5·30	3·4			3·6	
Zinc	0·88				2·5	
Technetium	11·2					

Critical temperatures and energy gaps at $T = 0$ for several elements (after Lynton, *Superconductivity*, John Wiley, New York, 1961). The values of the gap corresponding to different kinds of measuring technics are listed as follows: (a) infrared absorption (b) infrared transmission, (c) microwave absorption, (d) specific heat, (e) tunnelling.

Tunnelling experiments yield the density of states directly.

Table 1.1 gives the energy gap values for several superconductors.

Another relation for ξ_0 can be deduced from the knowledge of the gap Δ. To create an electron-hole pair, the forbidden values of momenta are given by

$$E_F - \Delta \leq \frac{\hbar^2 p^2}{2m} \leq E_F + \Delta,$$

so that the range of momentum δp is

$$\delta p \simeq \frac{p_F \Delta}{E_F}.$$

Thus the coherence length ξ_0 can be defined by the relation:

$$\xi_0 = \frac{\hbar v_F}{\pi \Delta}. \tag{1.22}$$

The factor $1/\pi$ is arbitrary and chosen for convenience.

It must be pointed out, however, that the existence of a gap in the low energy excitation spectrum is not a necessary condition for the occurrence of the typical superconducting properties, such as Meissner effect and permanent currents. "Gapless" superconductivity has been shown to exist in superconductors containing magnetic impurities, where the energy gap $\varDelta$ and the transition temperature T_0 vary with the impurity concentration, $\varDelta$ vanishing while T_0 is still finite. A similar situation also occurs in surface superconductivity where the absence of a gap in the low energy excitation spectrum does not prevent the existence of a superconducting sheath.†

The criterion for superconductivity seems to be the existence of a strong pairing correlation, and this does not necessarily imply the existence of a finite gap in the excitations. Nevertheless, the existence of a gap in most superconductors has been of a great importance for the establishment of the microscopic theories. The Bardeen, Cooper and Schrieffer theory predicts that the energy gap $\varDelta$ is $1.76\, k_B T_0$. Using this, the Faber–Pippard relation (1.20) may be written:

$$\xi_0 = 0.18 \frac{\hbar v_F}{k_B T_0}. \tag{1.23}$$

1.6. Classification of superconductors

The existence of the coherence length ξ_0 and of the London penetration depth λ_L leads to a natural classification of superconductors into two types.

(a) Type I superconductors for which $\lambda_L \ll \xi_0$. The London equation (1.17) is not valid. These superconductors exhibit a Meissner effect, but the penetration depth λ is given by a more complicated relation than (1.15) due to Pippard:

$$\lambda = \left(\frac{\sqrt{3}}{2\pi}\right)^{1/3} \xi_0^{2/3} \lambda_L^{1/3} \quad \text{for} \quad \xi_0 \gg \lambda_L \tag{1.24}$$

In this category are pure metals and low concentration alloys for which the effective mass m is close to the free electron mass, λ_L is short (~ 200 Å), and v_F is rather high ($\sim 10^8$ cm/sec).

(b) Type II superconductors for which $\lambda_L \gg \xi_0$. The London equation is valid only in low fields.

In this category are most chemical compounds such as Nb_3Sn, V_3Ga, etc., for which m is very high, $\lambda_L \sim 2000$ Å, v_F is small ($\sim 10^6$ cm/sec), the transition temperature T_0 (and consequently $\varDelta$) is rather high ($\sim 18°K$ for Nb_3Sn), and ξ_0 is thus small (~ 50 Å). It also includes higher concentration superconducting alloys in which the coherence length and the penetration depth are a function of the normal electron mean free path l. An increase in l decreases ξ_0 and increases λ_L.

The distinction between type I and type II superconductors is very important. Their properties in the presence of magnetic fields are strikingly different.

† See Chapters 4 and 6.

Introduction

Type I superconductors (or Pippard superconductors) behave roughly as "ideal" superconductors, which were outlined above. They are described not by the London theory but by the nonlocal Pippard theory. From the microscopic point of view, the properties of type I superconductors are well explained by the B.C.S. theory in its original form. Most of the initial experimental work was devoted to them. The B.C.S. theory is able to account for the specific heat dependence, the acoustic attenuation, the infrared absorption, the tunnelling effects, etc. These properties of type I super-

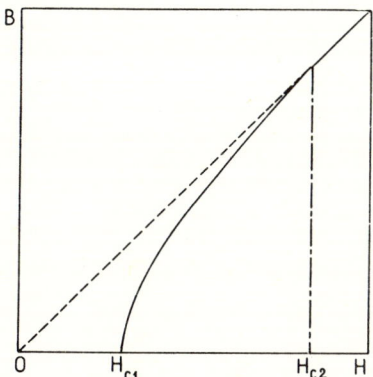

FIG. 1.6. Schematic variation of the induction B versus the field H in a type II superconductor

conductors have been extensively studied in a number of books to which we refer the reader.

This book is concerned with type II superconductors (or London superconductors). These were first investigated by Shubnikov *et al.* (1937) and have become increasingly important in the past few years. We shall begin with a short review of their main magnetic properties.

1.7. Magnetic and thermodynamic properties of type II superconductors

A macroscopic cylinder of type II superconductor placed in a field H parallel to its axis exhibits the properties shown in Figs 1.6 and 1.7. There are three regions of interest depending on the value of the external field.

(α) The exclusion of the field from the sample (Meissner effect) is total only for values of field H smaller than H_{c1}, where H_{c1} is a critical field smaller than the thermodynamic critical field H_c as defined by (1.5).

(β) For $H_{c2} > H > H_{c1}$ flux gradually penetrates the sample, but even at the thermodynamic equilibrium this flux is smaller than in the normal state. A new state appears in which a lattice of quantized flux-enclosing supercurrent vortices (or filaments) is formed: this state is commonly called

13

the "mixed state", and sometimes the "vortex state" or the "Shubnikov state".

(γ) For $H > H_{c2}$ the specimen becomes normal and the transition at $H = H_{c2}$ is of the second order.† The critical field H_{c2} is greater than H_c and can be very high. Table 1.2 gives the values of H_{c1} and H_{c2} for several materials.

H_{c1} and H_{c2} vary with temperature. This variation is shown on Fig. 1.8.

The curves $H_{c1}(T)$ and $H_{c2}(T)$ are the boundaries of three domains (α) for which $B = 0$ (Meissner effect), (β) for which $B < H$ (mixed state), (γ) for which $B = H$ (normal state).

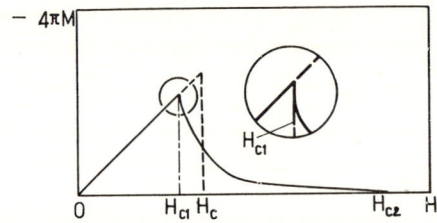

FIG. 1.7. Schematic variation of the magnetization versus the field H in a type II superconductor (an enlarged section of the vicinity of H_{c_1} is represented in the circle)

The magnetic properties of type II materials at a fixed temperature may be displayed as the induction inside the material (Fig. 1.6) or as the magnetization

$$-4\pi M = H - B \qquad (1.25)$$

(Fig. 1.7).

The concept of a "thermodynamic critical field" H_c for a type II superconductor can be introduced using (1.5):

$$F_n(T) - F_s(T) = \frac{H_c^2}{8\pi},$$

where $F_n(T)$ and $F_s(T)$ are respectively the free energy of the normal and superconducting states in zero field. It must be emphasized that for a type I superconductor, this equation defines an actual critical field, while for a type II superconductor it defines only a convenient concept. The magnetic properties of type I and type II superconductors having the same shape and the same value of H_c are compared in Figs 1.7 and 1.8 where the dashed lines give the variation of the magnetization versus H and of H_c versus T for the type I superconductor.

† The phenomenon of surface superconductivity is neglected here. It will be considered in detail later.

TABLE 1.2

Compound or element	H_{c1} (Oe)	H_c (Oe)	H_{c2} (Oe)	Temperature (°K)	Reference
Pb	550	550	550	4·2	(a)
0·85 Pb 0·15 Ir	250	650	3040	4·2	(a)
0·75 Pb 0·25 In	195~215	570	3500	4·2	(a)
0·70 Pb 0·30 Tl	145	430	2920	4·2	(a)
0·976 Pb 0·024 Hg	340	580	1460	4·2	(a)
0·912 Pb 0·088 Bi	245	675	3250	4·2	(a)
Nb	~1300	1608	2680	4·2	(b)
0·50 Nb 0·50 Ta	—	~252	1470	5·6	(b)

Some recent determinations of H_{c1}, H_{c2} and H_c in various compounds:
(a) DRUYVESTEYN, W. F., Thesis, Eindhoven (1965).
(b) MC CONVILLE, T. and SERIN, B., *Phys. Rev.* **140** (1965) A1169.

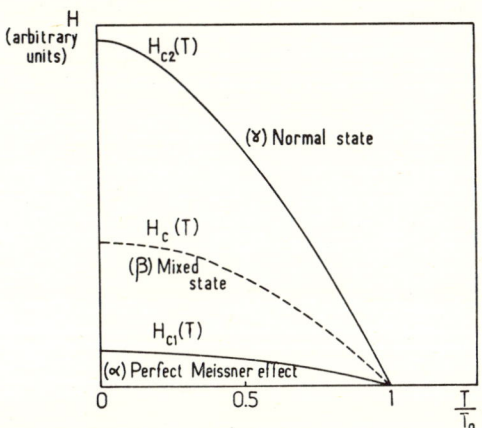

FIG. 1.8. Schematic variation of the critical fields H_{c1} and H_{c2} versus temperature in a type II superconductor

In the former, for $H < H_c$, $B = 0$ and it has the property that

$$\int_0^{H_c} M dH = -\frac{H_c^2}{8\pi}. \tag{1.26}$$

A type II superconductor has the analogous property

$$\int_0^{H_{c2}} M dH = -\frac{H_c^2}{8\pi}. \tag{1.27}$$

The Gibbs thermodynamic potential can be generally written as

$$G = F(B) - \frac{BH}{4\pi}, \tag{1.28}$$

Reversible properties

the equilibrium value $B(H)$ being obtained from

$$\left(\frac{\partial G}{\partial B}\right)_H = 0. \tag{1.29}$$

In the normal state where the induction B_n is equal to H, the suitable form for G is

$$G_n = F_n(0) + \frac{B^2}{8\pi} - \frac{BH}{4\pi} \tag{1.30}$$

while G_s is given by (1.28) with $F(B) = F_s(B_s)$. Then

$$\frac{\partial}{\partial H}(G_n - G_s) = \frac{B_s - B_n}{4\pi} = \frac{B_s - H}{4\pi}. \tag{1.31}$$

According to (1.29),

$$\frac{\partial}{\partial H}(G_n - G_s) = \frac{d}{dH}(G_n - G_s)$$

and

$$\int_0^{H_{c2}} M dH = |G_n - G_s|_0^{H_{c2}}.$$

For $H = H_{c2}$, $G_n = G_s$ and for $H = 0$, $G_n - G_s = F_n - F_s = -H_c^2/8\pi$ which proves (1.27). Hence, the thermodynamic critical field can be found from the reversible magnetization curve by evaluating the area under the curve $M = f(H)$.

Another interesting feature of type II superconductors is that the transitions at $H_{c1}(T)$ and $H_{c2}(T)$ are of the second order. There should be no latent heat and a discontinuity of the specific heat should exist. Several interesting thermodynamic relations have been deduced by Goodman (1962) for these transitions.

Defining the thermodynamic potential G_i corresponding to phase i [$i = (\alpha), (\beta)$ or (γ)], then

$$G_i = F_i(T, B_i) - \frac{B_i H}{4\pi}. \tag{1.32}$$

G_i must be a minimum for fixed H and T, so that

$$\left(\frac{\partial F}{\partial B_i}\right)_{H,T} = \frac{H}{4\pi}. \tag{1.33}$$

The corresponding entropy is

$$S_i = -\left(\frac{\partial G_i}{\partial T}\right)_H = -\frac{\partial F_i}{\partial T}. \tag{1.34}$$

Introduction

At the equilibrium between the two phases i and j, the field is equal to the critical field $H_{ij}(T)$, the thermodynamic potentials are equal, and if the transition is of the second order, there is no latent heat, and the entropies are also equal. As a result of this, it can be shown that the induction is continuous at the transition.

Calculating the variation of G_i for a variation dH of the field along the curve $H = H_{ij}(T)$ [i.e. for $dH = (dH_{ij}/dT)\,dT$]:

$$\frac{dG_i}{dT} = \frac{dF_i}{dT} - \frac{B_i}{4\pi}\frac{dH_{ij}}{dT} - \frac{H_{ij}}{4\pi}\frac{dB_i}{dT} = \frac{\partial F_i}{\partial T} + \frac{\partial F_i}{\partial B_i}\frac{dB_i}{dT}$$

$$- \frac{B_i}{4\pi}\frac{dH_{ij}}{dT} - \frac{H_{ij}}{4\pi}\frac{dB_i}{dT} = -S_i - \frac{B_i}{4\pi}\frac{dH_{ij}}{dT}.$$

(1.35)

In establishing (1.35), expressions (1.33) and (1.34) have been used. On the transition curve $H = H_{ij}(T)$. $G_i \equiv G_j$, i.e. $dG_i/dT \equiv dG_j/dT$. If there is no latent heat, $S_i \equiv S_j$, so that

$$B_i \equiv B \tag{1.36}$$

Let us now calculate the discontinuity of specific heat. For a fixed H in the phase i

$$C_i = T\left(\frac{\partial S_i}{\partial T}\right)_H. \tag{1.37}$$

Along the equilibrium curve $H = H_{ij}(T)$, the total derivative of the entropy is

$$\frac{\partial S_i}{\partial T} = \left(\frac{\partial S_i}{\partial T}\right)_H + \left(\frac{\partial S_i}{\partial H}\right)_T \frac{dH_{ij}}{dT}. \tag{1.38}$$

As there is no latent heat, $S_i \equiv S_j$ along the curve $H = H_{ij}(T)$ and

$$\frac{dS_i}{dT} = \frac{dS_j}{dT}. \tag{1.39}$$

The discontinuity in the specific heat is

$$C_j - C_i = T\left(\frac{\partial S_j}{\partial T}\right)_H - T\left(\frac{\partial S_i}{\partial T}\right)_H = T\frac{dH_{ij}}{dT}\left[\left(\frac{\partial S_i}{\partial H}\right)_T - \left(\frac{\partial S_j}{\partial H}\right)_T\right] \tag{1.40}$$

The derivative $\partial S_i/\partial H$ can be written as

$$\left(\frac{\partial S_i}{\partial H}\right)_T = \left(\frac{\partial S_i}{\partial B_i}\right)_T \left(\frac{\partial B_i}{\partial H}\right)_T = -\frac{\partial^2 F_i}{\partial B_i \partial T}\left(\frac{\partial B_i}{\partial H}\right)_T = -\frac{1}{4\pi}\left(\frac{\partial H}{\partial T}\right)_{B_i}\left(\frac{\partial B_i}{\partial H}\right)_T,$$

(1.41)

where (1.33) and (1.34) have been used.

Reversible Properties

On the other hand dH_{ij}/dT is given by

$$\frac{dH_{ij}}{dT} = \left(\frac{\partial H}{\partial T}\right)_{B_i} + \left(\frac{\partial H}{\partial B_i}\right)_T \frac{dB}{dT}, \qquad (1.42)$$

where $dB/dT = dB_i/dT = dB_j/dT$ is the variation of B along the equilibrium curve [see (1.36)].

The combination of expressions (1.41) and (1.42) yields

$$\left(\frac{\partial S_i}{\partial H}\right)_T = -\frac{1}{4\pi}\frac{dH_{ij}}{dT}\frac{\partial B_i}{\partial H} + \frac{1}{4\pi}\frac{dB}{dT} \qquad (1.43)$$

and the discontinuity of specific heat is given by

$$C_j - C_i = \frac{T}{4\pi}\left(\frac{dH_{ij}}{dT}\right)^2\left[\left(\frac{\partial B_j}{\partial H}\right)_T - \left(\frac{\partial B_i}{\partial H}\right)_T\right]. \qquad (1.44)$$

This relation is in fact a generalization of Ehrenfest's formula.

A number of conclusions can be drawn from (1.44):

for the transition $(\alpha) \to (\beta)$ (perfect Meissner effect $\to$ mixed state), $H = H_{c1}(T)$ and we expect from experiment $\partial B_\alpha/\partial H = 0$ and $\partial B_\beta/\partial H = \infty$ (see Fig. 1.6). One should find an infinite discontinuity of the specific heat. Actually, this is difficult to observe because the singularity is masked by hysteresis effects. However, a lambda transition for $H = H_{c_1}$ has recently been observed in niobium by Serin (1965);

for the transition $(\beta) \to (\gamma)$ (mixed state $\to$ normal state) $H = H_{c2}(T)$ and, from experiment, $\partial B_\beta/\partial H > 1$ and $\partial B_\gamma/\partial H = 1$. Knowing the value of dH_{c2}/dT, $C_\beta - C_\gamma$ can be calculated. It is first concluded that $C_\beta > C_\gamma$. However, the comparison of (1.44) with experiment is not always possible since not all the quantities of interest have been measured. In V_3Ga, it has been found that there is no latent heat and the discontinuity $C_\beta - C_\gamma$ has been measured as well as dH_{c2}/dT. An extrapolation of $\partial B/\partial H$ for $H = H_{c2}$ gives an agreement of the order of 10% with the value calculated from (1.44).

1.8. Ideal type II superconductors and hard superconductors

We have seen that in type II superconductors, superconductivity may exist in very high fields ($\sim 10^6$ Oe). Because of Silsbee's rule the critical current I_c might also be expected to be very high. Actually, for *thermodynamic equilibrium* this critical current is smaller than the critical current of a type I superconductor having the same critical field H_c.

This can be seen by calculating the theoretical current density I_c for a cylinder of radius a in which a current I flows parallel to the axis. At the surface of the cylinder, the field is

$$H(a) = \frac{2I}{ca}, \quad (1.45)$$

where c is the velocity of light in vacuum.

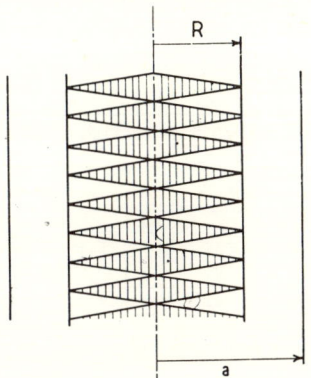

Fig. 1.9. The intermediate state of a cylinder of a type I superconductor with a current parallel to the cylinder axis

For a type I superconductor, when I exceeds the value

$$I_c(\text{I}) = \frac{ca}{2} H_c \quad (1.46)$$

the cylinder must become normal close to the surface,† and the flow of current is accompanied by heat dissipation.

For a type II superconductor, as long as $2I/ca < H_{c1}$, the Meissner effect is perfect and the current flows in a region close to the surface of width λ where λ is the penetration depth of the field. When $H(a)$ exceeds H_{c1}, dissipation of energy occurs, and the critical current $I_c(\text{II})$ is given by

$$I_c(\text{II}) = \frac{ca}{2} H_{c1}. \quad (1.47)$$

This critical current is smaller than $I_c(\text{I})$, as $H_{c1} < H_c$.

The fact that $I_c(\text{II})$ is given by (1.47) follows from the fact that for $H > H_{c1}$, type II superconductors are in the "mixed state" which can be *roughly* pictured as an array of normal regions of small radius (filaments) embedded in a superconducting matrix. As soon as I becomes greater than

† Actually, the cylinder divides into two parts. For $R < r < a$, there is a normal sheath, and for $0 > r > R$ the sample is in the intermediate state where normal and superconducting regions perpendicular to the cylinder axis alternate (Fig. 1.9).

Reversible Properties

$I_c(\text{II})$, filaments appear on the surface of the cylinder, which take the form of circles (along the lines of force). They penetrate into the cylinder and eventually annihilate in the centre, releasing heat.

To obtain currents higher than $I_c(\text{II})$ it is necessary to hinder the displacement of the vortex lines. Structural defects are used to pin the filaments.†

Type II superconductors with pinning centres are often called hard superconductors. They are of great technological interest because they can be used in the fabrication of superconducting magnets to produce fields of the order of several hundred thousand oersteds. The study of the reversible properties of ideal type II superconductors (with no pinning) remains, nevertheless, essential both from the point of view of theory and experiment. It has been the subject of numerous developments in the past 10 years. These are considered in the next chapters. The second part of this book will be devoted to the irreversible properties.

---

† This does not have a large effect on H_{c2}, but I_c (II) is strongly modified (Gorter, 1962).

Bibliography of Chapter 1

BARDEEN, J., COOPER, L. N. and SCHRIEFFER, J. R. (1957), *Phys. Rev.* **108**, 1175.
FABER, T. E. and PIPPARD, A. B. (1955), *Proc. Roy. Soc.* **A231**, 336.
GOODMAN, B. B. (1962), *Phys. Letters*, **1**, 215.
GORTER, C. J. (1962), *Phys. Letters*, **1**, 69; (1962), **2**, 26.
GORTER, C. J. and CASIMIR, H. B. G. (1934), *Physica*, **1**, 306.
KAMERLINGH—ONNES, H. (1911), *Leiden Comm.* 122b, 124c.
LONDON, F. and LONDON, H. (1935), *Proc. Roy. Soc.* **A149**, 71; (1935), *Physica*, **2**, 341.
MEISSNER, W. and OCHSENFELD, R. (1933), *Naturwissen.* **21**, 787.
PIPPARD, A. B. (1950), *Proc. Roy. Soc.* **A216**, 547.
QUINN, D. J. and ITTNER, W. B. (1962), *J. Appl. Phys.* **33**, 748.
SERIN, B. (1965), *Phys. Letters*, **16**, 112.
SILSBEE, F. B. (1916), *J. Wash. Acad. Sci.* **6**, 597.
SHUBNIKOV, L. W., KHOTKEVICH, V. I., SHEPELEV, J. D. and RJABININ J. N. (1937) *Zh. Eksperim. i Teor. Fiz.* **7**, 221.

CHAPTER 2

The Phenomenological Theory of Ginzburg and Landau

2.1. The free energy and the Ginzburg–Landau equations

In 1937 Landau proposed a general theory of second order phase transitions based on three fundamental assumptions (see Landau and Lifshitz, 1958):

(a) there exists an order parameter ψ which goes to zero at the transition;
(b) the free energy may be expanded in powers of ψ;
(c) the coefficients of the expansion are regular functions of T.

Thus, following Landau, the free energy per unit volume is written as†

$$F = F_n + \alpha(T)|\psi|^2 + \frac{\beta(T)}{2}|\psi|^4 + \ldots \qquad (2.1)$$

F_n is the energy of the normal state. The coefficients α and β, according to assumption (c), must have the following properties: $\alpha(T)$ is negative (for $T < T_0$) and vanishes at $T = T_0$, while the slope $d\alpha/dT$ remains finite at $T = T_0$. Close to T_0, $\alpha(T)$ may be written as

$$\alpha(T) = (T - T_0)\left(\frac{d\alpha}{dT}\right)_{T=T_0}. \qquad (2.2)$$

$\beta(T)$ on the other hand is a positive quantity and, close to T_0, may be written

$$\beta(T) \simeq \beta(T_0). \qquad (2.3)$$

For $T < T_0$ the free energy (2.1) presents an absolute minimum for

$$|\psi|^2 = -\frac{\alpha}{\beta}. \qquad (2.4)$$

† It can be shown from the microscopic theory that only even powers of $|\psi|$ appear in (2.1). See Chapter 5.

Theory of Ginzburg and Landau

Expression (2.4) ensures that $|\psi|$ vanishes for $T = T_0$. The free energy corresponding to (2.4) takes the form

$$F = F_n - \frac{\alpha^2}{2\beta}. \tag{2.5}$$

If the theory is to be applied to a superconductor, F is the free energy F_s of the superconductor in the absence of any field. It has been shown in Chapter 1 that

$$F_n - F_s = \frac{H_c^2(T)}{8\pi} \tag{2.6}$$

and from (2.5) it follows that close to T_0

$$H_c^2(T) = \frac{4\pi\alpha^2}{\beta}. \tag{2.7}$$

Equations (2.2) and (2.3) show that $F_n - F_s \sim (T - T_0)^2$, which is a general result of Landau's theory of second order phase transitions, and that $H_c(T)$ is proportional to $(T - T_0)$, which is in agreement with the formula (1.1) for the variation of H_c as a function of T

$$H_c(T) = H_c(0)\left[1 - \left(\frac{T}{T_0}\right)^2\right]. \tag{2.8}$$

The expression (2.1) is limited to the case where the order parameter ψ is a constant throughout the specimen. If ψ has a spatial variation, then the spatial derivatives of ψ must be added to (2.1). The first term can be expected to be proportional to $|\nabla\psi|^2$. The general Landau expression for the free energy in the vicinity of a second order transition should take the form

$$f = \int_\Omega F d\mathbf{r} \tag{2.9}$$

where the integral is extended over the volume Ω of the sample and $F(\mathbf{r})$ is given by

$$F = F_n + \alpha(T)|\psi|^2 + \frac{\beta(T)}{2}|\psi|^4 + \gamma|\nabla\psi|^2 + \ldots \tag{2.10}$$

The form (2.10) would not have been of a great help in the understanding of the properties of superconductors if Ginzburg and Landau (1950) had not proposed an extension of (2.10) to describe the superconductor in the presence of a magnetic field. With a great physical insight, they considered the order parameter ψ as a kind of a "wave function" for a "particle" of

Reversible Properties

charge e^* and mass m^*. In order to ensure gauge invariance they wrote the free energy as

$$F_s = F_n + \alpha |\psi|^2 + \frac{\beta}{2}|\psi|^4 + \frac{1}{2m^*}\left|\left(-i\hbar\nabla - \frac{e^*A}{c}\right)\psi\right|^2 + \frac{h^2}{8\pi}. \tag{2.11}$$

A is the vector potential for the field h, i.e.

$$h = \operatorname{curl} A \tag{2.12}$$

and $h^2/8\pi$ is the magnetic energy.

We shall now minimize the free energy $f = \int_\Omega F d\mathbf{r}$ for variations of the order parameter ψ and of the magnetic field h (i.e. of A):

$$\delta f = \int d\mathbf{r} \left\{ \alpha\psi \, \delta\psi^* + \beta|\psi|^2\psi\delta\psi^* + \frac{1}{2m^*}\left[i\hbar\nabla - \frac{e^*A}{c}\right]\delta\psi^* \right.$$
$$\left. \cdot \left[-i\hbar\nabla - \frac{e^*A}{c}\right]\psi + cc \right\} + \int d\mathbf{r} \left\{ \frac{h}{8\pi} \cdot \operatorname{curl}\delta A - \frac{e^*}{2m^*c}\psi^*\delta A \right.$$
$$\left. \cdot \left[-i\hbar\nabla - \frac{e^*}{c}A\right]\psi + cc \right\} \tag{2.13}$$

or

$$\delta f = \int d\mathbf{r} \left\{ \delta\psi^*\left[\alpha\psi + \beta|\psi|^2\psi + \frac{1}{2m^*}\left(-i\hbar\nabla - \frac{e^*A}{c}\right)^2\psi\right] + cc \right\}$$
$$+ \int d\mathbf{r} \left\{ \delta A \cdot \left(\frac{\operatorname{curl} h}{8\pi} - \frac{e^*}{2m^*c}\psi^*\left(-i\hbar\nabla - \frac{e^*}{c}A\right)\psi\right) + cc \right\}, \tag{2.14}$$

where cc means the complex conjugate quantity.

In deriving (2.14) the surface integral

$$\int \left(\delta\psi^*\left(-i\hbar\nabla - \frac{e^*}{c}A\right)\psi + cc\right) \cdot d\mathbf{s} \tag{2.15}$$

has been set equal to zero. There is no obvious reason for this integral to vanish and we will come back to this question later†. For f to be a minimum, i.e. $\delta f = 0$, (2.14) yields the two equations of Ginzburg and Landau:

$$\frac{1}{2m^*}\left[-i\hbar\nabla - \frac{e^*A}{c}\right]^2\psi + \alpha\psi + \beta|\psi|^2\psi = 0 \tag{2.16}$$

$$\frac{\operatorname{curl} h}{4\pi} = \frac{j}{c} = \frac{e^*\hbar}{2im^*c}(\psi^*\nabla\psi - \psi\nabla\psi^*) - \frac{e^{*2}}{m^*c^2}\psi^*\psi A. \tag{2.17}$$

† Actually one should have minimised the thermodynamic potential G. An extra surface integral then appears which must be set equal to zero. As usual this ensures the continiuty of the field component parallel to the surface at the sample boundary.

Theory of Ginzburg and Landau

The first equation allows the order parameter ψ to be calculated in the presence of the field, while (2.17) gives the distribution of the currents. It must be emphasized that, in these equations, $\mathbf{h}$ is the internal field, not the applied field, and that in principle, the two equations (2.16) and (2.17) plus (2.12) allow $\mathbf{h}$ and ψ to be determined.

Let us now investigate the problem of the charge e^* and the mass m^*. In their original paper Ginzburg and Landau found "no reason to consider e^* as different from the electronic charge". Consequently, they set $e^* = e$ and $m^* = m$. Actually, Gorkov (1959 a, b) was able to derive (2.16) and (2.17) from the microscopic theory. This point will be discussed in more detail in Chapter 5. However, some of the results obtained by Gorkov will be reviewed here.

In the microscopic theory the important physical quantity is the so-called pair potential† $\varDelta(r) = V \langle \psi_\uparrow(r) \psi_\downarrow(r) \rangle$. Gorkov obtained for this quantity two equations analogous to (2.16) and (2.17) showing that $\psi(r)$ is proportional to $\varDelta(r)$. At the same time, this shows also that $e^* = 2e$, a property related to the presence of the Cooper pairs. The mass m^*, on the contrary, occurs in the proportionality coefficient between ψ and $\varDelta$, and is thus left arbitrary. One could choose $2m$, for m^*, but the convention is to keep the mass m^* equal to the electronic mass m. The Ginzburg–Landau equations then take the form

$$\frac{1}{2m}\left[-i\hbar\nabla - \frac{2e\mathbf{A}}{c}\right]^2 \psi + \alpha\psi + \beta|\psi|^2\psi = 0, \qquad (2.18)$$

$$\frac{\operatorname{curl}\mathbf{h}}{4\pi} = \frac{\mathbf{j}}{c} = \frac{e\hbar}{imc}(\psi^*\nabla\psi - \psi\nabla\psi^*) - \frac{4e^2}{mc^2}|\psi|^2\mathbf{A}. \qquad (2.19)$$

α and β have also been deduced from the microscopic theory, and for pure materials‡ Gorkov obtained

$$\alpha = 1\cdot 83 \frac{\hbar^2}{2m}\frac{1}{\xi_0^2}\frac{T-T_0}{T_0}, \qquad (2.20)$$

$$\beta = 0\cdot 35 \frac{1}{N(0)}\left(\frac{\hbar^2}{2m}\frac{1}{\xi_0^2}\right)^2 \frac{1}{(k_B T_0)^2}, \qquad (2.21)$$

where ξ_0 is the coherence length and $N(0)$ the density of states at the Fermi level.

In order to define the problem completely, it is necessary to add a boundary condition to (2.16) and (2.17). For example, the current flowing through the surface of the sample can be taken to be zero. Equation (2.19), shows

---

† V is the interaction between the electrons, cf. Chapter 5.
‡ The concept of "pure" superconductors will be discussed in Chapter 5.

that this condition is equivalent to

$$\left(-i\hbar\nabla - \frac{2e\mathbf{A}}{c}\right)_n \psi = i\lambda\psi, \qquad (2.22)$$

where n indicates the component of $(-i\hbar\nabla - 2e\mathbf{A}/c)$ parallel to the normal of the surface. However, condition (2.22) does not ensure that the surface integral (2.15) is zero. Ginzburg and Landau in their original paper set $\lambda = 0$, i.e.

$$\left(-i\hbar\nabla - \frac{2e\mathbf{A}}{c}\right)_n \psi = 0 \qquad (2.23)$$

and (2.15) is then satisfied. In fact the vanishing of the surface integral (2.15) is not a necessary condition for the theory since the Ginzburg and Landau form for the free energy F is not valid within a distance ξ_0 of the surface. The microscopic theory (de Gennes, 1964) has shown that condition (2.23) holds for a superconductor–insulator† interface and (2.22) for a superconductor–normal metal interface. We shall see in Chapter 4 that conditions (2.22) and (2.23) have important implications in the onset of superconductivity.

2.2. The two characteristic lengths $\xi(T)$, $\lambda(T)$ and the validity of the Ginzburg—Landau scheme

(a) Range of variation of the order parameter. The temperature dependent coherence length

The two Ginzburg–Landau equations (2.16) and (2.17) have two special and obvious solutions:

(α) $\psi \equiv 0$, where $\mathbf{A}$ is determined only by $\mathbf{H} = \text{curl }\mathbf{A}$ and $\mathbf{H}$ is the applied field. This solution describes the normal state.

(β) $\psi = \psi_0 = (-\alpha/\beta)^{1/2}$ and $\mathbf{A} = 0$. This solution describes the ordinary superconducting state with perfect Meissner effect (neglecting surface effects). According to (2.4), ψ_0 corresponds to the lowest free energy when $\alpha < 0$, i.e. when $T < T_0$.

In the case of a very weak field, ψ is expected to vary very slowly, close to the value ψ_0. The range of variation of ψ can be deduced from the first Ginzburg–Landau equation by setting $\mathbf{A} = 0$. Let us introduce the new function

$$f = \frac{\psi}{\psi_0}, \qquad (2.24)$$

where f satisfies the following equation:

$$-\frac{\hbar^2}{2m}\nabla^2 f + \alpha f - \alpha f^3 = 0. \qquad (2.25)$$

† Or a superconductor-vacuum interface.

Theory of Ginzburg and Landau

It is then natural to introduce the length $\xi(T)$ such that

$$-\frac{\hbar^2}{2m\alpha} = \xi^2(T) \tag{2.26}$$

which gives the range of variation of f (i.e. of ψ).

This new characteristic length is called the temperature dependent coherence length.

(b) Penetration depth in low field. London equation

Let us now examine the equation for the current. To first order in h, $|\psi|^2$ can be replaced by ψ_0^2, the value of $|\psi|^2$ in the absence of the field; i.e.

$$\boldsymbol{j} = \frac{e\hbar}{im}(\psi^* \nabla \psi - \psi \nabla \psi^*) - \frac{4e^2}{mc}\psi_0^2 \boldsymbol{A}. \tag{2.27}$$

Taking the curl of $\boldsymbol{j}$, one obtains

$$\operatorname{curl} \boldsymbol{j} = -\frac{4e^2}{mc}\psi_0^2 \boldsymbol{h} \tag{2.28}$$

which is equivalent to the London equation with the penetration depth[†]

$$\lambda(T) = \left[\frac{mc^2}{16\pi e^2 \psi_0^2}\right]^{1/2}. \tag{2.29}$$

The above expression for the penetration depth is the London expression (1.15) where n_s the number of superconducting electrons, is replaced by $4\psi_0^2$.

We have thus introduced the second characteristic length $\lambda(T)$, the temperature dependent penetration depth, which determines the range of variation of the magnetic field, i.e. of $\boldsymbol{h}$ and $\boldsymbol{A}$.

(c) Validity of the Ginzburg–Landau theory

Four points need to be discussed in relation to the validity of the Ginzburg–Landau equations.

(α) The first point is related to the validity of Landau's general theory of second order transitions. Landau assumes that the free energy can be expanded in powers of ψ and that the coefficients α' and β have the form

---

[†] The problem of a superconductor filling the half plane $z > 0$, could have been solved using the boundary condition (2.23). In this case the problem is the same as in the introduction, cf. section 1.4.

Reversible Properties

(2.2) and (2.3). None of these assumptions are, in general, valid, especially in the presence of short range order effects. However, in the case of superconductors, it can be shown theoretically that the Landau expansion is valid.

(β) As will be shown later (Chapter 5), the discussion of the microscopic derivation of the Ginzburg–Landau equations leads to conditions of validity which are identical to the conditions that will be derived in (γ) and (δ).

(γ) ψ must be a slowly varying function over distances of the order of ξ_0. A necessary condition for the validity of the theory is, therefore,

$$\xi(T) \gg \xi_0 . \tag{2.30a}$$

Using (2.26), the expression (2.20) for α, obtained by Gorkov from the microscopic theory in the case of pure superconductors shows that

$$\xi(T) = 0{\cdot}74 \left[\frac{T_0}{T_0-T}\right]^{1/2} \xi_0 \tag{2.31}$$

and the condition (2.30) is equivalent to

$$\frac{T_0-T}{T_0} \ll 1 , \tag{2.30b}$$

i.e. the temperature must be close to T_0, the critical temperature in zero field.

(δ) Let us now examine the variation of the field $\mathbf{h}$.

The Ginzburg–Landau equations yield a local relation between the current and the vector potential. It has already been mentioned in the introduction (sections 1.5 and 1.6) that the general relation cannot be local. Actually for a constant $|\psi|$ and a small value of $\mathbf{h}$, the current $\mathbf{j}(\mathbf{r})$ depends on $\mathbf{A}(\mathbf{r}')$ for distances $|\mathbf{r}-\mathbf{r}'| \sim \xi_0$ in a pure superconductor. The local approximation will be valid only if $\mathbf{h}$ and $\mathbf{A}$ are slowly varying functions over distances of the order of ξ_0. In order for this to be true, it is necessary for

$$\lambda(T) \gg \xi_0 . \tag{2.32a}$$

Comparing expressions (2.29) for $\lambda(T)$, (1.19) for $\lambda_L(0)$ (the London penetration depth at $T=0$), and the value of α and β obtained by Gorkov we get†

$$\lambda(T) = \frac{1}{\sqrt{2}} \lambda_L(0) \left(\frac{T_0}{T_0-T}\right)^{1/2} \gg \xi_0 \tag{2.32b}$$

which is equivalent to

$$\frac{T_0-T}{T_0} \ll \left[\frac{\lambda_L(0)}{\xi_0}\right]^2 . \tag{2.32c}$$

† For the derivation of this expression see Chapter 5.

This again indicates that the temperature must be close to T_0.

There are two cases to be considered:

If $\lambda_L(0) \ll \xi_0$, the superconductor belongs to type I, and the stronger limitation is given by (2.32c).

If $\lambda_L(0) \gg \xi_0$, the superconductor belongs to type II, and the stronger limitation is now given by (2.30b). In this case, the temperature range over which the theory is valid is much wider than for type I superconductors.

Nevertheless, the Ginzburg and Landau theory can be used to describe type I as well as type II materials (in the temperature range close to T_0) and in the next paragraph it will be used to clarify the distinction between the two types of superconductivity.

2.3. The Ginzburg–Landau parameter $\kappa(T)$

In the foregoing section we have defined the two characteristic lengths $\xi(T)$ and $\lambda(T)$. According to Gorkov's derivation they vary as $(T_0 - T)^{-1/2}$ for $T \to T_0$. However, their ratio

$$\kappa(T) = \frac{\lambda(T)}{\xi(T)} \tag{2.33}$$

remains finite for $T = T_0$.

The dimensionless parameter $\kappa(T)$ may be related to the critical field $H_c(T)$ using the relations $\psi_0^2 = -\alpha/\beta$ and $H_c^2 = 4\pi\alpha^2/\beta$.

It is easily found that

$$\kappa(T) = 2\sqrt{2}\,\frac{e}{\hbar c}\,\lambda^2(T)\,H_c(T). \tag{2.34}$$

According to (2.34) $\kappa(T)$ can be determined by measuring the penetration depth in low field $\lambda(T)$ and the thermodynamic critical field $H_c(T)$. In practice, however, the best values of $\kappa(T)$ are obtained from supercooling data as will be shown in the next chapter.

Using the values of Gorkov for $\lambda(T)$ and $\xi(T)$

$$\kappa(T) \simeq \frac{\lambda_L(0)}{\xi_0}. \tag{2.35}$$

Hence one is led to expect that the Ginzburg–Landau parameter κ should be related in some way to the distinction between type II and type I superconductors. It must be emphasized that Ginzburg and Landau, in their original paper, were not aware of the existence of the coherence length ξ_0, and they defined the parameter κ by the expression†

$$\kappa(T) = \frac{mc}{2e\hbar}\left(\frac{\beta}{2\pi}\right)^{1/2}. \tag{2.36}$$

† Of course, with e instead of $2e$.

Reversible Properties

The Ginzburg–Landau equations are very often written in the literature in a form which introduces only the following dimensionless quantities:

$$\psi = \psi_0 f; \quad \mathbf{r} = \lambda(T)\boldsymbol{\rho}; \quad \frac{2e}{\hbar c}\xi(T)\mathbf{A} = \mathcal{A}; \quad \frac{2e}{\hbar c}\xi(T)\lambda(T)\mathbf{h} = \mathbf{h}.$$
(2.37)

In this notation it is seen that $f = 1$ in the superconducting state and zero in the normal state and that the reduced critical field is

$$h_c = \left(\frac{4\pi\alpha^2}{\beta}\right)^{1/2} \frac{2e}{\hbar c}\xi(T)\lambda(T) = \frac{1}{\sqrt{2}}.$$
(2.38)

Equations (2.18) and (2.19) are then transformed into

$$\left(i\frac{\nabla}{\kappa} + \mathcal{A}\right)^2 f = f - |f|^2 f$$
(2.39)

and

$$-\operatorname{curl}\operatorname{curl}\mathcal{A} = -\operatorname{curl}\mathbf{h} = |f|^2\mathcal{A} + \frac{i}{2\kappa}(f^*\nabla f - f\nabla f^*)$$
(2.40)

to which is added the Maxwell equation†

$$\mathbf{h} = \operatorname{curl}\mathcal{A}.$$
(2.41)

From these three expressions two coupled equations can be deduced introducing only f and $\mathbf{h}$. This can be most simply done by writing f as

$$f = f_0 e^{i\phi},$$
(2.42)

where f_0 is the modulus and ϕ the phase of f, and the vector potential as

$$\mathcal{A} = \mathcal{A}_0 + \frac{\nabla\phi}{\kappa}.$$
(2.43)

The gradient of any scalar function can be added to a vector potential. However, in this case it is the gradient of the phase which is being added so that this transformation is only valid for simply connected geometries. Equations (2.39) – (2.41) become

$$\left(i\frac{\nabla}{\kappa} + \mathcal{A}_0\right)^2 f_0 = f_0 - f_0^3,$$
(2.44)

$$-\operatorname{curl}\operatorname{curl}\mathcal{A}_0 = -\operatorname{curl}\mathbf{h} = f_0^2 \mathcal{A}_0,$$
(2.45)

$$\mathbf{h} = \operatorname{curl}\mathcal{A}_0.$$
(2.46)

† In (2.39) and (2.40) the derivatives are taken with respect to $\boldsymbol{\rho}$.

Actually, the expression (2.44) depends only on $\mathcal{A}_0^2$ because the cross-term is

$$\frac{i}{f_0}\frac{\nabla}{\kappa} \cdot (f_0^2 \mathcal{A}_0) = -\frac{i}{f_0}\frac{\nabla}{\kappa} \cdot (\text{curl curl } \mathcal{A}_0) = 0.$$

The elimination of $\mathcal{A}_0$ is then very simple and the equations become

$$-\frac{\nabla^2}{\kappa^2}f_0 + \frac{1}{f_0^3}(\text{curl } h)^2 = f_0 - f_0^3, \tag{2.47}$$

$$f_0^2 h = \frac{2}{f_0} \text{grad} f_0 \times \text{curl } h - \text{curl curl } h. \tag{2.48}$$

In deriving (2.47) and (2.48) use is made of the well-known properties:

$$\text{div} \cdot \text{curl} = 0 \quad \text{and} \quad \text{curl} \times \text{grad} = 0.$$

Expressions (2.47) and (2.48) are two coupled equations which depend only on the field and the modulus of the order parameter. They are particularly suitable for one-dimensional problems and we shall use them in the derivation of the surface energy.

2.4. The surface energy

In principle the Ginzburg–Landau equations allow the order parameter the field and hence the currents to be calculated. However, these equations are non-linear and the calculations are rather tedious and in general purely numerical. Nevertheless, it is possible to draw from them a number of qualitative conclusions without calculating ψ or h explicitly. In some other cases, i.e. in the case of linear problems and for some special values of the parameter $\kappa(T)$ the complete solution can be found relatively easily. One of these special cases is related to the concept of surface energy. This concept was introduced by London who noted that the total exclusion of the field (Meissner effect) would not lead to a state of lowest energy unless a boundary energy exists. Indeed, the exclusion of the external field H increases the energy of the superconductor by $H^2/8\pi$ per unit volume. It is then expected that, for a suitably shaped superconductor, it will be energetically favourable for the bulk of the material to divide up into alternate normal and superconducting layers. As the field is expelled from the superconducting part the creation of such layers needs an interface surface energy σ_{ns} per unit surface whose magnitude is such that its contribution exceeds the gain in magnetic energy.

There are two important different behaviours of superconductors with respect to the surface energy, and this difference is the basis for the classification of superconductors into the two main types that have been mentioned.

Reversible Properties

In the Ginzburg–Landau theory the concept of surface energy is simply introduced by solving the following problem.

Consider an infinite sample (see Fig. 2.1) which has the following boundary conditions:

for $z = -\infty$ we want the sample to be normal and we impose

$\psi = 0$ and $h = H_c$,

for $z = +\infty$ the sample is superconducting so that

$\psi = \psi_0$ and $h = 0$.

The order parameter ψ and the field h will vary gradually, and a barrier is therefore produced. The surface energy will then be the difference between the energy F_p of the above situation and the condensation energy $\int(H_c^2/8\pi)dv$ that would appear if the sample were entirely superconducting. According to (2.11) the free energy F_p is given by

$$F_p = \int dv \left\{ \alpha |\psi|^2 + \frac{\beta}{2}|\psi|^4 + \frac{1}{2m}\left|\left(-i\hbar\nabla - \frac{2eA}{c}\right)\psi\right|^2 + \frac{h^2}{8\pi} \right\}.$$

(2.49)

This expression does not take into account the reduction of magnetic energy due to the penetration of the field. As this is a one-dimensional problem the contribution may be written as

$$-MH_c = -\frac{h - H_c}{4\pi}H_c,$$

(2.50)

where h is a function of the variable z only.

The surface energy per unit area will then be

$$\sigma_{ns} = \frac{F_p - \int(H_c^2/8\pi)dv}{S} = \int_{-\infty}^{+\infty} dz \left\{ \alpha|\psi|^2 + \frac{\beta}{2}|\psi|^4 + \right.$$

$$\left. + \frac{1}{2m}\left|\left(-i\hbar\nabla - \frac{2eA}{c}\right)_z\psi\right|^2 + \frac{h^2}{8\pi} - \frac{hH_c}{4\pi} + \frac{H_c^2}{4\pi} - \frac{H_c^2}{8\pi} \right\} =$$

$$= \int_{-\infty}^{+\infty} dz \left\{ \alpha|\psi|^2 + \frac{\beta}{2}|\psi|^4 + \frac{\hbar^2}{2m}\left(\frac{d\psi}{dz}\right)^2 + \right.$$

$$\left. + \frac{2e^2}{mc^2}A^2|\psi|^2 + \frac{h^2}{8\pi} + \frac{H_c^2}{8\pi} - \frac{hH_c}{4\pi} \right\}.$$

(2.51)

Theory of Ginzburg and Landau

If the reduced variables introduced in the last paragraph are used [see (2.37)], (2.47) and (2.48) can be written in the form

$$-\frac{1}{\kappa^2}\frac{d^2 f}{dz^2} + \frac{1}{f^3}\left(\frac{dh}{dz}\right)^2 = f - f^3, \tag{2.52}$$

$$f^2 h = -\frac{2}{f}\frac{df}{dz}\frac{dh}{dz} + \frac{d^2 h}{dz^2}, \tag{2.53}$$

$$\mathscr{A} f^2 = \frac{dh}{dz}, \tag{2.54}$$

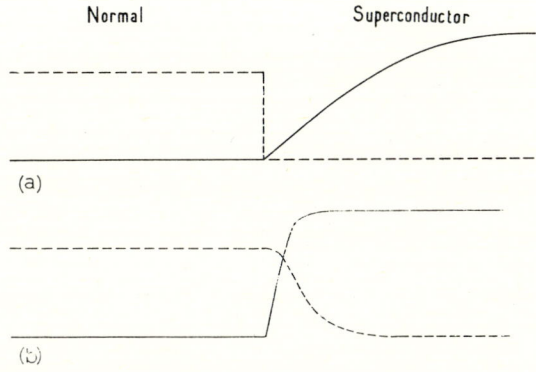

Fig. 2.1. Variation of f (full line) and h (dashed line) with position: (a) for $\kappa \ll 1$, (b) for $\kappa \gg 1$.

where h and $\mathscr{A}$ are the moduli of the reduced field and of the reduced vector potential and f is the modulus of the reduced order parameter. It is clear that these quantities are functions of z alone and that f varies from zero to 1 while h varies from $h_c = 1/\sqrt{2}$ to zero. The surface energy will take the following form:

$$\sigma_{ns} = \frac{H_c^2}{4\pi}\left(\frac{mc^2\beta}{16\pi e^2|\alpha|}\right)^{1/2}\int_{-\infty}^{+\infty} dz\left\{\frac{1}{2}(1-f^2)^2 + \right.$$

$$\left. + \frac{1}{f^2}\left(\frac{dh}{dz}\right)^2 + \frac{1}{\kappa^2}\left(\frac{df}{dz}\right)^2 + h^2 - \sqrt{2}h\right\}. \tag{2.55}$$

Using (2.52) this can be written as

$$\sigma_{ns} = \frac{H_c^2}{8\pi} 2\lambda(T)\int_{-\infty}^{+\infty} dz\left\{\frac{1}{2}(1-f^4) + h^2 - \sqrt{2}h + \frac{1}{\kappa^2}\frac{d}{dz}\left(f\frac{df}{dz}\right)\right\}$$

$$= 2\lambda(T)\frac{H_c^2}{8\pi}\int_{-\infty}^{+\infty} dz\left\{\frac{1}{2}(1-f^4) + h^2 - \sqrt{2}h\right\}, \tag{2.56}$$

Reversible Properties

where account is taken of the boundary conditions

$$f = 0 \text{ for } z = -\infty$$

$$\frac{df}{dz} = 0, \ f = 1 \text{ for } z = +\infty. \tag{2.57}$$

It is convenient, and customary, to introduce the length

$$\delta = 2I\lambda(T), \tag{2.58}$$

where I is the value of the integral

$$I = \int_{-\infty}^{+\infty} dz \left\{ \frac{1}{2}(1 - f^4) + h^2 - \sqrt{2}h \right\} \tag{2.59}$$

so that the surface energy is given by

$$\sigma_{ns} = \delta \frac{H_c^2}{8\pi}. \tag{2.60}$$

The literature usually quotes the values of σ_{ns} in terms of the parameter δ. The values of σ_{ns} (i.e. of I) will now be calculated in two limiting cases.
(a) $\kappa \ll 1$

If $\kappa \ll 1$, the penetration depth $\lambda(T)$ is very small in comparison to $\xi(T)$ and it can be assumed that h is zero as soon as f is different from zero. Equation (2.52) yields

$$-\frac{1}{\kappa^2}\frac{d^2f}{dz^2} = f - f^3. \tag{2.61}$$

i.e.

$$-\frac{1}{2\kappa^2}\left(\frac{df}{dz}\right)^2 = \frac{f^2}{2} - \frac{f^4}{4} + c. \tag{2.62}$$

The constant c is obviously equal to $-\frac{1}{4}$ since, for $f = 1$, we impose $df/dz = 0$ so that

$$\frac{1}{\kappa^2}\left(\frac{df}{dz}\right)^2 = \frac{1}{2}(1 - f^2)^2. \tag{2.63}$$

The convenient solution is

$$f = 0 \qquad h = \frac{1}{\sqrt{2}} \quad \text{for } z < 0$$

$$f = \tanh\frac{z\kappa}{\sqrt{2}} \quad h = 0 \qquad \text{for } z > 0 \tag{2.64}$$

(see Fig 2.1a) and the integral is seen to be

$$I = \frac{1}{2}\int_0^{+\infty}(1-f^4)\,dz = \frac{\sqrt{2}}{2\kappa}\int_0^1(1+f^2)\,df = \frac{2}{3}\frac{\sqrt{2}}{\kappa} \qquad (2.65)$$

and the surface energy is

$$\sigma_{ns} = \frac{4\sqrt{2}}{3}\frac{\lambda(T)}{\kappa(T)}\frac{H_c^2}{8\pi} = \frac{4\sqrt{2}}{3}\xi(T)\frac{H_c^2}{8\pi}. \qquad (2.66)$$

Hence

$$\delta = 1\cdot 89\,\xi(T) \qquad (2.67)$$

a result first obtained by Ginzburg and Landau.
Thus for $\kappa \ll 1$ the surface energy is positive.
(b) $\kappa \gg 1$
In this case the term $(1/\kappa^2)(d^2f/dz^2)$ in (2.52) can be neglected so that

$$\left(\frac{dh}{dz}\right)^2 = f^4(1-f^2). \qquad (2.68)$$

As h must decrease with increasing z,

$$\frac{dh}{dz} = -f^2(1-f^2)^{1/2}. \qquad (2.69)$$

From (2.53) we obtain

$$h = \frac{d}{dz}\frac{1}{f^2}\frac{dh}{dz} = -\frac{d}{dz}(1-f^2)^{1/2}. \qquad (2.70)$$

Here f obeys the following differential equation:

$$(1-f^2)^{1/2} - (1-f^2)^{3/2} = \frac{d^2}{dz^2}(1-f^2)^{1/2} \qquad (2.71)$$

or

$$\frac{d^2u}{dz^2} = u - u^3. \qquad (2.72)$$

where $u = (1-f^2)^{1/2}$, $\qquad (2.73)$

i.e.

$$\left(\frac{du}{dz}\right)^2 = u^2\left(1 - \frac{u^2}{2}\right) + c. \qquad (2.74)$$

The constant of integration is obviously equal to zero and this can be written:

$$\frac{du}{dz} = -u\left(1 - \frac{u^2}{2}\right)^{1/2} \qquad (2.75)$$

since du/dz must be negative.

Reversible Properties

The integral is readily seen to be

$$I = \int_{-\infty}^{+\infty} dz \left\{ 2u^2 \left(1 - \frac{u^2}{2}\right) - \sqrt{2}u \left(1 - \frac{u^2}{2}\right)^{1/2} \right\}$$

$$= \int_0^1 \left[2u \left(1 - \frac{u^2}{2}\right)^{1/2} - \sqrt{2} \right] \frac{du}{dz} dz = -\frac{4}{3}(\sqrt{2} - 1) \quad (2.76)$$

and the surface energy is†

$$\sigma_{ns} \simeq -\frac{8}{3}(\sqrt{2} - 1)\lambda(T)\frac{H_c^2}{8\pi} \simeq -\lambda(T)\frac{H_c^2(T)}{8\pi}. \quad (2.77)$$

Thus for $\kappa(T) \gg 1$ the surface energy is negative.

The two preceding calculations yield the interesting result that for type I superconductors $[\kappa \ll 1, \lambda(T) \ll \xi(T)]$ the surface energy is positive, while for type II superconductors $[\kappa \gg 1, \lambda(T) \gg \xi(T)]$, this energy is negative. One therefore expects very different behaviour of the two kinds of superconductors. The study of these behaviours, and more particularly of type II, will be the subject of the next chapter. The boundary between type II and I superconductivity can be defined by finding the value of κ which corresponds to a surface energy equal to zero.

The surface integral is given by

$$I = \int_{-\infty}^{+\infty} dz \left[\frac{1}{2}(1 - f^4) + h^2 - \sqrt{2}h \right] \quad (2.59)$$

and it is clear that this integral is zero if

$$h = \frac{1}{\sqrt{2}}(1 - f^2) \quad (2.78)$$

In this case, (2.52) and (2.53) take the form

$$-\frac{1}{\kappa^2}\frac{d^2f}{dz^2} + \frac{2}{f}\left(\frac{df}{dz}\right)^2 = f - f^3, \quad (2.79)$$

$$-\frac{2}{\sqrt{2}}f\frac{d^2f}{dz^2} + \frac{2}{\sqrt{2}}\left(\frac{df}{dz}\right)^2 = \frac{f^2}{\sqrt{2}}(1 - f^2). \quad (2.80)$$

These two equations are identical for

$$\kappa = \frac{1}{\sqrt{2}}. \quad (2.81)$$

† Using the London equation instead of the Ginzburg–Landau equation it is found that $\sigma_{ns} = -\lambda_L H_c^2/8\pi$ where λ_L is the London penetration depth.

This is the condition for zero surface energy.

The relation between the parameter κ and the classification of superconductors is clear and, following Abrikosov (1952) the distinction between type I and type II may be made as follows:

Type I superconductivity: $\kappa < 1/\sqrt{2}$, positive surface energy.
Type II superconductivity: $\kappa > 1/\sqrt{2}$, negative surface energy.

2.5. Laminar model and filamentary structure

For the London superconductors the surface energy associated with the interface between a normal and a superconducting region is negative. It is therefore energetically favourable for a superconductor in the presence of a field to divide up into a large number of normal and superconducting regions (the mixed state). Thus the contribution of the surface energy to the free energy of the system is important.

In type I superconductors, on the contrary, the division of the sample into normal and superconducting regions (intermediate state) is less favourable, because the surface energy is positive, and the contribution of the interface energy to the free energy of the system is negligible. In type I superconductors this division occurs only for a suitably shaped specimen with a demagnetization coefficient D, for fields H such that

$$(1-D)H_c < H < H_c \tag{2.82}$$

The simplest picture which was used to describe the mixed state of type II superconductors, by analogy with type I superconductivity, is to consider a periodic array of parallel alternately normal and superconducting *laminae* (Goodman, 1961). In the low induction limit, the volume of the normal region will be small and in order to increase the surface/volume ratio of the normal region a model in which normal laminae of width $2\xi(T)$ separated by a distance equal to $d - 2\xi(T)$ is used (Fig. 2.2).

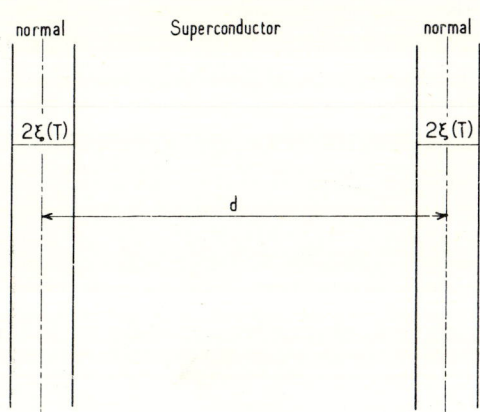

Fig. 2.2.

Reversible Properties

The case $\lambda(T) \gg \xi(T)$ (high κ) and field $h(x)$ parallel to $0z$, the laminae being perpendicular to the x axis, will be considered here.

In the superconducting region the field obeys the London equation:

$$\lambda^2(T)\frac{d^2h}{dx^2} = h, \qquad (2.83)$$

i.e.

$$h = H_n \frac{\cosh x/\lambda_L}{\cosh d/2\lambda_L}, \qquad (2.84)$$

where H_n is the field in the normal region.

The free energy will consist of three contributions: the magnetic energy; the kinetic energy of the superducting electrons; and the energy of formation of the normal regions.

The first two terms are (per unit volume):

$$F_1 = \frac{1}{\Omega}\int d\mathbf{r}\left(\frac{h^2}{8\pi} + \frac{1}{2}mn_sv^2\right) = \frac{1}{\Omega}\int\frac{d\mathbf{r}}{8\pi}[h^2 + \lambda^2(T)(\mathrm{curl}\,\mathbf{h})^2]$$

$$= \frac{2}{d}\int_0^{d/2}dx\,\frac{h^2 + \lambda^2(T)(dh/dx)^2}{8\pi}. \qquad (2.85)$$

The third term is

$$F_2 = \frac{2\xi(T)}{d}\frac{H_c^2}{8\pi} \qquad (2.86)$$

so that

$$F = F_1 + F_2 = \frac{H_n^2}{8\pi}\frac{\tanh d/2\lambda_L}{d/2\lambda_L} + \frac{H_c^2}{8\pi}\frac{2\lambda_L}{\kappa d}, \qquad (2.87)$$

where the Ginzburg–Landau parameter

$$\kappa = \frac{\lambda(T)}{\xi(T)}$$

has been introduced.

The thermodynamic potential is

$$G = F - \frac{BH}{4\pi}, \qquad (2.88)$$

where B is the induction, i.e. the mean value of the field h, and H the applied field.

$$B = \frac{2}{d}H_n\int_0^{d/2}dx\,\frac{\cosh x/\lambda_L}{\cosh d/2\lambda_L} = H_n\frac{\tanh d/2\lambda_L}{d/2\lambda_L}. \qquad (2.89)$$

Theory of Ginzburg and Landau

In equilibrium the thermodynamic potential must be a minimum with respect to H_n,

$$\frac{\partial G}{\partial H_n} = 0 \qquad (2.90)$$

so that

$$H_n = H \qquad (2.91)$$

and

$$G = \frac{1}{4\pi d/\lambda_L}\left[-H^2 \tanh\frac{d}{2\lambda_L} + \frac{H_c^2}{\kappa}\right]. \qquad (2.92)$$

for $H < H_c/\kappa^{1/2}$ G is a minimum for $d = \infty$ (Meissner effect: the field is expelled);
for $H > H_c/\kappa^{1/2}$ G is a minimum for a finite value of d.

Thus using this model a field $H_{c1}^L = H_c/\kappa^{1/2}$ is predicted at which the fields begins to penetrate.

Another way of increasing the surface/volume ratio is to consider a periodic two-dimensional set of normal *filaments* of diameter 2ξ separated by distances of order d. It will be seen in the next chapter that in this case the Meissner effect is perfect for fields H smaller than

$$H_{c_1}^F \simeq \frac{H_c}{\kappa}\log\kappa: \qquad (2.93)$$

for $\kappa \gg 1$

$$H_{c1}^F < H_{c1}^L;$$

for

$$H_{c1}^F < H < H_{c1}^L;$$

$$G_F < G_{\text{Meissner}},$$

$$G_L > G_{\text{Meissner}}.$$

Hence the filamentary structure is more favourable than the laminar one.

It can be shown that this conclusion is true for all values of the field up to H_{c2}. The laminar model has thus been abandoned in favour of the filamentary structure which gives a better account of the properties of type II superconductors and which has recently received direct experimental confirmation (Cribier *et al.*, 1964).

Bibliography of Chapter 2

ABRIKOSOV, A. A. (1952), *Dokl. Akad. Nauk. SSSR*, **86** 489.
CRIBIER, D., JACROT, B., RAO, L. M. and FARNOUX, B. (1964), *Phys. Letters*, **9,** 106 and Private Communications.
DE GENNES, P. G., (1964) International Conference on the Science of Superconductivity, Hamilton, *Rev. Mod. Phys.* **36,** 225.
GINZBURG, V. L. and LANDAU, L. D. (1950), *Zh. Eksperim. i Teor. Fiz.* **20,** 1064.
GOODMAN, B. B. (1961), *Phys. Rev. Letters*, **6,** 597.
GORKOV, L. P. (1959a), *Zh. Eksperim. i Teor. Fiz.* **36,** 1918 (English translation: (1959), *Soviet Phys. JETP*, **9,** 1364): (1959b), **37,** 1407 (English translation: (1960), *Soviet Phys. JETP*, **10,** 998).
LANDAU, L. D. (1937), *Phys. Z. Sowjet U.* **11,** 545.
LANDAU, L. D. and LIFSHITZ, E. M. (1958), *Course of Theoretical Physics*, Vol. 5, *Statistical Physics*, Pergamon Press, p. 430 and ff.

CHAPTER 3

Magnetic Properties of Type II Superconductors the Vortex Line Structure

3.1. Introduction

In 1952, Abrikosov, using the Ginzburg–Landau theory of superconductivity, conceived a theory explaining the magnetic properties of type II superconductors. This theory proposed a structure of the mixed state consisting of a two-dimensional periodic array of vortex lines (filaments), i.e. schematically, the core of each filament contains magnetic flux and is surrounded by a vortex of superconducting electrons. This was a completely new concept and the publication of Abrikosov's paper was delayed until 1957. It must be emphasized that, even in 1957, the importance of Abrikosov's contribution was not fully recognized, mainly because of the lack of experimental data.

The recent developments in the study of type II superconductors have, in general, confirmed this theory: the vortex concept is used extensively to account for the properties. Cribier *et al.* (1964) have observed the periodic array of filaments directly, using neutron diffraction techniques, in niobium and in lead–bismuth alloys.

This chapter will be devoted to the study of infinite samples, leaving aside the problem of surface superconductivity. It will follow Abrikosov's paper in its main developments, and will, thus, make extensive use of Ginzburg–Landau equations. The principal features of the vortex structure will be emphasized, and a short account of the recent experimental work will be given.

3.2. Onset of superconductivity. The critical field H_{c2}

Let us consider a superconductor in a uniform external field $\boldsymbol{H}$. If the magnitude of $\boldsymbol{H}$ is sufficiently high, superconductivity is destroyed and the microscopic field $\boldsymbol{h}$ is constant and equal to $\boldsymbol{H}$ throughout the specimen. Let us now decrease the magnitude of $\boldsymbol{H}$. At a certain value H_{c2} superconductiv-

Reversible Properties

ity begins to appear in the sample. This critical field H_{c2} is different from the thermodynamic critical field H_c. In order to calculate this field, Ginzburg and Landau noted that, close to H_{c2} the magnitude $|\psi|$ of the order parameter ψ is very small so that the first Ginzburg–Landau equation (2.18) may be linearized and written as

$$\frac{1}{2m}\left[-i\hbar\nabla - \frac{2e}{c}A\right]^2 \psi = -\alpha\psi. \qquad (3.1)$$

Moreover, (2.19) shows that the correction due to the supercurrents is of the order of $|\psi|^2$, so that, to a first approximation the microscopic field h can be taken equal to the applied field H, i.e.

$$\operatorname{curl} A = H. \qquad (3.2)$$

Equation (3.1) is formally identical to a Schrödinger equation for the motion of a particle of charge $2e$ and mass m in a constant magnetic field H. It is well known (Landau, 1930) that such a particle moves along the direction of the field H with a constant velocity V_z (V_z can take any value) and has a circular motion in the plane perpendicular to H with the frequency

$$\omega = \frac{2eH}{mc}. \qquad (3.3)$$

If $|\psi|$ is taken to be a bounded function, the eigenvalues of (3.1) (which are to be identified with $-\alpha$) are:

$$-\alpha_n = \frac{mV_z^2}{2} + \left(n + \frac{1}{2}\right)\hbar\omega \qquad (3.4)$$

so that the critical field H is given by

$$(2n+1)\frac{\hbar eH}{mc} = -\alpha_n - \frac{mV_z^2}{2}. \qquad (3.5)$$

Obviously, the field of interest is the highest field for which superconductivity begins to occur. This corresponds to $n = 0$ and $V_z = 0$ in (3.5) and the field, obtained by setting $\alpha = \alpha_0$, is

$$H_{c2} = -\alpha\frac{mc}{\hbar e}. \qquad (3.6)$$

Using the expressions (2.7)

$$\frac{\alpha^2}{2\beta} = \frac{H_c^2}{8\pi}$$

and (2.36)

$$\kappa = \frac{mc}{2e\hbar}\left(\frac{\beta}{2\pi}\right)^{1/2},$$

a relation is obtained between the critical field H_{c2} and the thermodynamic critical field H_c:

$$H_{c2} = \kappa \sqrt{2} H_c(T). \qquad (3.7)$$

The preceding derivation of the expression for H_{c2} is valid whatever the value of κ, and it is seen that type I and type II superconductors behave quite differently:

(1) for type II superconductors ($\kappa > 1/\sqrt{2}$), superconductivity appears at a field $H_{c2} > H_c$. Actually, below H_{c2}, the specimen is in the mixed state;

(2) for type I superconductors ($\kappa < 1/\sqrt{2}$), the specimen may remain normal, in decreasing fields, down to a field $H_{c2} < H_c$: the normal phase is supercooled (metastable state);

(3) for $\kappa = 1/\sqrt{2}$, expression (3.7) yields, as expected, $H_{c2} = H_c$.

Expression (3.7) suggests a method of measuring the parameter κ, simply by determining the nucleation field H_n. However, this method does not yield the correct value of κ if one identifies H_n with H_{c2}. Expression (3.7) is derived for an infinite medium and does not take into account surface effects. It will be shown in Chapter 4 that, in a finite sample, the nucleation takes place on the surfaces parallel to the applied field at a nucleation field $H_{c3} > H_{c2}$; a superconducting sheath exists close to these surfaces, the rest of the specimen being normal. H_{c2} corresponds to the restoration of superconductivity in the whole sample.

The linearized equation (3.1) is a homogeneous equation. The norm of $|\psi|$ is not determined. In order to obtain the value of the maximum of $|\psi|$, just below H_{c2} it is necessary to use the complete equations. This problem will be investigated in the next section.

Each solution ψ_n of (3.1) corresponding to an eigenvalue $-\alpha_n$ is highly degenerate.

This property, which is shown in Appendix 3.1, will be used to obtain the Abrikosov vortex structure close to H_{c2} (see section 3.3c).

3.3. Structure of the mixed state

Let us now investigate the structure of a type II superconductor for applied fields H smaller than H_{c2}. The magnitude of the order parameter $|\psi|$ is expected to increase and the problem now involves the solution of the two complete Ginzburg–Landau equations and not only of the linearized form

Reversible Properties

(3.1). This is a difficult mathematical task, even on a purely numerical basis.[†] A number of conclusions can be drawn, however, in two limiting cases: the vicinity of H_{c2} and the vicinity of H_{c1}.

(a) Flux quantization. The flux quantum Φ_0

In 1950 F. London made the suggestion that the flux trapped by a superconducting ring is quantized. This is an important prediction because it foretells a situation in which one of the characteristic effects of quantum mechanics occurs on a macroscopic scale. For the sake of simplicity the theoretical prediction of the flux quantization will be made here within the Ginzburg–Landau frame. It must be kept in mind, however, that flux quantization is a general phenomenon not restricted to the vicinity of T_0.[‡]

It has already been pointed out in Chapter 2 that the order parameter ψ is not necessarily real. If one introduces the magnitude $|\psi|$ and the phase ϕ of this parameter, the second Ginzburg–Landau equation becomes

$$\boldsymbol{j} = \frac{2e\hbar}{m}|\psi|^2 \nabla\phi - \frac{4e^2}{mc}|\psi|^2 \boldsymbol{A} \tag{3.8a}$$

or

$$\boldsymbol{A} = -\frac{mc}{4e^2|\psi|^2}\boldsymbol{j} + \frac{c\hbar}{2e}\nabla\phi. \tag{3.8b}$$

Let us now calculate the line integral of $\boldsymbol{A}$ around a closed loop Γ; i.e.

$$\int_\Gamma \boldsymbol{A}\cdot d\boldsymbol{l} = \int_\Sigma \operatorname{curl}\boldsymbol{A}\cdot d\boldsymbol{\sigma} = \int_\Sigma \boldsymbol{h}\cdot d\boldsymbol{\sigma} = \Phi. \tag{3.9}$$

This line integral is equal to the flux of $\boldsymbol{h}$ through the loop. Using (3.8b), it is seen that

$$\Phi = \int_\Gamma \boldsymbol{A}\cdot d\boldsymbol{l} = -\frac{mc}{4e^2}\int_\Gamma \frac{\boldsymbol{j}}{|\psi|^2}\cdot d\boldsymbol{l} + \frac{c\hbar}{2e}\int_\Gamma \nabla\phi\cdot d\boldsymbol{l}. \tag{3.10}$$

The line integral $\int_\Gamma \nabla\phi\cdot d\boldsymbol{l}$ does not necessarily vanish, because the only general requirement is that the modulus of the order parameter ψ is a single valued function, i.e. the phase of ψ varies by $2\pi n$, where n is an integer (or zero), when we make a complete turn around Γ. Consequently, (3.10) predicts the *fluxoid* quantization

$$\Phi = n\frac{ch}{2e} - \frac{mc}{4e^2}\int_\Gamma \frac{\boldsymbol{j}}{|\psi|^2}\cdot d\boldsymbol{l}. \tag{3.11}$$

[†] Several attempts have been made in this direction by Marcus (1964) using a computer.

[‡] Flux quantization may be deduced from the microscopic theory. See, for example, De Gennes (1965).

Properties of Type II Superconductors

If the path of integration Γ is chosen to be a line where $\boldsymbol{j} \equiv 0$, or such that $\boldsymbol{j}$ is orthogonal to $d\boldsymbol{l}$, the relation becomes

$$\Phi = \int_\Sigma \boldsymbol{h} \cdot d\boldsymbol{\sigma} = n\Phi_0, \qquad (3.12)$$

where

$$\Phi_0 = \frac{ch}{2e} = 2 \times 10^{-7} \text{ gauss-cm}^2 \qquad (3.13)$$

is the flux quantum.† It is seen that the flux trapped in a superconductor is quantized and equal to $n\Phi_0$. This quantization is a general feature of superconductivity.

In a singly connected type I superconductor the Meissner effect is perfect and n is equal to zero. Flux quantization can be observed only in a multiply connected geometry, such as a superconducting ring. When the applied field is removed, flux equal to $n\Phi_0$ remains trapped. It must be emphasized that London (1950) predicted a flux quantum of 4×10^{-7} gauss-cm² (i.e. $2\Phi_0$). The actual value given by (3.13) is a consequence of the existence of the Cooper pairs. Deaver and Fairbank (1961), as well as Doll and Nabauer (1961), have measured Φ_0 in a ring shaped layer, thus providing experimental evidence for the existence of Cooper pairs.‡

In type II superconductors the situation is somewhat different. Flux quantization is expected to exist even in singly connected geometries because, in the mixed state, superconducting regions surround lines of force (where the field penetrates) and form a multiconnected system of filaments. This is a situation somewhat equivalent to the annulus case and the total flux enclosed can take only discrete values of the form (3.13).

Moreover, in the vortex structure each filament consists of a core, the radius of which is of the order of $\xi(T)$, while annular currents $\boldsymbol{j}$ encircle the filament and screen out the field for $r \geq \lambda(T)$. The preceding arguments lead to flux quantization inside *each* filament. The problem remains: how many quanta are there in each filament? In Appendix III.4 it is demonstrated that close to H_{c2} each filament carries one flux quantum whatever the value of κ. This is true also for an isolated filament, i.e. close to H_{c1}. It is then reasonable to assume that the situation will remain the same all over the field range (i.e. from H_{c1} to H_{c2}), for all values of $\kappa > 1/\sqrt{2}$.

This conclusion is plausible because, if each filament carries only one quantum of flux Φ_0, this leads to the state of maximum subdivision and this is favourable from the point of view of the surface energy. However this argument is not entirely convincing because one should make an energy balance between the formation energy of the filament and the gain in surface energy. That is to say, one should solve the Ginzburg–Landau equations

† Here $h = 2\pi\hbar$ is Planck's constant.
‡ At the time of these experiments it was expected that the flux quantum would be $2\Phi_0$ as predicted by London.

and calculate the free energy to get a definitive answer.† The experiments by Cribier *et al.* (1964) have shown that in a large range of fields and for κ values of the order of 1, there is still only one flux quantum per vortex line. It must also be emphasized that these experiments were performed at temperatures rather far from T_0, where the Ginzburg–Landau scheme is no longer valid.‡

It is interesting to write the fields H_c and H_{c2} in terms of Φ_0. Using expressions (2.7) for H_c and (3.6) for H_{c2} and expressions (2.26) and (2.36) for $\xi(T)$ and $\kappa(T)$, it is seen that

$$H_c = \frac{1}{2\pi\sqrt{2}} \frac{\Phi_0}{\lambda(T)\xi(T)}, \tag{3.14a}$$

$$H_{c2} = \frac{1}{2\pi} \frac{\Phi_0}{\xi^2(T)}. \tag{3.14b}$$

These expressions are often used in the literature.

Moreover, the flux quantization allows the induction B to be calculated in a type II superconductor. It is clear that the microscopic field $\boldsymbol{h}$ is parallel to the applied field $\boldsymbol{H}$, and if $0z$ is the axis parallel to $\boldsymbol{H}$, is independent of z. The mean value of $\boldsymbol{h}$, i. e. the induction, will be

$$B = \frac{1}{S}\int_S \boldsymbol{h}\cdot d\boldsymbol{\sigma} = \frac{n}{S}\Phi_0, \tag{3.15}$$

where S is the cross-section of the sample perpendicular to the applied field, and n the number of vortex lines.

(b) Isolated filament. The field for first penetration: H_{c1}

The field H_{c1}, the lowest field in which the mixed state exists, may be identified with the field at which the first filament appears. If $\boldsymbol{H}$ is the applied field and if $0z$ is parallel to $\boldsymbol{H}$, the problem has cylindrical symmetry and the microscopic field $\boldsymbol{h}$ as well as the magnitude of the order parameter $|\psi|$ will depend only on $|\boldsymbol{r}|$.

Using the dimensionless variables introduced in (2.37), the two Ginzburg–Landau equations may be written as

$$-\frac{1}{\kappa^2}\left(\frac{1}{\rho}\frac{d}{d\rho}\rho\frac{d}{d\rho}f_0\right) - \frac{1}{f_0^3}\left(\frac{d}{d\rho}h\right)^2 + f_0(1-f_0^2) = 0, \tag{3.16}$$

$$\frac{1}{\rho}\frac{d}{d\rho}\frac{\rho}{f_0^2}\frac{d}{d\rho}h = h. \tag{3.17}$$

---

† We have already mentioned that this is a formidable task. One can consider that the experimental determination by Cribier *et al.* is as good a proof as the numerical solution of the Ginzburg–Landau equations.

‡ These experiments will be discussed in more detail in section 3.5.

Properties of Type II Superconductors

These equations are obtained from (2.47) and (2.48), or by direct elimination of $\mathscr{A}_0$. h is the magnitude of the reduced field.

For an infinitely large superconductor, the following boundary conditions are added to (3.16) and (3.17). For $\rho \to \infty$, superconductivity is entirely restored so that

$$\rho \to \infty ; \quad f_0 = 1, \quad h = 0, \quad j = 0, \quad (3.18)$$

where $j = \text{curl } h$ is the current. This last condition reads:

$$-\frac{dh}{d\rho} = 0 \quad \text{for } \rho \to \infty \quad (3.19)$$

The boundary conditions (3.18) do not specify the solutions of (3.16) and (3.17) completely. Four conditions are needed and two extra conditions arise from the flux quantization. With the above reduced variables $\Phi_0 = 2\pi/\kappa$ and the flux quantization condition reads

$$\Phi = 2\pi \int_0^\infty h\rho \, d\rho = \frac{2\pi p}{\kappa}, \quad (3.20)$$

where p is an integer.

Using (3.17) it is readily seen that†

$$\Phi = 2\pi \left| \frac{\rho}{f_0^2} \frac{dh}{d\rho} \right|_0^\infty = -2\pi \left| \frac{\rho}{f_0^2} \frac{dh}{d\rho} \right|_{\rho=0} \quad (3.21)$$

so that three of the four boundary conditions are

$$\rho = 0; \quad \frac{1}{f_0^2} \frac{dh}{d\rho} = -\frac{p}{\kappa\rho} ; \quad p \text{ integer} \quad (3.22)$$

$$\rho = \infty ; \quad f_0 = 1 ; \quad h = 0 ; \quad \frac{dh}{d\rho} = 0. \quad (3.23)$$

It should be noted that $-(1/f_0^2)(dh/d\rho)$ is the modulus of the vector potential $\mathscr{A}_0$. For $\rho \to 0$, (3.16) takes the form

$$\frac{1}{\kappa^2} \left(\frac{1}{\rho} \frac{d}{d\rho} \rho \frac{d}{d\rho} f_0 \right) - \frac{p^2}{\kappa^2 \rho^2} f_0 + f_0(1 - f_0^2) = 0. \quad (3.24)$$

† It is shown in equation (3.27) (see below) that h varies as $K_0(\rho)$ the Hankel function of zero order, for $\rho \to \infty$, i.e. as $e^{-\rho}/\sqrt{\rho}$.

Reversible Properties

This equation is satisfied by regular solutions of the form

$$f_0 = c_p \rho^p + \ldots \tag{3.25}$$

$$h = h_p(0) - \frac{c_p \rho^{2p}}{2\kappa} + \ldots \tag{3.26}$$

The fourth boundary condition is thus $f_0 = 0$ for $\rho = 0$.

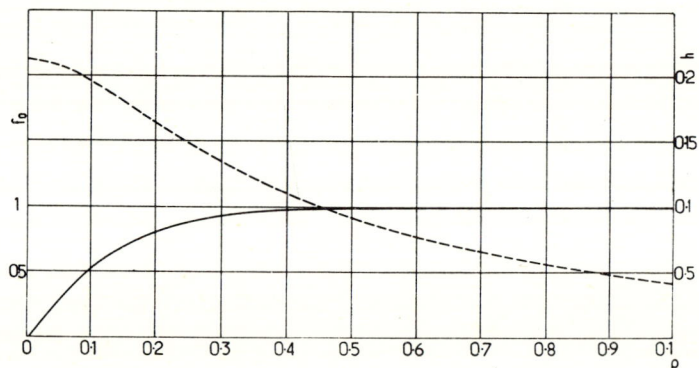

FIG. 3.1. The variation of the order parameter f_0 (full line) and of the field (dashed line) for an isolated vortex ($\kappa = 10$). The field reaches the $K_0(\rho)$ variation for $\rho \simeq 0.5$

c_p and $h_p(0)$ are constants to be determined by complete integration of (3.16) and (3.17). It is seen that for $\rho \to 0$, the order parameter for a vortex line with p flux quanta behaves as ρ^p. In particular, for a one quantum line f_0 depends linearly on ρ for $\rho \to 0$ (Fig. 3.1).

For $\rho \to \infty$, (3.17) yields

$$h = \frac{1}{\rho}\frac{d}{d\rho}\rho\frac{dh}{d\rho}. \tag{3.27}$$

The convenient solution is the Hankel function of imaginary argument of zero order $K_0(\rho)$ so that†

$$h = \alpha K_0(\rho) \tag{3.28}$$

where α is a constant which must also be determined by complete integration of the equations.

† See, for example, Morse and Feshbach, *Methods of Theoretical Physics* (McGraw-Hill, 1953), Chap. 10.

Properties of Type II Superconductors

Similarly,

$$\frac{dh}{d\rho} = \alpha \frac{dK_0(\rho)}{d\rho} = -\alpha K_1(\rho) \tag{3.29}$$

$K_1(\rho)$ is the Hankel function of imaginary argument of order 1.

For $\kappa \gg 1/\sqrt{2}$, the constant α can easily be determined, since the range of variation of h is much greater than the range of variation of f_0. (If δ is the range of variation of f_0, the range of variation of h is $\kappa\delta \gg \delta$.) The coefficient α will be determined by matching expressions (3.29) and (3.22).†

For $\rho \to \infty$:

$$\frac{1}{f_0^2} \frac{dh}{d\rho} = -\alpha K_1(\rho). \tag{3.30}$$

For $\rho \to 0$:

$$\frac{1}{f_0^2} \frac{dh}{d\rho} = -\frac{p}{\kappa\rho}. \tag{3.22}$$

As

$$K_1(\rho) \to \frac{1}{\rho} + \ldots \quad \text{for} \quad \rho \to 0,$$

it is easily seen that

$$\alpha = \frac{p}{\kappa}. \tag{3.31}$$

In order to find out the most stable situation, one must calculate the free energy. If $\mathscr{F}_p$ is the energy of the single line with p flux quanta, and if $\mathscr{F}_M$ is the magnetic energy, the formation of a filament will be favourable when

$$\mathscr{F}_p - \mathscr{F}_M \leq 0. \tag{3.32}$$

The magnetic energy is clearly given by‡

$$\mathscr{F}_M = 2h_e B = \frac{4\pi p}{\kappa} h_e, \tag{3.33}$$

where h_e is the applied field, and B has been calculated with the help of (3.15). The field of first appearance of a filament with p flux quanta is then

$$h_{c1}(p) = \frac{\kappa}{4\pi} \frac{\mathscr{F}_p}{p} \tag{3.34}$$

It is seen that the formation of a line with one quantum will be favourable

---

† This procedure is clearly valid only for high values of κ.
‡ All the energies are reduced and are calculated for a volume of height equal to 1. The actual energies are $F = (H_c^2/8\pi)\mathscr{F}$.

49

Reversible Properties

if $\mathscr{F}_p$ increases faster than p. The theoretical solution of this problem involves the solution of (3.16) and (3.17). This has not yet been done. However, for $\kappa \gg 1/\sqrt{2}$ the calculation of the energy $\mathscr{F}_p$ can be performed to the lowest order in κ^{-1}. Let us first write down the expression for the free energy of the line. The general Ginzburg–Landau energy is given by (2.48), i.e.

$$\mathscr{F} = \int dv \left\{ \frac{1}{2} - f_0^2 + f_0^4 + h^2 + \left| \left(-i\frac{\nabla}{\kappa} - \mathscr{A} \right) f \right|^2 \right\} \quad (3.35)$$

$$= \int dv \left\{ \frac{1}{2}(1 - f_0^4) + h^2 \right\}. \quad (3.36)$$

Expression (3.36) has been obtained by using the Ginzburg–Landau equations (2.39) and (2.40) and neglecting the surface integral which appears in the transformation. This is a general result not restricted to the case of cylindrical symmetry. For a single vortex line a further simplification of (3.36) may be obtained. It is shown in Appendix 3.2 that the free energy can be written as

$$\mathscr{F} = 2\pi \int_0^\infty \rho(1 - f_0^2)\, d\rho. \quad (3.37)$$

For high κ values the integral (3.37) can be calculated explicitly to first order in κ^{-1}. In the region in which f_0 is varying $(1/f_0^2)(dh/d\rho)$ is still equal to its boundary value $-p/\kappa\rho$, so that f_0 obeys (3.24). In the region $p/\kappa < \rho < p$, the convenient solution can be written as

$$f_0^2 = 1 - \frac{p^2}{\kappa^2 \rho^2}. \quad (3.38)$$

Provided that $\log \kappa \gg 1$, the main contribution to $\mathscr{F}$ comes from this domain; the free energy is thus

$$\mathscr{F}_p \simeq 2\pi \int_{p/\kappa}^{p} \frac{p^2}{\kappa^2 \rho}\, d\rho = \frac{2\pi}{\kappa^2} p^2 \log \kappa \quad (3.39)$$

and

$$h_{c1}(p) = \frac{p}{2\kappa} \log \kappa. \quad (3.40)$$

It is seen that the lowest field is obtained for $p = 1$, i.e. the most stable solution for $\kappa \gg 1/\sqrt{2}$ is the formation of vortex lines with one flux quantum per line. A numerical integration by Abrikosov (1957) has shown that in the high κ limit, the field h_{c1} is given by

$$h_{c1} = \frac{1}{2\kappa}(\log \kappa + 0{\cdot}08). \quad (3.41)$$

Properties of Type II Superconductors

and the actual first penetration field is

$$H_{c1} = \frac{H_c}{\sqrt{2}\kappa}(\log \kappa + 0.08) = \frac{\Phi_0}{4\pi\lambda^2(T)}\left[\log\left(\frac{\lambda(T)}{\xi(T)}\right) + 0.08\right]$$

(3.42)

The field in the centre of the filament can easily be computed.

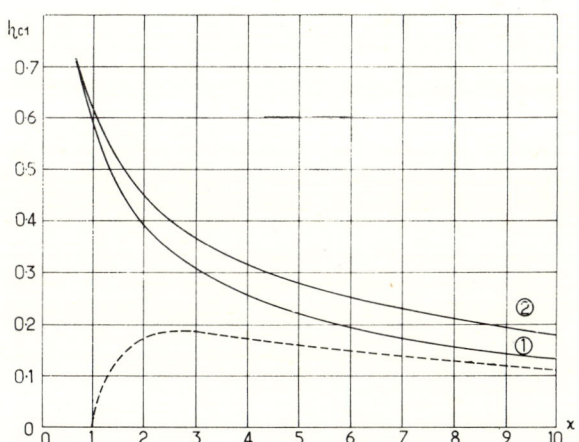

FIG. 3.2. h_{c1} as function of κ for one quantum ① and two quanta per line ②. The curve labelled ② was also obtained by Harden and Arp (1963). The dashed curve corresponds to $\log \kappa/2\kappa$

In Appendix 3.3, it is shown that the line free energy for high values of κ is, to lowest order in κ^{-1},

$$\mathscr{F} \simeq \frac{2\pi}{\kappa} h(0).$$

(3.43)

By comparing (3.43) and (3.34) one can see that the field $h(0)$ is about twice the field h_{c1}. The numerical integration by Abrikosov had given

$$h(0) = \frac{1}{\kappa}(\log \kappa - 0.18).$$

(3.44)

In conclusion, it is seen that for high values of κ the field varies in a distance of order $\lambda(T)$ while the order parameter varies over a range of the order of $\xi(T)$.

For smaller values of κ it is necessary to use a numerical integration of (3.16) and (3.17). This has recently been done by Matricon (1966). The energy $\mathscr{F}_p$ increases faster than p, so from (3.34) we conclude that the most

Reversible Properties

stable situation is obtained with one flux quantum per line. The variation for H_{c1} as a function of κ is shown in Fig. 3.2. It is seen that H_{c1} decreases with κ. It reaches the variation predicted by (3.39) only for values of $\kappa > 20$.

(c) The vicinity of H_{c2}. The vortex structure

It has already been remarked that, below H_{c2}, the magnitude of the order parameter ψ is expected to increase, so that the description of the behaviour of the superconductor involves the solution of the complete Ginzburg–Landau equations. The only slight simplification, which occurs here, comes from the fact that the microscopic field $\mathbf{h}$ is expected to be parallel to the applied field $\mathbf{H}$. If the direction of the field $\mathbf{H}$ is taken to be the z direction, the equations can be reduced to a two-dimensional form. Using the dimensionless quantities (2.37), the equations for h and f_0 read:

$$\frac{\nabla^2}{\kappa^2} f_0 - \frac{1}{f_0^3}(\nabla h)^2 + f_0 - f_0^3 = 0, \qquad (3.45)$$

$$\nabla^2 h - \frac{2}{f_0} \nabla h \cdot \nabla f_0 - f_0^2 h = 0. \qquad (3.46)$$

In these equations h is the modulus of the reduced field $\mathbf{h}$ which is parallel to $0z$. h is, of course, a function of x and y. The two-dimensional equations (3.45) and (3.46) are easily deduced from (2.47) and (2.48) by setting $\mathbf{h}$ parallel to $0z$.

Close to H_{c2}, as Abrikosov (1957) noted, the solution for f_0 must keep some features of the solution for H_{c2}. This will be shown below using a perturbation method.

At H_{c2}, (3.45) may be linearized and written as

$$\frac{\nabla^2}{\kappa^2} f_0 - \frac{1}{f_0^3}(\nabla h)^2 + f_0 = 0. \qquad (3.47)$$

It is seen that (3.46) and (3.47) allow the solution

$$h = \kappa - \frac{f_0^2}{2\kappa}, \qquad (3.48)$$

where f_0 obeys the following differential equation

$$\frac{\nabla^2}{\kappa^2} f_0 - \frac{(\nabla f_0)^2}{\kappa^2 f_0} + f_0 = 0. \qquad (3.49)$$

Expression (3.46) reduces to $h = \kappa$ (i.e. $H = H_{c2}$) for $f_0 = 0$, while expression (3.49) is equivalent to the linearized equation (3.1).† Departing

† Or (A. 3.1) of the Appendix.

Properties of Type II Superconductors

from H_{c2}, f_0 increases but is still small. Thus we look for a solution of the form

$$h = \kappa + \varepsilon_2 + \varepsilon_4 + \cdots \quad (3.50)$$

where $\varepsilon_2, \varepsilon_4, \ldots$ are quantities of order $f_0^2, f_0^4 \ldots$

It is clear that the expansion (3.50) introduces only even orders since (3.45) and (3.46) contain only even powers of f_0. The expression for h is reported in (3.45) and (3.46) and the resulting conditions to each order in f_0 are:

$$\begin{cases} \dfrac{\nabla^2 f_0}{\kappa^2} - \dfrac{(\nabla \varepsilon_2)^2}{f_0^3} + f_0 = 0, & (3.51a) \\[1em] \nabla^2 \varepsilon_2 - \dfrac{2}{f_0} \nabla \varepsilon_2 \cdot \nabla f_0 - \kappa f_0^2 = 0, & (3.51b) \end{cases}$$

$$\begin{cases} \dfrac{2}{f_0^3} \nabla \varepsilon_2 \cdot \nabla \varepsilon_4 + f_0^3 = 0, & (3.52a) \\[1em] \nabla^2 \varepsilon_4 - \dfrac{2}{f_0} \nabla \varepsilon_4 \cdot \nabla f_0 - \varepsilon_2 f_0^2 = 0. & (3.52b) \end{cases}$$

Let us write $\nabla \varepsilon_2$ as

$$\nabla \varepsilon_2 = -\dfrac{f_0}{\kappa} \nabla f_0 + \nabla \phi_2 \quad (3.53)$$

by analogy with the solution (3.48), ϕ_2 being a function of x and y of order f_0^2. This yields for the equations (3.51):

$$\begin{cases} \dfrac{\nabla^2 f_0}{\kappa^2} - \dfrac{(\nabla f_0)^2}{\kappa^2 f_0} + \dfrac{2}{\kappa f_0^2} \nabla f_0 \cdot \nabla \phi_2 - \dfrac{(\nabla \phi_2)^2}{f_0^3} + f_0 = 0, & (3.54a) \\[1em] \nabla^2 \phi_2 - f_0 \dfrac{\nabla^2 f_0}{\kappa} + \dfrac{(\nabla f_0)^2}{\kappa} - \dfrac{2}{f_0} \nabla f_0 \cdot \nabla \phi_2 - \kappa f_0^2 = 0. & (3.54b) \end{cases}$$

Multiplying (3.54b) by $1/\kappa f_0$ and adding to (3.54a) yields

$$\dfrac{\nabla^2 \phi_2}{\kappa f_0} - \dfrac{(\nabla \phi_2)^2}{f_0^3} = 0. \quad (3.55)$$

ϕ_2 and f_0 must be bounded functions. Multiplying (3.55) by f_0 and integrating over the volume gives

$$\dfrac{1}{\kappa} \int \nabla^2 \phi_2 \, dv = \int \dfrac{(\nabla \phi_2)^2}{f_0^2} \, dv. \quad (3.56)$$

Reversible Properties

The left-hand integral can be transformed into a surface integral. Since ϕ_2 is bounded, this surface integral is negligible in comparison with the volume integral of the right-hand side. It must therefore be zero showing that the only bounded solution of (3.55) is

$$\nabla \phi_2 \equiv 0. \tag{3.57}$$

Thus ϕ_2 is a constant. Therefore the solution of (3.15) is given by

$$\varepsilon_2 = \phi_2 - \frac{f_0^2}{2\kappa}, \tag{3.58}$$

where f_0 obeys the same equation which was obtained for $H = H_{c2}$, i.e.

$$\frac{\nabla^2 f_0}{\kappa^2} - \frac{(\nabla f_0)^2}{\kappa^2 f_0} + f_0 = 0. \tag{3.49}$$

Since (3.49) is linear the norm of f_0 is not determined. This norm will be obtained by considering the fourth order correction to h. Equations (3.52) become

$$-\frac{2\nabla f_0 \cdot \nabla \varepsilon_4}{\kappa f_0^2} + f_0^3 = 0, \tag{3.59a}$$

$$\nabla^2 \varepsilon_4 - \frac{2}{f_0} \nabla \varepsilon_4 \cdot \nabla f_0 + \frac{f_0^4}{2\kappa} - \phi_2 f_0^2 = 0. \tag{3.59b}$$

We are not interested in the calculation of ε_4, but these two equations yield an interesting relation for the constant ϕ_2 and the norm of f_0, which was first obtained by Abrikosov. Multiplying (3.59a) by κf_0, subtracting from (3.59b) and integrating over the volume yields

$$\int \nabla^2 \varepsilon_4 \, dv = \phi_2 \int f_0^2 \, dV - \kappa \left(\frac{1}{2\kappa^2} - 1 \right) \int f_0^4 \, dV. \tag{3.60}$$

The left-hand integral can be transformed into a surface integral and is therefore negligible and one obtains the following relation:

$$\phi_2 \overline{f_0^2} - \kappa \left[\frac{1}{2\kappa^2} - 1 \right] \overline{f_0^4} = 0, \tag{3.61}$$

where the bar denotes an average over the volume.

Equation (3.61) shows that ϕ_2 is negative, and of order 2 in f_0 as it should be. It will be shown that the value of the applied field is $(\kappa + \phi_2)$ and (3.61) determines the norm of f_0 as a function of the applied field.

Properties of Type II Superconductors

To second order in f_0 the microscopic field is seen to be

$$h = \kappa + \varepsilon_2 = \kappa + \phi_2 - \frac{f_0^2}{2\kappa}. \qquad (3.62)$$

This shows that the lines of constant f_0 are also the lines of constant h, i.e. the lines of current.

The induction and the free energy can be easily computed in terms of f_0^2 if one introduces the ratio†

$$\beta_A = \frac{\overline{f_0^4}}{(\overline{f_0^2})^2}. \qquad (3.63)$$

β_A is obviously larger than 1 (Schwartz inequality).

The induction $\mathscr{B}$ is obtained by taking the mean value of the microscopic field h:

$$\mathscr{B} = \kappa + \phi_2 - \frac{\overline{f_0^2}}{2\kappa} = \kappa + \phi_2 + \frac{\phi_2}{(2\kappa^2 - 1)\beta_A} \qquad (3.64)$$

while the free energy is obtained from (3.36) and is

$$\mathscr{F} = \overline{h^2} + \frac{1}{2} - \frac{\overline{f_0^4}}{2} \qquad (3.65\text{a})$$

or using (3.61), (3.62) and (3.64):

$$\mathscr{F} = \frac{1}{2} + \mathscr{B}^2 - \frac{(\kappa - \mathscr{B})^2}{1 + (2\kappa^2 - 1)\beta_A}. \qquad (3.65\text{b})$$

The applied field strength h_e is given by

$$h_e = \frac{1}{2}\frac{\partial \mathscr{F}}{\partial \mathscr{B}} = \mathscr{B} + \frac{(\kappa - \mathscr{B})}{1 + (2\kappa^2 - 1)\beta_A} = \kappa + \phi_2. \qquad (3.66)$$

So the applied field h_e is equal to $(\kappa + \phi_2)$ as previously mentioned and

$$\frac{\phi_2}{\kappa} = \frac{H - H_{c2}}{H_{c2}}. \qquad (3.67)$$

In terms of the applied field h_e the induction is

$$\mathscr{B} = h_e - \frac{\kappa - h_e}{(2\kappa^2 - 1)\beta_A}. \qquad (3.68)$$

† β_A is essentially a geometrical factor which will be shown to be independent of κ.

Reversible Properties

The magnetization is then

$$\mu = \frac{1}{4\pi} \frac{\kappa - h_e}{(2\kappa^2 - 1)\beta_A}. \tag{3.69}$$

At this stage, the remaining problem is to find the convenient solution for f_0 which obeys (3.49). This equation is readily transformed into

$$\nabla^2 (\log f_0) + \kappa^2 = 0. \tag{3.70}$$

It is clear from the solution of this equation that the geometrical factor β_A is independent of κ.† The most general solution of (3.70) may be written as

$$\log f_0 = -\frac{\kappa^2 y^2}{2} + \alpha(x, y), \tag{3.71}$$

where $\alpha(x,y)$ is the general solution of the Laplace equation

$$\nabla^2 \alpha = 0. \tag{3.72}$$

So $\alpha(x,y)$ is the real part of any analytical function of $z = x + iy$. Since

$$e^{Re\,(a+ib)} = e^a = |e^{a+ib}|, \tag{3.73}$$

the most general solution of (3.70) is

$$f_0 = e^{-\kappa^2 y^2/2} \; |g(z)|, \tag{3.74}$$

where $g(z)$ is any analytical function of $z = x + iy$.

For a given $\mathscr{B}$, the free energy (3.65) is an increasing function of β_A ($\kappa \geqslant 1/\sqrt{2}$), and the most favourable f_0 corresponds to the lowest β_A. It has already been stated that, in order to increase the contribution of the surface energy, it is favourable to consider a periodic two-dimensional set of filaments. Let us suppose that the unit cell is the parallelogram shown in Fig. 3.3. One can use the new coordinate system X, Y such that

$$z = x + iy = X + Y e^{i\alpha} \tag{3.75a}$$

$$x = X + Y \cos \alpha \tag{3.75b}$$

$$y = Y \sin \alpha \tag{3.75c}$$

and f_0 reads:

$$f_0 = |g(X + Y e^{i\alpha})| \, e^{-(\kappa^2 Y^2 \sin^2 \alpha)/2}. \tag{3.76}$$

f_0 is a periodic function of X with period a and of Y with period b. It is

† For different values of κ the solutions of this equation are homothetic.

assumed that a and b are the smallest possible periods in X and Y. $g(X + Y e^{i\alpha})$ may be expanded as a Fourier series in X, i.e.

$$g(X + Y e^{i\alpha}) = \sum_{n=-\infty}^{+\infty} \gamma_n \exp\left\{\frac{2\pi ni}{a}(X + Y e^{i\alpha})\right\}, \qquad (3.77)$$

where γ_n are constants and f_0 becomes

$$f_0 = \left| \sum_{n=-\infty}^{+\infty} C_n \exp\left\{\frac{2\pi ni}{a}(X + Y \cos\alpha)\right\} \right.$$

$$\left. \exp\left\{-\frac{\kappa^2}{2}\sin^2\alpha\left(Y + \frac{2\pi n}{a\kappa^2 \sin\alpha}\right)^2\right\} \right|, \qquad (3.78)$$

where

$$C_n = \gamma_n \exp\left(\frac{2\pi^2 n^2}{a^2 \kappa^2}\right). \qquad (3.79)$$

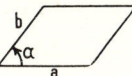

FIG. 3.3.

This function is periodic in Y with period b if

$$\frac{bn}{p} = \frac{2\pi n}{a\kappa^2 \sin\alpha} \qquad (3.80a)$$

and if

$$C_{n+p} = C_n \exp\left(\frac{2\pi ibn}{a}\cos\alpha\right) \qquad (3.80b)$$

where p is an integer.

Equation (3.80a) reads

$$ab \sin\alpha = \frac{2\pi p}{\kappa^2}. \qquad (3.81)$$

It is seen that this is the surface of the unit cell. If the applied field is equal to H_{c2}, the flux of the reduced field h in the unit cell is

$$\int h \cdot d\sigma = \kappa\, ab \sin\alpha = \frac{2\pi p}{\kappa}. \qquad (3.82)$$

According to the definition of the reduced field, the flux of the actual field in the unit cell is

$$\Phi_{H_{c2}} = p\Phi_0. \qquad (3.83)$$

Reversible Properties

So p represents the number of flux quanta per line.†

It is shown in Appendix 3.4 that the most favourable solutions correspond to $p = 1$, that is to one flux quantum per line. [The argument presented here is a simplified version of the proof given by Lasher (1965)].

It is thus sufficient to calculate β_A for $p = 1$. According to (3.80b), which reduces to

$$C_{n+p} = C_n \exp\left(\frac{2\pi i b}{a} n \cos \alpha\right),$$

f_0 depends only on one coefficient C_0 and reads

$$f_0 = |C_0| \left| \sum_{n=-\infty}^{+\infty} \exp\left\{2\pi i \frac{b}{a} \cos \alpha \frac{n(n-1)}{2}\right\} \right.$$
$$\left. \times \exp\left\{\frac{2\pi n i}{a}(X + Y \cos \alpha)\right\} \exp\left\{-\frac{\kappa^2 \sin^2 \alpha}{2}(Y + bn)^2\right\} \right|.$$
(3.84)

After some tedious algebra, β_A is obtained as

$$\beta_A = \frac{\langle f_0^4 \rangle}{\langle f_0^2 \rangle^2} = \frac{1}{\sqrt{(2\pi)}} |\kappa b \sin \alpha| \left\{ \left| \sum_{n=-\infty}^{+\infty} \exp\left\{\frac{4\pi^2}{a^2 \kappa^2} \frac{i e^{i\alpha}}{\sin \alpha} n^2\right\} \right|^2 \right.$$
$$\left. + \left| \sum_{n=-\infty}^{+\infty} \exp\left\{\frac{4\pi^2}{a^2 \kappa^2} \frac{i e^{i\alpha}}{\sin \alpha}\left(n + \frac{1}{2}\right)^2\right\} \right|^2 \right\}.$$
(3.85)

Introducing the complex variable

$$\zeta = \frac{b}{a} e^{i\alpha} = \rho + i\sigma \tag{3.86}$$

and using (3.81) with $p = 1$, β_A reads

$$\beta_A = \sigma^{1/2} \left\{ \left| \sum_{n=-\infty}^{+\infty} \exp(2\pi i n^2 \zeta) \right|^2 + \left| \sum_{n=-\infty}^{+\infty} \exp\left[2\pi i \left(n + \frac{1}{2}\right)^2 \zeta\right] \right|^2 \right\}.$$
(3.87)

As already stated β_A, which is a geometrical factor, is independent of κ. In the plane of the complex variable ζ, the function β_A has several symmetries.

(i) β_A is periodic of period 1 in ρ [i.e. $\beta_A(\rho) = \beta_A(\rho + 1)$],

(ii) β_A is conserved in a symmetry with respect to the axis $\rho = \frac{1}{2}$ [i.e. $\beta_A(\rho) = \beta_A(1 - \rho)$],

† For fields smaller than H_{c_2}, in first approximation, the unit cell is not modified, and only the amplitude of f_0 increases. It might seem that according to expression (3.62) for the field, the flux quantization condition is no longer obtained. This point is discussed below.

Properties of Type II Superconductors

(iii) it can be shown by using the Poisson's sum formula that $\beta_A(\zeta) = \beta_A(1/\zeta^*)$. β_A is conserved in an inversion, where the inversion circle is of radius 1 and centred at the origin.

In order to interpret these symmetry properties, it should be noticed that for a given ζ the unit cell of f_0 is homothetic to the parallelogram shown in Fig. 3.4a.†

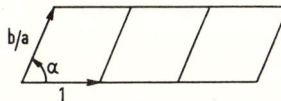

FIG. 3.4a.

It is clear that the translation $\rho \to \rho + 1$ is a new choice of the unit cell for the same lattice as shown on Fig. 3.4b.

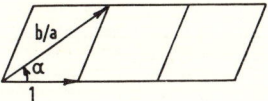

FIG. 3.4b.

The symmetry with respect to the axis $\rho = \frac{1}{2}$ (i.e. $\rho \to 1 - \rho$) yields a lattice obtained from the original lattice by the same symmetry (Fig. 3.4c).

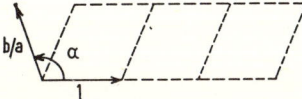

FIG. 3.4c.

The inversion consists in interchanging a and b and of course the lattice is unchanged. Thus none of the above symmetries change the lattice. Due to these symmetries, it is sufficient to study the behaviour of β_A for ζ located in the hatched region in Fig. 3.5.

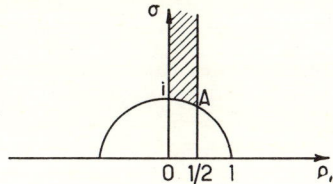

FIG. 3.5.

† The actual values of a and b depend on κ and are obtained from (3.79) with $p = 1$.

Reversible Properties

Using expression (3.85), for a given σ, β_A is a minimum for $\rho = \frac{1}{2}$ while on the line $\rho = \frac{1}{2}$, β_A is an increasing function of σ for $\sigma \geq \sqrt{3}/2$.

The absolute minimum of β_A is reached at the isolated point A where

$$\zeta = e^{i\pi/3}.$$

This point corresponds to an equilateral triangular lattice. Of course, in the ζ plane β_A has the same value at an infinite number of isolated points which can be deduced from A by the symmetry operations already mentioned, but such points correspond to the same lattice as shown above. The value of β_A has been calculated by Kleiner et al. (1964) and was found to be

$$\beta_A = 1\cdot 1596.$$

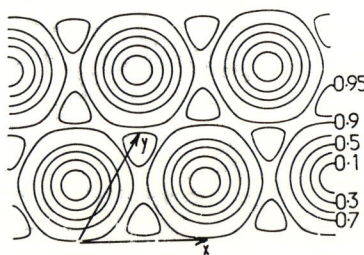

FIG. 3.6. Contour diagram for f_0^2 as computed by Kleiner et al. The maximum of f_0^2 is normalized to unity

In his original paper Abrikosov predicted the square lattice corresponding to $\zeta = i$, $\beta_A = 1\cdot 18$. This point is indeed a turning point of β_A, but corresponds to a saddle point and not to an absolute minimum.

In order to investigate the properties of the field [eqn. (3.62)], it is interesting to write f_0^2 as a double Fourier series. A straightforward calculation yields

$$f_0^2 = |C_0|^2\, 3^{-1/4} \sum_{m,n} (-)^{mn}\, e^{-i\pi n/2}\, e^{-\pi(m^2+n^2-mn)/\sqrt{3}}\, e^{2\pi i(nX+mY)/a}. \quad (3.88)$$

f_0^2 vanishes at the points translationally equivalent to† $X = \frac{3}{4}a$, $Y = \frac{a}{2}$, and behaves like r^2 in the vicinity of these points, which correspond to the minima of the field H. H therefore has hexagonal symmetry (symmetry group p_6 mm). A contour diagram for f_0^2 has been given by Kleiner et al. and is reproduced on Fig. 3.6. The distance between two maxima of the field is obviously equal to the lattice parameter $a = (2\sqrt{\pi/\kappa})\, 3^{-1/4}$ [obtained from (3.79)]. The actual distance between two vortex lines is thus

$$L = a\lambda(T) = 2\sqrt{\pi}\, 3^{-1/4}\, \xi(T) \simeq 2\cdot 7\, \xi(T). \quad (3.89)$$

† This is most easily seen when f_0 is written in the form (3.84) with $\alpha = \pi/3$ and $b = a$.

Properties of Type II Superconductors

The widths and the distances of the lines are both of the order of $\xi(T)$ and there is a strong overlap of the lines. The concept of individual vortex lines is meaningless in the vicinity of H_{c2} and is replaced by the concept of a periodic structure of the field.

The coefficient $|C_0|^2$ appearing in f_0^2 is easily computed from (3.61) and is, as expected, proportional to $(1 - h_e/\kappa)$. The convergence of the series for f_0^2 is very rapid so that in most calculations it is sufficient to write f_0^2†:

$$f_0^2 = |C_0|^2 \, 3^{-1/4} \left\{ 1 - 2e^{-\pi/\sqrt{3}} \left(\cos \frac{2\pi}{a} X + \cos \frac{2\pi}{a} Y + \cos \frac{2\pi}{a} (X+Y) \right) \right\} \quad (3.90)$$

This will be useful in the discussion of the neutron diffraction experiment (see p. 70).

Let us now return to the problem of flux quantization in a vortex line, i.e. in one unit cell. It has been shown in section 3.3a that the fluxoid quantization reads‡

$$\int_\Sigma \mathbf{h} \cdot d\boldsymbol{\sigma} = \frac{2\pi p}{\kappa} + \int_\Gamma \frac{\mathbf{j} \cdot d\mathbf{l}}{f_0^2}. \quad (3.91)$$

However, as stated above, the lines of current are the lines of constant f_0^2. Hence due to the translational invariance of the vortex lattice,

$$\int_\Gamma \frac{\mathbf{j} \cdot d\mathbf{l}}{f_0^2} = 0$$

when integrated along the sides of the unit cell. So the flux of the reduced field in the unit cell must be $2\pi/\kappa$ for one flux quantum per line. From (3.62) it is seen that

$$\int_\Sigma \mathbf{h} \cdot d\boldsymbol{\sigma} = h_e S_{he} - \int_\Sigma \frac{f_0^2}{2\kappa} d\boldsymbol{\sigma} = \frac{2\pi}{\kappa} = \kappa S_{hc_2}, \quad (3.92)$$

where S_{he} is the surface of the unit cell corresponding to the applied field h_e. This equation gives the variation in size of the unit cell below H_{c2} and yields

$$\frac{S_{H_{c_2}}}{S_{H_e}} = \frac{S_{hc_2}}{S_{he}} = \frac{H_e}{H_{c2}} - \frac{\overline{f_0^2}}{2\kappa^2}, \quad (3.93)$$

The above ratio is equal to 1 for $H_e = H_{c2}$, and for $H_e < H_{c2}$ the increase of the unit cell surface is of the order of $\overline{f_0^2}$.

This conclusion is nevertheless consistent with the preceeding calculation of the vortex structure. Indeed when the free energy is written in the form

---

† In this expression the origin has been taken as one of the vanishing points of f_0^2.
‡ This equation is obtained by using the dimensionless quantities (2.37) in (3.11).

(3.65b), the induction $\mathcal{B}$ is written to order f_0^2 and the lattice parameter is not altered.

If terms of higher order had been taken into account in $\mathcal{B}$ as in $\mathcal{F}$ the minimization of $\mathcal{F}$ would have yielded a lattice dependent on h_e. It is clear that if one takes into account the variation of the lattice in the calculation of $\mathcal{B}$, this yields a correction of the order of $\overline{f_0^4}$.

(d) The special case $\kappa = 1/\sqrt{2}$

For $\kappa = 1/\sqrt{2}$, the nucleation field H_{c2} is evidently equal to H_c. It will be shown below that the field H_{c1} is also equal to H_c. Let us consider (3.16) and (3.17). For $\kappa = 1/\sqrt{2}$, these two equations are identical if

$$h = \frac{1}{\sqrt{2}}(1 - f_0^2). \tag{3.94}$$

This form for h is correct since it satisfies the boundary conditions (3.18). The energy of the line can be expressed in terms of the flux of h. For the line with p flux quanta, (3.34) reads

$$\mathcal{F}_p = 2\pi \int_0^\infty (1 - f_0^2)\rho d\rho = 2\pi\sqrt{2}\int_0^\infty h\rho\, d\rho = 2\pi\sqrt{2}\,\frac{\phi}{\kappa}. \tag{3.95}$$

$\mathcal{F}_p$ is proportional to p, and consequently the field of first penetration H_{c1} will be independent of p.

$$h_{c1}(p) = \frac{\kappa \mathcal{F}_p}{4\pi p} = \frac{1}{\sqrt{2}} = h_c = h_{c2}. \tag{3.96}$$

This result is a consequence of the vanishing of the surface energy.

3.4. General properties of the vortex line structure. The magnetization curve

Using the results of the preceding section, it is natural to assume that a periodic structure of vortex lines with one quantum of flux per line will hold over the whole range of the applied field. Moreover, Matricon (1964), using a simplified model based on the generalized London equation (see later), has shown that the triangular arrangement is, for high κ values, the most favourable whatever the value of H ($H_{c1} < H < H_{c2}$).[†] For this pattern, the concentration of lines, and hence the induction, is easily computed as a function of the lattice parameter L.

† See also the work of Fetter, Hohenberg and Pincus (1966).

The line concentration per unit cross-sectional area is:

$$\rho_L = \frac{n}{S} = \frac{2}{\sqrt{3}\,L^2} \tag{3.97}$$

and the induction is

$$B = \frac{n}{S}\Phi_0 = \frac{2}{\sqrt{3}}\frac{\Phi_0}{L^2}. \tag{3.98}$$

In section 3.3b it has been shown that for high κ values the concept of an isolated vortex line leads one to consider a core of width $\xi(T)$ where the order parameter is highly perturbed while the field varies over the distance $\lambda(T) \gg \xi(T)$ (Friedel et al. 1963). It is then convenient to consider the properties of a type II superconductor in an applied magnetic field in three field regions.

(a) Near H_{c2}, where the concentration of lines is high. It has been shown that $L \sim \xi(T)$ and $\rho_L \simeq 1/\xi^2(T)$. There is a strong overlapping of the cores and the concept of isolated lines must be abandoned. However, the calculation of section 3.3c yields the main results needed.

(b) Near H_{c1}. This is the region where the concentration of lines is low $[\rho_L \simeq 1/\lambda^2(T)]$. The lines can be considered as isolated, or as weakly interacting.

(c) In the intermediate range $1/\lambda^2(T) \ll \rho_L \ll 1/\xi^2(T)$. Here the concentration is rather high, but the concept of interacting lines can be still used.

These three field regions will be considered in more detail below.

(a) *Near H_{c2}.*

Close to H_{c2}, the free energy has the form (3.65b):

$$F = \frac{H_c^2}{8\pi} + \frac{B^2}{8\pi} - \frac{(H_{c2} - B)^2}{1 + (2\kappa^2 - 1)\beta_A}. \tag{3.99}$$

It is seen, as expected, that the applied field is

$$H = \frac{1}{4\pi}\frac{dF}{dB}. \tag{3.100}$$

It has already been mentioned that the magnetization is of order $|\psi|^2$ and may be obtained from (3.69):

$$M = \frac{B - H}{4\pi} = \frac{H - H_{c2}}{4\pi\beta_A(2\kappa^2 - 1)}; \quad \left(\frac{H_{c2} - H}{H_{c2}} \ll 1\right) \tag{3.101}$$

The magnetization vanishes at $H = H_{c2}$ and the transition is of the second order. This is in agreement with the prediction made in section 1.7. The slope

Reversible Properties

of the magnetization is finite:

$$\frac{dM}{dH} = \frac{1}{4\pi\beta_A (2\kappa^2 - 1)} \tag{3.102}$$

(and will be very large when $\kappa \simeq 1/\sqrt{2}$).

Expression (3.102) may be used to determine κ (knowing β_A) or β_A (knowing κ). Clearly one needs a magnetization curve as reversible as possible. It is for this reason that Kinsel, Lynton and Serin (1962) studied the alloy In–2·5% Bi. There remains some uncertainty on the slope, but nevertheless the measurement of H_{c2} and H_c close to T_0 has yielded $\kappa = (1/\sqrt{2})(H_{c2}/H_c) = 1·80$, while the slope of the magnetization has yielded $\kappa = 1·81$ (with $\beta_A = 1·16$).† It should be noted that the preceding results are valid whatever the value of κ.

(b) *Near H_{c1}*

For high values of κ the concept of isolated lines can be described using the London equation to calculate the field and the current. The core is well represented, in the limit $\lambda(T) \gg \xi(T)$, by a two-dimensional delta function. For an isolated line the convenient "London equation" will be

$$\mathbf{h} + \lambda^2(T) \operatorname{curl} \operatorname{curl} \mathbf{h} = \boldsymbol{\varphi}_0 \delta_2(\mathbf{r}). \tag{3.103}$$

($\boldsymbol{\varphi}_0$ is parallel to z and has the magnitude Φ_0 of the flux quantum.)

This form yields the correct fluxoid quantization. By taking into account the Maxwell equation

$$\operatorname{div} \mathbf{h} = 0, \tag{3.104}$$

one can calculate the behaviour of the field and determine H_{c1}. The results are analogous to those of section 3.3b. For example, the microscopic field is given by

$$h = \frac{\Phi_0}{2\pi\lambda^2(T)} K_0\left(\frac{r}{\lambda(T)}\right), \quad r \gg \xi(T) \tag{3.105}$$

and the free energy, which is obtained from (1.14), reads (per unit length of line):

$$F = \left[\frac{\Phi_0}{4\pi\lambda(T)}\right]^2 \left[\log \frac{\lambda(T)}{\xi(T)} + \varepsilon\right]. \tag{3.106}$$

The numerical constant ε includes the effect of the core. The above value of F yields the expression (3.40) for H_{c1}.

† The comparison has to be made close to T_0. It will be shown in Chapter 5 that for $T < T_0$ it is necessary to introduce three parameters $\kappa_1(T)$, $\kappa_2(T)$, $\kappa_3(T)$ related to $H_{c2}(T)$, $(dM/dH)_{H=H_{c2}}$, $H_{c1}(T)$ which are equal only for $T = T_0$.

Properties of Type II Superconductors

Provided that the $\xi(T)$ and $\lambda(T)$ values used are those appropriate to the temperature T, the above results constitute in a certain way a generalization of the Abrikosov picture to any temperature below T_0.

If the applied field is raised slightly above H_{c1}, the concentration of lines will increase but it can be assumed that their distance apart is still greater than the radius of the core. The convenient "London equation" can now be written as

$$\boldsymbol{h} + \lambda^2(T) \operatorname{curl}\operatorname{curl}\boldsymbol{h} = \varphi_0 \sum_i \delta_2(\boldsymbol{r} - \boldsymbol{r}_i); \quad [|\boldsymbol{r}_i - \boldsymbol{r}_j| \gg \xi(T)], \tag{3.107}$$

where $\boldsymbol{r}_i$ is the position of the core of the line i.

Let us first examine the case of two lines. If the condition $|\boldsymbol{r}_1 - \boldsymbol{r}_2| \gg \xi(T)$ is fulfilled, then the field $\boldsymbol{h}$ will be the superposition of $\boldsymbol{h}_1$ and $\boldsymbol{h}_2$ due to the separate filaments.

$$\boldsymbol{h} = \boldsymbol{h}_1(\boldsymbol{r}) + \boldsymbol{h}_2(\boldsymbol{r}). \tag{3.108}$$

$$\boldsymbol{h}_1(\boldsymbol{r}) = \frac{\varphi_0}{2\pi\lambda^2(T)} K_0\left[\frac{|\boldsymbol{r} - \boldsymbol{r}_1|}{\lambda(T)}\right]. \tag{3.109}$$

The free energy per unit length of line is

$$\frac{1}{8\pi}\int [\boldsymbol{h}^2 + \lambda^2(T)|\operatorname{curl}\boldsymbol{h}|^2]\,dv = \frac{\lambda^2(T)}{8\pi}\int [\boldsymbol{h} \times \operatorname{curl}\boldsymbol{h}]\cdot d\boldsymbol{\sigma}. \tag{3.110}$$

The surface integral is taken over the surface of the two cores $|\boldsymbol{r} - \boldsymbol{r}_1| \simeq \xi(T)$, $|\boldsymbol{r} - \boldsymbol{r}_2| \simeq \xi(T)$.

The energy can be written as

$$F = \frac{\lambda^2(T)}{8\pi}\int (d\boldsymbol{\sigma}_1 + d\boldsymbol{\sigma}_2)\cdot[(\boldsymbol{h}_1 + \boldsymbol{h}_2)\times(\operatorname{curl}\boldsymbol{h}_1 + \operatorname{curl}\boldsymbol{h}_2)] \tag{3.111}$$

and divided into three contributions.

(α) The energy of the individual filaments

$$J = F_{11} + F_{22} = \frac{\lambda^2(T)}{8\pi}\left[\int d\boldsymbol{\sigma}_1\cdot(\boldsymbol{h}_1\times\operatorname{curl}\boldsymbol{h}_1) + \int d\boldsymbol{\sigma}_2\cdot(\boldsymbol{h}_2\times\operatorname{curl}\boldsymbol{h}_2)\right] \tag{3.112}$$

(β) The term

$$\int (\boldsymbol{h}_1 + \boldsymbol{h}_2)\cdot(\operatorname{curl}\boldsymbol{h}_1 \times d\boldsymbol{\sigma}_2 + \operatorname{curl}\boldsymbol{h}_2 \times d\boldsymbol{\sigma}_1) \tag{3.113}$$

Reversible Properties

which tends toward zero for $\xi(T) \ll \lambda(T)$ since $\boldsymbol{h}_1$ and curl $\boldsymbol{h}_1$ are finite over the domain of integration $d\boldsymbol{\sigma}_2$.

(γ) The interaction energy

$$F_{12} = \frac{\lambda^2(T)}{8\pi} \int \left\{ [\boldsymbol{h}_1 \times \text{curl } \boldsymbol{h}_2] \cdot d\boldsymbol{\sigma}_2 + [\boldsymbol{h}_2 \times \text{curl } \boldsymbol{h}_1] \cdot d\boldsymbol{\sigma}_1 \right\}. \quad (3.114)$$

This energy is finite since, for $|\boldsymbol{r} - \boldsymbol{r}_2| \ll \lambda(T)$, $|\text{curl } \boldsymbol{h}_2|$ is proportional to $1/|\boldsymbol{r}_1 - \boldsymbol{r}_2|$. From expression (3.105) for $\boldsymbol{h}$ it is seen that

$$F_{12} = \frac{\Phi_0^2}{16\pi^2\lambda^2(T)} K_0\left[\frac{r_{12}}{\lambda(T)}\right] \quad (\boldsymbol{r}_{12} = \boldsymbol{r}_1 - \boldsymbol{r}_2). \quad (3.115)$$

This interaction energy is repulsive and decreases as

$$\frac{1}{\sqrt{r_{12}}} \exp\left\{-r_{12}/\lambda(T)\right\} \quad (3.116)$$

at large distances, and varies as

$$\log \frac{\lambda(T)}{r_{12}} \quad (3.117)$$

at short distances.

The various thermodynamic quantities of interest can now be computed using the Gibbs free energy:

$$G = F - \frac{BH}{4\pi} \quad (3.118)$$

which may be written as

$$G = \sum_i F_i + \sum_{ij} F_{ij} - \frac{BH}{4\pi}. \quad (3.119)$$

F_i is the individual energy of the line i, as given by (3.106) and F_{ij} is the interaction energy. For H slightly larger than H_{c1}, the density of lines ρ_L is still small and one need only consider the interaction with the z nearest neighbours. Thus the Gibbs energy per unit surface area is†

$$G = \frac{B}{4\pi}\left[H_{c1} - H + \frac{1}{2}z\frac{\Phi_0}{2\pi\lambda^2(T)}K_0\left(\frac{L}{\lambda(T)}\right)\right] \quad (3.120)$$

In deriving (3.120) use has been made of (3.34) and of the relation

$$B = \rho_L\Phi_0.$$

† Note that the number of lines is proportional to B

For a triangular lattice, L is related to B by (3.98). The variation of G as a function of B is shown in Fig. 3.7. The initial slope is negative ($H > H_{c1}$) and the interaction energy is negligible for B small. When B increases the interaction term $K_0 \{1{\cdot}07 \sqrt{[\Phi_0/B\lambda^2(T)]}\}$ increases and B presents a minimum $B = B_M(H)$ which is the equilibrium induction for the field H.

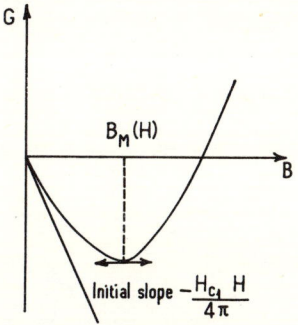

FIG. 3.7. Schematic variation of G as a function of B

$B_M(H)$, and hence $M(H)$, were first computed by Goodman (1964) (Fig. 3.8). Due to the particular form of the interaction term, it is seen that $(\partial M/\partial H)_{H=H_{c1}}$ is infinite. This follows from the fact that the interaction energy is of the form (3.116) for $L \gg \lambda(T)$. It is possible to form many lines

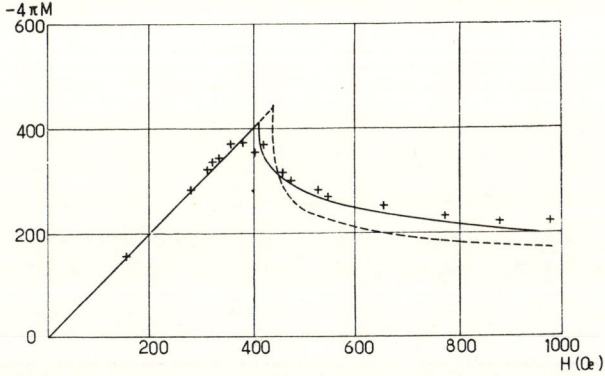

FIG. 3.8. Magnetization curve for a molybdenum–rhenium alloy at $T = 0{\cdot}52\, T_0$. The two theoretical curves (dashed line; laminar model; full line; vortex line model) have been calculated by Goodman. The experimental points are due to Joiner and Blaugher (1964)

without increasing the interaction energy drastically. The experimental curves (in Fig. 3.8 the experimental points correspond to a Mo–Re alloy) do not show the infinite slope at $H = H_{c1}$. This is in agreement with the low interaction between the lines, which is too small to prevent the pinning of the filaments by structural defects (see Chapter 8).

Reversible Properties

A similar calculation can be performed for the laminar model. The repulsion between the lines is still proportional to $e^{-L/\lambda(T)}$, but B is proportional to $1/L$ (instead of $1/L^2$) and $\partial M(H)/\partial H$ is greater for the laminar model than for the vortex lines (Goodman, 1964).

It is interesting to determine the condition of static equilibrium of the lines (De Gennes, 1965). Using the expression (3.115) for F_{12} it is seen that the force acting on line 2 is

$$f_2 = -\nabla F_{12}. \tag{3.121}$$

Static equilibrium is obtained when $f_2 = 0$. f_2 may be written in terms of the superfluid velocity at the point r_2. The current j which would exist at r_2 in the presence of line 1 alone is

$$j = n_1 e v = -\frac{e}{4\pi} (\operatorname{curl} h_1)_{(r=r_2)}. \tag{3.122}$$

Using (3.111) and (3.116) it is found that

$$f_{2x} = \frac{\hbar}{2\pi} v_y, \tag{3.123}$$

$$f_{2y} = -\frac{\hbar}{2\pi} v_x. \tag{3.124}$$

The line is in static equilibrium when the superfluid velocity is zero at any point in the line.

(c) The intermediate region

In this region the vortex lines form a dense lattice. However, some predictions can be made if one assumes that the distance between the lines L, is still greater than the core radius $\xi(T)$. This region will thus be characterised by $1/\xi^2(T) \gg \rho_L \gg 1/\lambda^2(T)$.† In the expression (3.119) for the Gibbs energy, the summation for the interaction energy must be extended to distant neighbours. In order to calculate this energy it is better to transform the summation to a sum over the reciprocal lattice. If one introduces the Fourier transform of the field h,

$$h_q = \rho_L \int_{\text{cell}} h(r) e^{iq \cdot r} d_2 r, \tag{3.125}$$

† It should be noticed that the concept of the intermediate region is restricted to high values of κ ($\kappa > 10$) since the smallest value of L is of the order of $2 \cdot 7 \, \xi(T)$ [see (3.89)].

it is seen that $h_q = 0$ unless q is equal to a reciprocal lattice vector M. h_M is obtained from (3.107) for the field and reads

$$h_M = \frac{\rho_L \Phi_0}{1 + \lambda^2(T) M^2}. \tag{3.126}$$

The free energy becomes

$$F = \frac{1}{8\pi} \int (h^2 + \lambda^2(T) |\operatorname{curl} h|^2) \, dr = \frac{B^2}{8\pi} \sum_M \frac{1}{1 + \lambda^2(T) M^2}. \tag{3.127}$$

The first reciprocal translation is of order $1/L$ so that for $M \neq 0$, $\lambda^2(T) q^2 \gg 1$ [as $\rho_L \gg 1/\lambda^2(T)$]. The free energy takes the form

$$F = \frac{B^2}{8\pi} + \frac{B^2}{8\pi} \sum_{M \neq 0} \frac{1}{\lambda^2(T) M^2}. \tag{3.128}$$

This energy depends on the particular form of the lattice. However, qualitative conclusions can be obtained by replacing the summation by an integral:

$$\sum_M \frac{1}{M^2} = \frac{1}{(2\pi)^2} \frac{1}{\rho_L} \int \frac{dM}{M^2} = \frac{1}{2\pi \rho_L} \int_{M_{min}}^{M_{max}} \frac{M dM}{M^2} = \frac{1}{2\pi \rho_L} \log \frac{M_{max}}{M_{min}}. \tag{3.129}$$

M_{min} is clearly of the order $1/L$, while M_{max} is of the order $1/\xi(T)$ as the Fourier components of the core have to be excluded. Hence F is seen to be

$$F = \frac{B^2}{8\pi} + \frac{B}{4\pi} H_{c1} \frac{\log \alpha L/\xi(T)}{\log \lambda(T)/\xi(T)}. \tag{3.130}$$

α is a numerical constant of the order of unity. The Gibbs function:

$$G = F - \frac{BH}{4\pi}$$

is a minimum at

$$H = B + H_{c1} \frac{\log \alpha' L/\xi(T)}{\log \lambda(T)/\xi(T)} \quad (\alpha' = \alpha e^{-1/2}). \tag{3.131}$$

The magnetization M is thus given by

$$M = \frac{B - H}{4\pi} = \frac{-H_{c1}}{4\pi} \frac{\log\{[\alpha'/\xi(T)](2\Phi_0/\sqrt{3B})^{1/2}\}}{\log\{\lambda(T)/\xi(T)\}}. \tag{3.132}$$

The logarithmic dependence of M on B is in quite good agreement with experimental data on reversible magnetization curves.

The behaviour of the magnetization is shown in Fig. 1.7 and in Fig. 3.8.

Reversible Properties

3.5. Experimental observation of the vortex line structure

(a) Neutron diffraction

In 1964, De Gennes and Matricon suggested that the periodic structure of the vortex lines could be detected by neutron diffraction since the periodic variation of the field h will result in Bragg peaks. The position of these peaks determines the lattice parameter L and the symmetry of the lattice.

The interaction energy between the neutron and the magnetic field is $\mu_N h(r)$ where $\mu_N = (1 \cdot 91 \; e\hbar/M_N c)$ is the neutron momentum, M_N being the neutron mass. In the scattering process, the neutron undergoes a change of momentum from $\hbar k_0$ to $\hbar(k_0 + q)$. The scattering amplitude is obtained, in the Born approximation, by evaluating the integral

$$a = \frac{M_N}{2\pi\hbar^2} \int \mu_N h(r) \, e^{iq \cdot r} \, dr \, . \qquad (3.133)$$

As expected, this integral is zero, unless q is equal to M, a reciprocal lattice vector (Bragg scattering), and the scattering amplitude is proportional to the corresponding coefficient of the Fourier series for h.

In the region $\rho_L \ll 1/\xi^2(T)$, and for high values of κ, the expansion (3.126) can be used and

$$a_M = \frac{1 \cdot 91}{2} \frac{\rho_L V}{1 + \lambda^2(T) M^2} \, . \qquad (3.134)$$

V is the sample volume. As the familiar quantity used by experimentalists is the cross-section per atom, we use for V an atomic volume of order 30Å^3. The first Bragg reflection occurs, with a triangular lattice, for $M_1 = 4\pi/\sqrt{3} \, L$. If $B = 2000$ gauss, $L = 10^3$ Å and $M \sim 7 \times 10^5$ cm^{-1}, for a penetration depth $\lambda(T) \sim 1000$ Å, $\lambda^2 M^2 \sim 50 \gg 1$ (as expected). The cross-section is then

$$\sigma_1 = 4\pi a_{M_1}^2 \simeq 5 \times 10^{-28} \; \text{cm}^2 = 0 \cdot 5 \; \text{millibarn} \, .$$

This is a small value close to the present limits of neutron techniques.

The cross-section σ_2 for the second Bragg peak is smaller. It is clear that the ratio σ_2/σ_1 is equal to $(h_{M_1}/h_{M_2})^2$, where h_{M_1} and h_{M_2} are the corresponding coefficient of the Fourier series, since the number of second neighbours is equal to the number of first neighbours (in triangular as in square lattices).

In the region $\rho_L \ll 1/\xi^2(T)$ and for high κ values the ratio $\sigma_2/\sigma_1 = M_1^4/M_2^4$ [see (3.126)], i.e.

$$\frac{\sigma_2}{\sigma_1} = \frac{1}{9} \simeq 10^{-1} \; \text{for the triangular lattice,}$$

$$\frac{\sigma_2}{\sigma_1} = \frac{1}{4} \simeq 2 \times 10^{-1} \; \text{for the square lattice.}$$

Close to H_{c2} this ratio is even smaller as, according to (3.88), the Fourier coefficients are proportional to $e^{-\pi(m^2+n^2-mn)/\sqrt{3}}$ in the triangular lattice and to $e^{-\pi(m^2+n^2)/2}$ in the square one. The ratio σ_2/σ_1 is thus

$$\frac{\sigma_2}{\sigma_1} = e^{-4\pi/\sqrt{3}} = 7 \times 10^{-4} \text{ for the triangular lattice}$$

$$\frac{\sigma_2}{\sigma_1} = e^{-\pi} \quad = 4 \times 10^{-2} \text{ for the square lattice.}$$

The observation of the second Bragg peak is expected to be very difficult, particularly in the case of a triangular arrangement. It must be emphasized, however, that the observation of this peak, if desirable, is not necessary to determine the structure of the lattice. Indeed, if one measures B (by conventional magnetic techniques and L (by neutron diffraction) the relation $B = \alpha_p \Phi_0/L^2$ will give the value of α.

For a square lattice and one flux quantum per line $\alpha_1^s = 1$, while for a triangular lattice with one quantum $\alpha_1^t = 2/\sqrt{3}$. Sufficient precision in the determination of B and L is enough to choose between the two types of lattice. It is easier to determine the number p of flux quanta, as the corresponding coefficient α_p is $p\alpha_1$.

The determination of L is not simple because the Bragg angle, for the first reflection, is very small. The Bragg condition yields

$$\lambda_N = 2\alpha_1 \sin\theta = \sqrt{3}\, L \sin\theta, \qquad (3.135)$$

where λ_N is the neutron wavelength and α_1 the distance between the reticular lines. In order to have a sufficient flux of incoming neutron λ_N cannot exceed 5 Å. This leads to $\theta \simeq 6 \times 10^{-3}$ radian (20′).

The experiments were performed at Saclay by Cribier, Jacrot, Madhav Rao and Farnoux (1964 and to be published). We quote here their more recent results: niobium was chosen as the substance for investigation since in this element the penetration depth is not too large. The sample consisted of a bunch of single crystals insulated from each other. The field is applied parallel to the axis. Figure 3.9 shows the magnetization curve obtained for this sample at 4·2°K. It is seen that the magnetization is irreversible. Neutron diffraction experiments were performed in the region 1 and 2 as well as in region 3 where trapped flux is present. Figure 3.10 gives the diffraction pattern for different values of the applied field, and for incoming neutrons with an average wavelength of 4·2 Å and a dispersion of about 1·3 Å. A peak is observed which moves to larger angles when the field is increased (L decreases). Moreover, an experiment made with a better resolution shows that the width of the peak is of the order of magnitude expected for a Bragg peak, taking into account the various experimental uncertainties for the angle and the incoming neutron wavelength. This shows that there is long range ordering of the vortex lines. To account for the experimental results this ordering must exist over several hundred lattice spacings. Figure 3.11 represents the variation of $(1/d_1^2)$ $[d_1 = (\sqrt{3}/2)L]$ with the induction B. It is seen that, except in the vicinity of H_{c1}, the relation between $1/d_1^2$ and B is linear as expected from the flux quantization con-

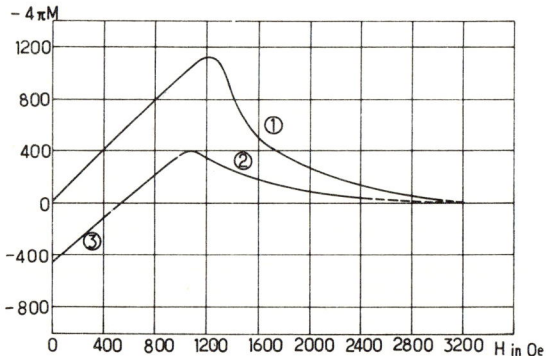

FIG. 3.9. Magnetization curve of the Nb sample used by Cribier et al. in the neutron diffraction experiment (private communication)

dition. The slope α is $2/\sqrt{3}\phi_0$ which is the value corresponding to the triangular lattice with one quantum per line.

The departure from this linear relation just above H_{c1} is due to the fact that, in the large sample used, only a part of the volume is occupied by the vortex lines and consequently the value of the induction is underestimated. This explanation is confirmed by the behaviour of the intensity of the scattered neutrons when the applied field is increased. This intensity increases in

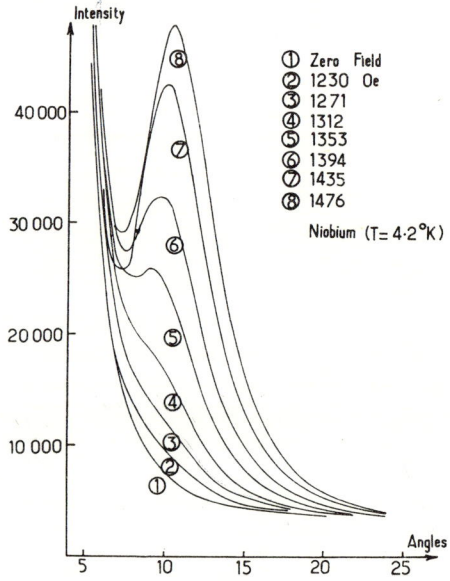

FIG. 3.10. Neutron diffraction pattern in Nb for various values of the applied field (private communication)

the domain where the relation $1/d_1^2 = f(B)$ departs from the straight line because the volume occupied by the line pattern increases. When the whole volume of the sample is occupied, the relation $1/d_1^2 = f(B)$ is linear and the intensity decreases as expected from the cross-section formulae.

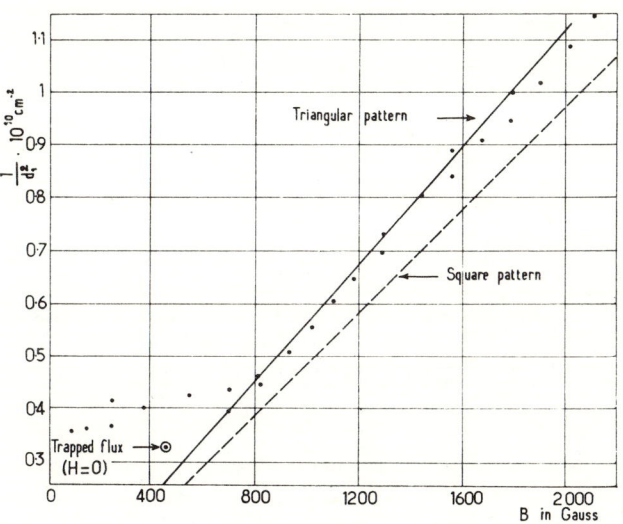

FIG. 3.11. Variation of $1/d_1^2$ versus the induction B in Nb (after Cribier et al.: private communication)

It was possible to observe only the first Bragg peak, as expected in the case of a triangular lattice. The variation of the field in space is almost sinusoidal as predicted by (3.88).

Even in the trapped flux region (region 3 in Fig. 3.9), the experiments show that a regular triangular array exists.

It is to be emphasized that the above experiments were performed at a temperature rather far from $T_0 (T_0 \simeq 8.2°\text{K})$ and with a low value of κ ($\kappa \sim 1.4$). Moreover another experiment has been performed on a Pb–2% Bi alloy for which κ is even smaller ($\kappa = 0.9$). The scattering angle is, of course, smaller than in niobium and the accuracy is not sufficient to allow the triangular and square lattices to be distinguished. However, one can infer from the data that there is only one quantum per line. It should be noted that these experiments have also shown that the laminar model is not correct.†

† Very recently Essmann and Trauble [Phys. Letters, 24A, 526 (1967)] have observed the flux line pattern direcly in Pb — 4% In and in niobium (Fig. 3.12). The pattern is obtained by depositing small ferromagnetic particles on the specimen and observing these with an electron microscope using a replica technique (see frontispiece).

The experiments were performed in the remanent state. The observed patterns confirm entirely the neutron diffraction results. In particular, in niobium for an induction smaller than 500 gauss, the flux line pattern is seen only near the axis of the sample. This confirms the fact that the induction is not homogeneous and explains why the relation $1/d^2 \propto B$ is not obtained close to H_{c1} (Fig. 3.11).

Reversible Properties

(b) Nuclear magnetic resonance

Other methods may be used to investigate the vortex line structure by measuring the field inhomogeneities. It is possible to use the Mössbauer effect (Sarma, 1964) and also nuclear magnetic resonance. This last technique has been used by Gossard *et al.* (1964) who have measured the line broadening, due to the field inhomogeneities, in the resonance of V^{51} in V_3Ga and V_3Si samples. Indeed the NMR line shape is determined by the moments of the magnetic field. For example, the second moment of the resonance line is

$$(\overline{\nabla H^2}) = \overline{h^2} - \bar{h}^2. \tag{3.136}$$

If the profile of the line is gaussian and if the broadening is caused by the field inhomogeneities alone, the above quantity determines the width of the line. The spatial average of the field $\bar{h}$ is, of course, the induction B. In the region where the density of lines is $\rho_L \ll 1/\xi^2(T)$, the average of the square of the field $\overline{h^2}$ is easily computed from (3.126) and

$$\overline{h^2} = \frac{1}{S}\int h^2 d\sigma = \sum_M h_M h_{-M} = \rho_L \Phi_0^2 \sum_M \frac{2}{[1+\lambda^2(T)M^2]^2}. \tag{3.137}$$

In (3.137) the summation can be replaced by an integral and the width becomes

$$\overline{\Delta H^2} = \frac{B^2 L^4}{4\pi\lambda^2(T)[L^2 + 4\pi^2\lambda^2(T)]}. \tag{3.138}$$

This result has been deduced by Gossard *et al.* (1964) for a square lattice. In the region $\xi(T) \ll L \ll \lambda(T)$ the width is approximatively equal to

$$(\overline{\Delta H^2})^{1/2} = \frac{\Phi_0}{\sqrt{(8\pi^3)}\lambda^2(T)}. \tag{3.139}$$

The broadening is of order H_{c1} (110 Oe for V_3Si and V_3Ga).

The comparison with experiment must be made cautiously because the observed line width arises from several other effects (nuclear dipolar broadening, quadrupolar broadening, inhomogeneous Knight shift and a broadening proportional to the magnitude of the field modulation employed) which is of the order of 10–20 Oe. It is interesting to note that for the laminar model:

$$(\overline{\Delta H^2})^{1/2} = \frac{\sqrt{5}}{60} B \left[\frac{L}{\lambda(T)}\right]^2 = \frac{\sqrt{5}}{60} B \left(\frac{3}{2}\frac{H_c^2}{B^2}\right)^{2/3}, \tag{3.140}$$

i.e. the laminar model predicts a variation of the width proportional to $B^{-1/3}$ while for the vortex line structure this width should be independent of B. Experiments are at present being made to check this dependency.

Delrieu and Winter (1966) have observed the nuclear magnetic resonance of ^{93}Nb nuclei in the superconducting state near H_{c2} They have also calculated the line shape as a function of the field distribution for a triangular and a square lattice. The experimental results confirm the triangular arrangement and also give the value of the magnetization near H_{c2} allowing the various κ values to be determined. Fite and Redfield (1966), using a relaxation time technique, arrive at a similar conclusion for fields varying from H_{c1} to H_{c2} in a vanadium sample.

Appendix to Chapter 3

3.1. In order to show that the solutions ψ_n corresponding to an eigenvalue $-\alpha_n$ of (3.1) are degenerate, let us calculate ψ_n explicity in a special gauge. Using the dimensionless quantities (2.37), eqn. (3.1) reads

$$\left(i\frac{\nabla}{\kappa} + \mathcal{A}\right)^2 f = f. \quad (A.3.1)$$

For $\mathcal{A}$ one can use the following gauge:

$$\mathcal{A} = (0, hx, 0), \quad (A.3.2)$$

where h is the reduced field corresponding to the applied field H. Equation (A.3.1) then becomes

$$-\frac{1}{\kappa^2}\frac{\partial^2}{\partial x^2}f + \left(\frac{i}{\kappa}\frac{\partial}{\partial y} + hx\right)^2 f - \frac{1}{\kappa^2}\frac{\partial^2}{\partial z^2}f = f. \quad (A.3.3)$$

The general solution of (A.3.3) has the form

$$f = e^{ik_z z} e^{ik_y y} \phi(x). \quad (A.3.4)$$

$\phi(x)$ satisfies the following second order differential equation:

$$-\frac{1}{\kappa^2}\frac{\partial^2}{\partial x^2}\phi(x) + \left(-\frac{k_y}{\kappa} + hx\right)^2 \phi(x) = \left(1 - \frac{k_z^2}{\kappa^2}\right)\phi(x). \quad (A.3.5)$$

This equation gives bounded solutions for $\phi(x)$ only for discrete values of h. These values are independent of k_y as can be seen by performing the translation $X = x - k_y/\kappa h$, for which the equation changes into the familiar equation for the harmonic oscillator. Each value of h is given by

$$h = \frac{\kappa^2 - k_z^2}{(2n+1)\kappa} \quad (n \text{ integer}), \quad (A.3.6)$$

For each value of h (i.e. of n and k_z) there exists an infinity of functions of the form (A.3.4). The highest value of h is given by $k_z = 0$, $n = 0$ and

$$h = \kappa \quad (\text{i.e. } H_{c2} = \kappa\sqrt{2}\,H_c), \quad (A.3.7)$$

Properties of Type II Superconductors

the corresponding function being

$$f_k = e^{-ik_y y} e^{(-\kappa^2/2)(x - k_y/\kappa^2)^2}. \qquad (A.3.8)$$

3.2. Proof of equation (3.37)

By multiplying (3.16) by $df_0/d\rho$ and (3.17) by $dh/d\rho$, subtracting and regrouping the various terms, one gets

$$h\frac{dh}{d\rho} = \frac{1}{2\rho^2}\frac{d}{d\rho}\frac{\rho^2}{f_0^2}\left(\frac{dh}{d\rho}\right)^2 - \frac{1}{2\kappa^2\rho^2}\frac{d}{d\rho}\rho^2\left(\frac{df_0}{d\rho}\right)^2 - f_0(1 - f_0^2)\frac{df_0}{d\rho}. \qquad (A.3.9)$$

The free energy can be written as

$$\mathscr{F} = 2\pi \int_0^\infty \left[h^2 + \frac{1}{2}(1 - f_0^4)\right]\rho\,d\rho$$

$$= \pi |h^2\rho^2|_0^\infty - 2\pi \int_0^\infty \rho^2 h \frac{dh}{d\rho} d\rho + \pi \int_0^\infty (1 - f_0^4)\rho\,d\rho.$$

Using (A.3.9)

$$\mathscr{F} = -\pi \left|\frac{\rho^2}{f_0^2}\left(\frac{dh}{d\rho}\right)^2\right|_0^\infty + \frac{\pi}{\kappa^2}\left|\rho^2\left(\frac{df_0}{d\rho}\right)^2\right|_0^\infty +$$

$$+ \pi \int_0^\infty \left[2\rho f_0(1 - f_0^2)\frac{df_0}{d\rho} + 1 - f_0^4\right]\rho\,d\rho$$

$$= -\frac{\pi}{2}|\rho^2(1 - f_0^2)^2|_0^\infty + 2\pi \int_0^\infty (1 - f_0^2)\rho\,d\rho$$

$$= 2\pi \int_0^\infty \rho(1 - f_0^2)\,d\rho. \qquad (A.3.10)$$

In deriving this expression, the boundary conditions (3.22) and (3.23) have been used.
The above expression is valid whatever the value of κ.

3.3. Proof of equation (3.43)

The line free energy is

$$\mathscr{F} = 2\pi \int_0^\infty \left[h^2 + \frac{1}{2}(1 - f_0^4)\right]\rho\,d\rho. \qquad (A.3.11)$$

Reversible Properties

Using (3.17) $\mathscr{F}$ becomes

$$\mathscr{F} = 2\pi \int_0^\infty \left[\frac{h}{\rho} \frac{d}{d\rho} \frac{\rho}{f_0^2} \frac{dh}{d\rho} + \frac{1}{2}(1-f_0^4) \right] \rho \, d\rho$$

$$= 2\pi \left| \frac{h\rho}{f_0^2} \frac{dh}{d\rho} \right|_0^\infty - 2\pi \int_0^\infty \frac{\rho}{f_0^2} \left(\frac{dh}{d\rho} \right)^2 d\rho + \pi \int_0^\infty (1-f_0^4) \rho \, d\rho.$$

(A.3.12)

Or, using the boundary condition (3.22) and (3.16),

$$\mathscr{F} = \frac{2\pi}{\kappa} h(0) - 2\pi \int_0^\infty \frac{f_0}{\kappa^2} \frac{d}{d\rho} \rho \frac{df_0}{d\rho} + \pi \int_0^\infty (1-f_0^2)^2 \rho \, d\rho.$$

(A.3.13)

Using the expression (3.36) for f_0 it is seen that the last integrals are of order $1/\kappa^2$ so that

$$\mathscr{F} \simeq \frac{2\pi}{\kappa} h(0).$$

(A.3.14)

3.4. *Proof of the one quantum per line condition*

It is obvious that, if $f_0(x, y)$ is a solution of (3.68) corresponding to p flux quanta per line, $f_0^{1/p}[(\sqrt{p})x, (\sqrt{p})y]$ is also a solution of this equation. The periods of this new function are $a/\sqrt{p}$, $b/\sqrt{p}$ and the area of the unit cell is $2\pi/\kappa^2$. The function $f_0^{1/p}[(\sqrt{p})x, (\sqrt{p})y]$ corresponds to one quantum per line. To show that there is only one quantum per line it is sufficient to prove that the β_A value for this last function is smaller than the value corresponding to $f_0(x, y)$, i.e.

$$\frac{\langle f_0^4 \rangle}{\langle f_0^2 \rangle^2} \geq \frac{\langle f_0^{4/p} \rangle}{\langle f_0^{2/p} \rangle^2}$$

(A.3.15)

or

$$\frac{\langle \phi^{2p} \rangle}{\langle \phi^p \rangle^2} \geq \frac{\langle \phi^2 \rangle}{\langle \phi \rangle^2},$$

(A.3.16)

where

$$\phi = f_0^{2/p} = |f|^{2/p}$$

(A.3.17)

is a positive definite function of x and y.
Equation (A.3.16) reads:

$$\langle \phi^{2p}(\mathbf{r}_1) \phi(\mathbf{r}_2) \phi(\mathbf{r}_3) - \phi^2(\mathbf{r}_1) \phi^p(\mathbf{r}_2) \phi^p(\mathbf{r}_3) \rangle \geq 0$$

(A.3.18)

or, equivalently,

$$\frac{1}{3} \langle \phi^{2p}(r_1) \phi(r_2) \phi(r_3) - \phi^2(r_1) \phi^p(r_2) \phi^p(r_3)$$

$$+ \phi^{2p}(r_2) \phi(r_3) \phi(r_1) - \phi^2(r_2) \phi^p(r_3) \phi^p(r_1)$$

$$+ \phi^{2p}(r_3) \phi(r_1) \phi(r_2) - \phi^2(r_3) \phi^p(r_1) \phi^p(r_2) \rangle \geq 0. \quad (A.3.19)$$

It is easily seen that the quantity

$$ABC \{A(A^{2q} - B^q C^q) + B(B^{2q} - C^q A^q) + C(C^{2q} - A^q B^q)\}, \quad (A.3.20)$$

where q is an integer and A, B, C are positive quantities, is positive and vanishes only for $A = B = C$.

Thus the integrand of (A.3.19) is positive and (A.3.16) is verified. The equal sign holds only when $\phi(r)$ is reduced to a constant. This last case, of course, never happens in the problem.

Bibliography of Chapter 3

ABRIKOSOV, A. A. (1957), *Zh. Eksperim. i Teor. Fiz.* **32**, 1442. (English translation: *Soviet Phys. JETP*, **5**, (1957) 1174.)

CRIBIER, D., JACROT, B., MADHAV RAO, L. and FARNOUX B. (1964), *Phys. Letters*, **9**, 106. To be published in *Progr. Low Temp. Phys.* vol. **5**.

CRIBIER, D., FARNOUX, B., JACROT, B., MADHAV RAO, L., VIVET, B. and ANTONINI, M. (1964), *Proceedings of the IXth International Conference on Low Temperature Physics*, Colombus, Ohio. In this paper it was concluded wrongly that the lattice is a square one.

DEAVER, B. S. and FAIRBANK, W. M. (1961), *Phys. Rev. Letters*, **7**, 43.

DELRIEU, J. H. and WINTER, J. H. (1966) *Solid State Comm.*, **4**, 545.

DOLL, R. and NABAUER, M. (1961), *Phys. Rev. Letters*, **7**, 51.

FETTER, A. L., HOHENBERG, P. C. and PINCUS, P. (1966), *Phys. Rev.* **147 A**, 140.

FITE, W. and REDFIELD, A. G. (1966), *Phys. Rev. Letters*, **7**, 381.

FRIEDEL, J., DE GENNES, P. G. and MATRICON, J. (1963), *Appl. Phys. Letters*, **2**, 119.

DE GENNES, P. G. (1963–4–5), *Métaux et Alliages Supraconducteurs*. Cours professe à la Faculté d'Orsay.

DE GENNES, P. G. and MATRICON, J. (1964), *Rev. Mod. Phys.* **36**, 45.

GOODMAN, B. B. (1964), *Compt. Rend. Acad. Sci. Paris*, **258**, 5175.

GOSSARD, A. C., JACCARINO, V., PINCUS, P. and WERNICK, J. H. (1964), *Communication to the Conference on the Physics of Type II Superconductivity*. Western Reserve University, Cleveland, Ohio, U.S.A.

HARDEN, J. L. and ARP, V. (1963), *Cryogenics* **3**, 105.

JOINER, W. C. H. and BLAUGHER, R. D. (1964), *Rev. Mod. Phys.* **36**, 67.

KINSEL, T., LYNTON, E. A. and SERIN, B. (1963), *Phys. Letters*, **3**, 30.

KLEINER, W. M., ROTH, L. M. and AUTLER, S. H. (1964), *Phys. Rev.* **133**, A, 1226.

LANDAU, L. (1930), *Z. Phys.* **64**, 629.

LASHER, G. (1965), *Phys. Rev.* **140 A**, 523.

LONDON, F. (1950), *Superfluids*, John Wiley, vol. I, p. 152.

MARCUS, P. M. (1964), *Communication to the Conference on the Physics of Type II Superconductivity*. Western Reserve University, Cleveland, Ohio, U.S.A.

MATRICON, J. (1964), *Phys. Letters*, **9**, 289; (1966), Thèse, Université d'Orsay.

SARMA, G. (1964), *Compt. Rend. Acad. Sci. Paris*, **258**, 1461.

CHAPTER 4

Surface Superconductivity

4.1. Nucleation in a semi-infinite medium. The field H_{c3}

In Chapter 3, it has been shown that, for an infinite sample, superconductivity is destroyed when the applied field H is larger than $H_{c2} = \kappa\sqrt{2}H_c$. Conversely, when the field H is reduced, superconducting regions appear spontaneously at $H = H_{c2}$. This conclusion was obtained for an infinite sample, i.e. by neglecting surface effects. The value of H_{c2} was obtained from the linearized Ginzburg–Landau equation (3.1) using the condition that the order parameter ψ is a bounded function. It will be shown in this chapter that, for an ideal material, nucleation always takes place first on the surface and that this has a drastic influence on the nucleation field (Saint-James and de Gennes, 1963).

It has already been mentioned in Chapter 2 that the two Ginzburg–Landau equations must be supplemented by a boundary condition to ensure that there is no current flowing through the surface of the sample. For a boundary separating a superconductor from an insulator or from vacuum, the microscopic theory shows that the convenient boundary condition is†

$$\left(-i\hbar\nabla - \frac{2e}{c}\mathbf{A}\right)_n \psi = 0, \qquad (4.1)$$

where n indicates the component of $\left[-i\hbar\nabla - (2e/c)\mathbf{A}\right]$ parallel to the normal of the surface. Condition (4.1) implies that the radius of curvature of the surface is much larger than $\xi(T)$, a criterion which is related to the general validity of the Ginzburg–Landau scheme (see section 2.1).

In order to investigate the influence of the boundary condition (4.1), we shall thus consider the problem of a semi-infinite superconductor bounded by the plane $x = 0$ and filling the half-plane $x > 0$. The half-plane $x < 0$ is filled by an insulator (or is vacuum) (Fig. 4.1).

Close to the nucleation field the magnitude of the order parameter is small and the first Ginzburg–Landau equation may be linearized and written as

$$\frac{1}{2m}\left(-i\hbar\nabla - \frac{2e}{c}\mathbf{A}\right)^2 \psi + \alpha\psi = 0. \qquad (4.2)$$

† For a more detailed discussion on the boundary condition see Caroli *et al.* (1962).

Reversible Properties

In section 3.1, it was shown that the vector potential A corresponding to the applied field H is consistent with this linearization, i.e.

$$H = \operatorname{curl} A \tag{4.3}$$

The difference with the treatment of section 3.1 arises from the fact that (4.2) and (4.3) are now supplemented by the boundary condition (4.1). Two cases must be investigated.

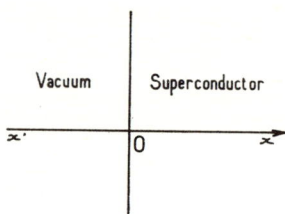

Fig. 4.1.

(a) The applied field perpendicular to the surface of the sample

If the z axis is taken parallel to the field, and if A is chosen in the gauge

$$A_x = 0, \quad A_y = Hx, \quad A_z = 0, \tag{4.4}$$

(4.2) reads

$$-\frac{\hbar^2}{2m}\frac{\partial^2}{\partial x^2}\psi + \frac{1}{2m}\left[-i\hbar\frac{\partial}{\partial y} - \frac{2e}{c}Hx\right]^2\psi - \frac{\hbar^2}{2m}\frac{\partial^2}{\partial z^2}\psi = -\alpha\psi \tag{4.5}$$

while the boundary condition becomes

$$\frac{\partial}{\partial z}\psi = 0 \quad \text{for } z = 0. \tag{4.6}$$

The general solution of this problem is given by:

$$\psi(x, y, z) = \psi(x, y) \cos k_z z, \tag{4.7}$$

where $\psi(x, y)$ must be a bounded function.

It is clear that the problem is now equivalent to the problem of section 3.1 and that the highest value of the field is obtained for the lowest eigenvalue, i.e. for

$$-\alpha = \frac{e\hbar}{mc} H \tag{4.8}$$

which yields

$$H = H_{c2}.$$

The corresponding eigenfunction is

$$\psi = \psi_k(x, y) = e^{ik_y y} e^{-(x-x_0)^2/2\,\xi^2(T)}, \qquad (4.9)$$

where

$$x_0 = \frac{\hbar c}{2eH} k_y \qquad (4.10)$$

and $\xi(T)$ is the temperature dependent coherence length.

Hence, for an applied field perpendicular to the boundary, the presence of the surface does not modify the nucleation field which is equal to H_{c2}.

(b) The applied field parallel to the surface

If the z axis is parallel to the applied field H, (4.2) takes the form (4.5) but the boundary condition becomes

$$\frac{\partial \psi}{\partial x} = 0 \quad \text{for } x = 0. \qquad (4.11)$$

The solution of (4.5) may be written as

$$\psi = e^{ik_z y} e^{ik_y z} f(x). \qquad (4.12)$$

As we are interested in the lowest eigenvalue we consider only the case $k_z = 0$ and (4.5) reads

$$-\frac{\hbar^2}{2m}\frac{d^2 f}{dx^2} + \frac{2e^2 H^2}{mc^2}(x - x_0)^2 f = -\alpha f \qquad (4.13)$$

the boundary condition being

$$\frac{df}{dx} = 0 \quad \text{for } x = 0 \qquad (4.14)$$

and x_0 is given by (4.10).

Equation (4.13) has the form of the Schrödinger equation for an harmonic oscillator of frequency $\omega = 2eH/mc$, the minimum of the potential, i.e. the equilibrium position, being located at x_0.

The boundary condition (4.14) shows that the eigenvalue will be given by

$$-\alpha = g_0 \frac{e\hbar}{mc} H, \qquad (4.15)$$

where g_0 is a number depending on the value of x_0. As the nucleation field

Reversible Properties

corresponds to the highest field H obtainable from (4.15), an extra condition is obtained by writing

$$\frac{dg_0}{dx_0} = 0. \tag{4.16}$$

The problem is equivalent to finding the lowest value of g_0.

In order to investigate the behaviour of g_0 with x_0, one can first consider the two special cases $x_0 \gg \xi(T)$ and $x_0 = 0$.

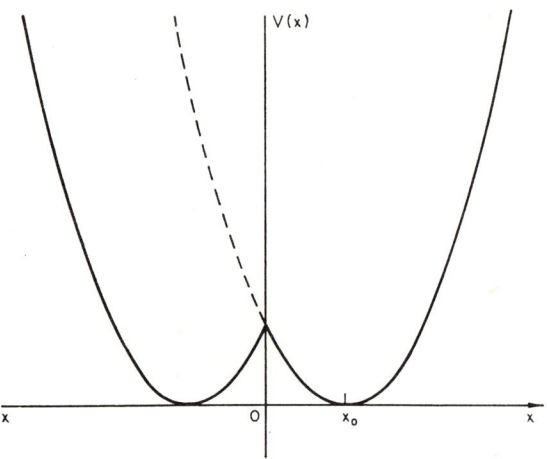

FIG. 4.2. Form of the potential (4.17)

For $x_0 \gg \xi(T)$, the harmonic oscillator function (4.9) is nearly zero for $x = 0$ and the boundary condition is fulfilled. The solution is still given by (4.9) the eigenvalue being (4.8), i.e. $g_0 = 1$.

For $x_0 = 0$, the harmonic oscillator function (4.9) is still the solution required as ψ is an extremum for $x = x_0$ and $g_0 = 1$.

For intermediate values, the eigenvalue is lowered (i.e. $g_0 < 1$) as can be seen from the following argument.

The eigenvalue problem (4.13) and eqn. (4.14) may be replaced by another Schrödinger equation where the potential $V(x)$ is symmetrized (Fig. 4.2):

$$\left. \begin{array}{ll} V(x) = \dfrac{2e^2H^2}{mc^2}(x-x_0)^2 & \text{for } x > 0, \\[2mm] V(x) = V(-x) & \text{for } x < 0, \end{array} \right\} \tag{4.17}$$

The eigenfunction corresponding to the lowest eigenvalue associated to the potential $V(x)$ has no node and is even so that it satisfies the boundary conditions (4.14) automatically.† For $x < 0$, $V(x)$ is always smaller

† The only condition to be imposed on this eigenfunction is that it should be bounded in the whole domain $-\infty < x < +\infty$.

than the potential $(2e^2H^2/mc^2)(x - x_0)^2$. Consequently, the eigenvalue associated with $V(x)$ is smaller than the eigenvalue (4.1) associated with $(2e^2H^2/mc^2)(x - x_0)^2$ and the boundary conditions $f(x) = 0$ for $x = \pm \infty$, i.e. g_0 is smaller than 1. A particular value of x_0 exists close to the surface for which g_0 is a minimum and this means that the nucleation will be easier in the vicinity of the surface.

To calculate the optimum value of x_0 and of g_0 it is convenient to adopt the following dimensionless quantities:

$$X = \frac{x}{\xi(T)}, \quad K = \xi(T)k_y, \quad h = \frac{2eH}{\hbar c}\xi^2(T), \quad \xi^2(T) = -\frac{\hbar^2}{2m\alpha}. \quad (4.18)$$

Equations (4.13) and (4.14) become

$$-\frac{d^2f}{dX^2} + (K - hX)^2 f = f, \quad (4.19)$$

$$\frac{df}{dX} = 0 \quad \text{at} \quad X = 0. \quad (4.20)$$

A further change of variable, writing $t = X\sqrt{h} - K/\sqrt{h}$, yields

$$-\frac{d^2f}{dt^2} + t^2 f = \frac{1}{h}f = g_0 f, \quad (4.21)$$

$$\frac{df}{dt} = 0 \quad \text{for} \quad t = -\mu = -\frac{K}{\sqrt{h}}, \quad (4.22)$$

since

$$g_0 = -\frac{mc\alpha}{e\hbar H} = \frac{1}{h}. \quad (4.23)$$

The most general solution of (4.21) is a linear combination of Weber functions (see, for example, Morse and Feshbach, *Methods of Theoretical Physics*, vol. 2, p. 1403).

As g_0 is smaller than 1, the following integral representation can be used for these functions:

$$\Delta_{g_0}(t) = e^{t^2/2} \int_0^\infty u^{-(1+g_0)/2} e^{-(u-t)^2} du \quad (4.24)$$

and the solution of (4.21) is

$$f(t) = A\Delta_{g_0}(t) + B\Delta_{g_0}(-t). \quad (4.25)$$

The coefficients A and B are determined by the boundary conditions. It is clear that for $t \to \infty$, $\Delta_{g_0}(t) \to \infty$, and $\Delta_{g_0}(-t) \to 0$. Thus $A = 0$.

Reversible Properties

For $x = 0$, i.e. for $t = -\mu$, the boundary condition yields

$$\int_0^\infty (2u - \mu) u^{-(1+g_0)/2} e^{-(u-\mu)^2} du = 0, \qquad (4.26)$$

or

$$2I_1 - \mu I_0 = 0, \qquad (4.27)$$

where

$$I_\alpha(\mu) = \int_0^\infty u^\alpha u^{-(1+g_0)/2} e^{-(u-\mu)^2} du. \qquad (4.28)$$

Equation (4.25) [or (4.26)] yields an implicit relation between g_0 and μ (i.e. between the critical field H and x_0). The lowest eigenvalue is obtained for $dg_0/d\mu = 0$, i.e.

$$4I_2 - 6\mu I_1 + (2\mu^2 - 1)I_0 = 0. \qquad (4.29)$$

Using the recursion formula,

$$2I_2 = 2\mu I_1 + \frac{1 - g_0}{2} I_0, \qquad (4.30)$$

one easily obtains

$$g_0 = \mu^2, \qquad (4.31)$$

where μ is a number given by

$$\int_0^\infty (2u - \mu) u^{-(1+\mu^2)/2} e^{-(u-\mu)^2} du = 0. \qquad (4.32)$$

This equation yields

$$g_0 = \mu^2 = 0.59010, \qquad (4.33)$$

i.e.

$$h = \frac{1}{0.5901} = 1.6946. \qquad (4.34)$$

The corresponding nucleation field is

$$H_{c3} = 1.695 \frac{\hbar c}{2e} \frac{1}{\xi^2(T)} = 1.695 \, H_{c2} = 1.695 \sqrt{2}\kappa H_c. \qquad (4.35)$$

In order to investigate the physical situation corresponding to this nucleation, one must consider the corresponding variation of the order parameter, i.e. the eigenfunction ψ of (4.5). Equation (4.33) shows that $K = 1$, i.e. $k_y = 1/\xi(T)$, so that the minimum of the potential is located at

$$x_0 = -\frac{\hbar c}{2eH_{c3}} k_y = \mu^2 \xi(T). \qquad (4.36)$$

Surface Superconductivity

The eigenfunction (which is the same as the order parameter) is then

$$\psi(x, y) = e^{iy/\xi(T)} e^{1/2[(x-\mu^2\xi(T))/\mu\xi(T)]^2} \int_0^\infty u^{-(1+\mu^2)/2} e^{-[u+(x-\mu^2\xi(T))/\mu\xi(T)]^2} du.$$

(4.37)

Thus the order parameter varies over a distance of the order of $\mu\xi(T)$ from the surface (Fig. 4.3).

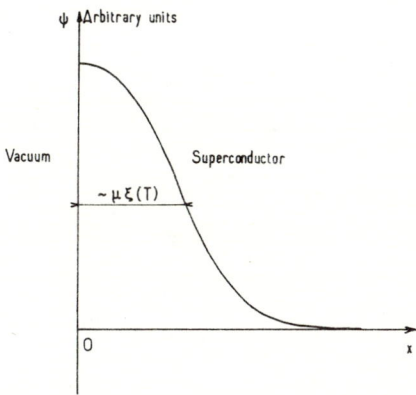

FIG. 4.3. Variation of the order parameter close to the surface

This result has important consequences in the onset of the superconductivity. They are reviewed below. Three cases must be distinguished according to the value of κ.

(α) $\kappa > 1/\sqrt{2}$ (Type II superconductor): $H_{c3} > H_{c2} > H_c$.

For type II superconductors the above calculation shows that superconductivity is not entirely destroyed for $H_{c2} < H < H_{c3}$. A superconducting sheath remains close to the surface parallel to the applied field. Conversely when the field is decreased below H_{c3}, a superconducting sheath appears at the surface before superconductivity is restored in the bulk at $H = H_{c2}$. If the sample is a long cylinder with the applied field parallel to the axis, the sheath will cover all the surface of the cylinder. If it is a sphere, the sheath will be restricted to a small zone near the equatorial plane when H is close to H_{c3}. When the field is decreased towards H_{c2} the sheath will progressively extend up to the poles.

The presence of this sheath explains several anomalies in the behaviour of type II superconductors which were very often attributed to metallurgical imperfections of the samples:

Reversible Properties

(i) the transition from the superconducting to the normal state (as a function of field) occurs over a wide fields range, since it takes place from H_{c2} to H_{c3} (Fig. 4.4);

(ii) the transition field is different when measured by magnetization (which usually gives H_{c2}) and by resistivity techniques (giving H_{c3}) (Joiner and Blaugher, 1964; Druyvesteyn *et al.*, 1964; Bon Mardion *et al.*, 1964). The superconducting sheath acts as a short circuit up to H_{c3};

(iii) the critical field depends on the orientation of the applied field relative to the surface. (This point will be discussed later. See section 4.3.)

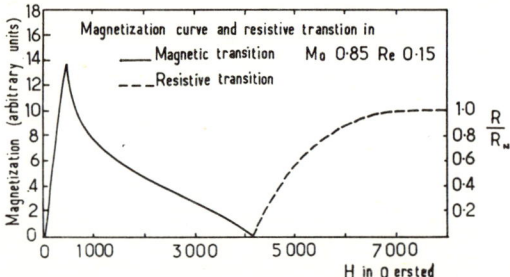

FIG. 4.4. Resistive transition from the superconducting to the normal state showing the broadening due to the existence of the surface sheath (after Joiner and Blaugher, 1964)

Many experiments have been performed to detect the surface sheath and to determine the ratio H_{c3}/H_{c2}. Table 4.1 surveys these experiments (listed by Serin, 1964). It is seen that the agreement between experiment and the predicted value for H_{c3}/H_{c2} is fairly good.

The presence of the superconducting sheath is shown by the following observations:

(i) magnetization measurements have shown that the anomaly in the nucleation field is not a bulk effect;

(ii) the anomaly is still seen when the impedance is measured at high frequencies (Cardona and Rosenblum, 1964);

(iii) the measurement of magnetic permeability in a parallel field does not show any variation close to H_{c2}. The superconducting sheath gives perfect screening for fields up to H_{c3}, and the permeability increases only at $H \sim H_{c3}$ (Burger *et al.*, 1965a);

(iv) tunnelling experiments (Tomash, 1964; Guyon *et al.*, 1965b) display a density of states which deviates significantly from the normal state value;

(v) the sheath acts as a small diamagnetic region which can be detected by torque measurements (Tomash and Joseph, 1964).

Surface Superconductivity

TABLE 4.1. SUMMARY OF EXPERIMENTAL RESULTS ON LIMITING FIELD FOR SURFACE CONDUCTIVITY, (AFTER SERIN. 1964)

Ref.	Method and Specimen Form	Material	Type	Results
1	R ; B	Pb + (0·03–6·6% In)	I II	Qualitative agreement
2	R ; B	In + Sn (Doidge) Pb + (15–65%Tl)	I II	Agreement to ± 20% $H_{c3}/H_{c2} = 1·64 \pm 0·08$
3	R ; B	50% Nb + 50% Ta Pb + (17% In) (Pb–In) + Cu-plate	II II II	$H_{c3}/H_{c2} = 1·71$ $H_{c3}/H_{c2} = 1·96$ $H_{c3}/H_{c2} = 1·15$
4	M ; Fi, Fo	Pb + (4–10%Tl)	II	$H_{c3}/H_{c2} = 1·69 \pm 0·02$
5	R, μ (600 c/s); Fi	Sn + (6% In)	II	$H_{c3}/H_{c2} = 1·67$
6	R ; B	In + (2–6% Pb)	II	$H_{c3}/H_{c2} = 1·86$
7	R (23 kMc/sec); B	50% Pb + 50% Tl 83% Pb + 17% In	II II	$H_{c3}/H_{c2} = 1·4–1·9$ $H_{c3}/H_{c2} = 1·6 \pm 0·1$
8	R ; B	In· + 6% Pb (In–Pb) + Cu-plate	II II	$H_{c3}/H_{c2} = 1·86$ $H_{c3}/H_{c2} = 1·15$
9	Tunnel, Fi	Pb + (5–14% Tl)	II	$H_{c3}/H_{c2} = 1·69$
10	R (23 kMc/sec); B	Pb + 1% Tl	I	$H_{c3}/\sqrt{2}\kappa H_c = 1·9$
11	χ, χ'' (18–100 c/s);B	Pb + (0·1–5% Bi)	I	$H_{c3}/\sqrt{2}\kappa H_c = 1·86$
12	μ (10⁴ c/s); B	Pb + (5–20% Tl)	II	$H_{c3}/H_{c2} = 1·75 \pm 0·02$

Legend: R = resistance; μ = permeability; χ, χ'' = real and imaginary part of susceptibility; B = bulk; Fi = film; and Fo = foil, M = Magnetization.

1. W. F. DRUYVESTEYN, D. J. VAN OOIJEN and T. J. BERBEN, *Rev. Mod. Phys.* **36**, 58 (1964).
2. G. BON MARDION, B. B. GOODMAN and A. LACAZE, *Phys. Letters*, **8**, 15 (1964).
3. C. F. HEMPSTEAD and Y. B. KIM, *Phys. Rev. Letters*, **12**, 145 (1964).
4. W. J. TOMASCH and A. S. JOSEPH, *Phys. Rev. Letters*, **12**, 148 (1964).
5. J. P. BURGER and E. GUYON, *Contribution to type II superconductors conference*, Cleveland, (Ohio), 1964.
6. S. GYGAX, J. L. OLSEN and R. H. KROPSCHOT, *Phys. Letters*, **8**, 228 (1964).
7. M. CARDONA and M. B. ROSENBLUM, *Phys. Letters*, **8**, 308 (1964).
8. S. GYGAX and R. H. KROPSCHOT, *Phys. Letters*, **9**, 91 (1964).
9. W. J. TOMASCH, *Phys. Letters*, **9**, 104 (1964).
10. B. ROSENBLUM and M. CARDONA, *Phys. Letters*, **9**, 220 (1964).
11. M. STRONGIN, A. PASKIN, D. G. SCHWEITZER, O. F. KAMMERER and P. P. CRAIG, *Phys. Rev. Letters*, **12**, 442 (1964).
12. H. R. HART, Jr. and P. S. SWARTZ, *Phys. Letters*, **10**, 40 (1964).

(β) $\qquad 0·418 \leq \kappa \leq \dfrac{1}{\sqrt{2}} \; ; H_{c2} < H_c < H_{c3} .$

For these type I superconductors in the field $H_c < H < H_{c3}$ a superconducting sheath still exists in the regions where $\boldsymbol{H}$ is parallel to the surface.

Reversible Properties

This situation can be observed in dilute alloys. For example, Seraphim (1962) has made a systematic study of In alloys. A broadening in the resistive transition was observed above the following concentrations in the alloys:

$$\left.\begin{array}{l} 0\cdot7 - 1\% \text{ Bi} \\ 6\cdot2 - 7\cdot4\% \text{Ta} \\ 1\cdot9\% \text{ Pb} \\ 3 - 4\% \text{ Cd} \end{array}\right\} \text{ in In}$$

The broadening in the magnetization curve, however, appeared only above concentrations approximately twice as large as these.

All the critical concentrations of the alloys listed above have about the same resistivity and κ is found to be of the order of $0\cdot4$ for each of them.

De Gennes (1964a) and Maki (1964), predicted that κ will vary with temperature, increasing by about 20% when T goes from T_0 to zero (see Chapter 5). It is then possible to have a situation in which $H_{c3} > H_c$ at $T = 0$ and $H_{c3} < H_c$ for $T = T_0$. For example, in lead (Strongin et al., 1964; Goldstein, 1964) tunnelling is observed in fields applied parallel to the sample up to $1\cdot22\, H_c$ at a temperature $T = 1\cdot4\,°K$ where κ is of the order of $0\cdot50$ but is not observed for fields greater than H_c for $T \sim T_0$ where $\kappa \sim 0\cdot38$.

(γ) $\qquad\qquad \kappa < 0\cdot418\,; \quad H_{c2} < H_{c3} < H_c$

In these type I superconductors H_{c2} and H_{c3} are supercooling fields. When the field is reduced below H_c nucleation is generally observed in localized regions of the sample, probably around defects. In very good samples, and close to T_0 where $\xi(T)$ is large (much larger than the radius of the impurity centres), the field may be lowered below H_c, while the sample remains in the normal state, in metastable equilibrium. However, at H_{c3} superconducting regions are created spontaneously near the sample surface and the normal state is strictly unstable for $H < H_{c2}$. This has been carefully studied by Faber (1954, 1957). He observed that:

(i) the nucleation field is proportional to H_c, and
(ii) the nucleation always starts close to the surface parallel to the applied field.

In calculating κ, Faber identified the nucleation field H_n with H_{c2}.

In fact it has to be identified with H_{c3} and if this is done one obtains the κ values given below.

However, it has to be kept in mind that this procedure is not rigorously correct since the experiments were performed far below T_0, and as stated in Chapter 2, the Ginzburg–Landau scheme is strictly correct only at tem-

	Al	In	Sn
κ_{obs}	0·015	0·07	0·10
κ_{calc}	0·01	0·05	0·15

peratures close to T_0. However the values obtained for κ are in rather good agreement with κ calculated using the relation:

$$\kappa = 0.96\, \lambda_L/\xi_0$$

4.2. Nucleation in a slab

The preceding calculation is valid for a semi-infinite medium. The case of a slab of thickness $2a$ will be investigated below. It will be shown that the nucleation field varies with the ratio $a/\xi(T)$ when the applied field is parallel to the surface. However the conclusions of section 4.1a are not modified and, for an applied field perpendicular to the surface, the nucleation field turns out to be $H_{c2} = \kappa \sqrt{2} H_c$, as above.

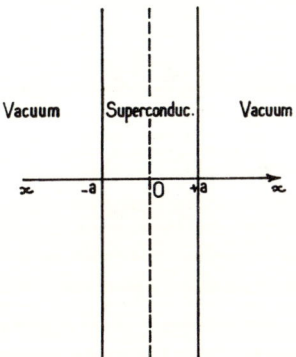

FIG. 4.5.

a) *The parallel nucleation field in a slab* (Fig. 4.5)

The presence of the two boundaries will modify the conclusion of the section 4.1b and the calculations have to be repeated in this new geometry. As several results obtained here will be useful in the discussion of the angular dependence of the nucleation field (see section 4.3), the calculation will be developed below.

The boundary condition (4.11) must be replaced by

$$\frac{\partial \psi}{\partial x} = 0 \quad \text{for} \quad x = \pm a. \tag{4.38}$$

Reversible Properties

If one introduces the following dimensionless quantities

$$\left. \begin{array}{l} X = \dfrac{x}{a}, \quad Y = \dfrac{y}{a}, \quad Z = \dfrac{z}{a}; \\[2mm] h = \dfrac{2eH}{\hbar c} \; a^2 = \dfrac{2\pi a^2}{\phi_0} H; \quad \varepsilon = -\dfrac{2m\alpha}{\hbar^2} a^2 = \dfrac{a^2}{\xi^2(T)} = \dfrac{2\pi a^2}{\phi_0} H_{c2} \end{array} \right\} \quad (4.39)$$

(4.5) and (4.38) become

$$-\frac{\partial^2 \psi}{\partial X^2} - \frac{\partial^2 \psi}{\partial Z^2} + \left[i\frac{\partial}{\partial Y} + hX \right]^2 \psi = \varepsilon \psi, \qquad (4.40)$$

$$\frac{\partial \psi}{\partial X} = 0 \quad \text{for} \quad X = \pm 1, \qquad (4.41)$$

ψ being a function of (X,Y,Z) bounded and normalizable. The solution of (4.40) may be written in the form

$$\psi(X, Y, Z) = e^{ik_z Z} e^{ik_y Y} f(X). \qquad (4.42)$$

As the highest nucleation field is obtained for the lowest value of ε/h one considers the case $k_z = 0$ and the equations read

$$-\frac{d^2 f}{dX^2} + (k_y - hX)^2 f = \varepsilon f \qquad (4.43)$$

$$\frac{df}{dX} = 0 \quad \text{for} \quad X = \pm 1. \qquad (4.44)$$

At this stage it is particularly convenient to introduce the new variable

$$t = \sqrt{(2h)} \left[X - \frac{k_y}{h} \right] \qquad (4.45)$$

which yields

$$-\frac{d^2 f}{dt^2} + \frac{t^2}{4} f = \frac{\varepsilon}{2h} f, \qquad (4.46)$$

$$\frac{df}{dt} = 0 \quad \text{for} \quad t = \sqrt{(2h)} \left[\pm 1 - \frac{k_y}{h} \right]. \qquad (4.47)$$

Equation (4.46) is the so called Weber equation. Its most general solution is

$$f(t) = \alpha_\nu D_\nu(t) + \beta_\nu D_\nu(-t), \qquad (4.48)$$

where α_ν and β_ν are constants and $D_\nu(t)$ is the Weber function.† ν is related to ε/h by

$$\frac{\varepsilon}{h} = 2\nu + 1. \qquad (4.49)$$

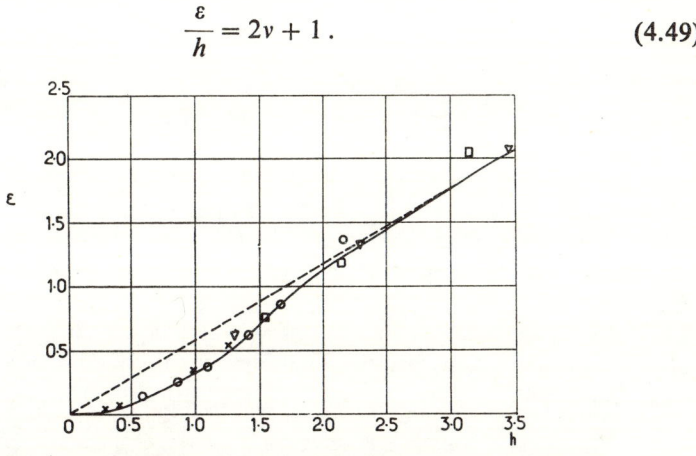

FIG. 4.6. Variation of ε versus h, i.e., of the nucleation field as a function of the slab thickness as calculated by Saint-James and De Gennes (1963). The experimental points are due to Burger et al., and refer to various samples (1965a)

The relation between ν and k_y is obtained by writing the two boundary conditions (4.47). This yields the following implicit relation:

$$D'_\nu\left[\sqrt{(2h)}\left(1 - \frac{k_y}{h}\right)\right] D'_\nu\left[\sqrt{(2h)}\left(1 + \frac{k_y}{h}\right)\right] =$$
$$= D'_\nu\left[-\sqrt{(2h)}\left(1 + \frac{k_y}{h}\right)\right] D'_\nu\left[-\sqrt{(2h)}\left(1 - \frac{k_y}{h}\right)\right], \qquad (4.50)$$

where D'_ν is the derivative of D_ν with respect to t.

For each value of h and k_y this equation gives a infinite but discrete, set of values of ν. Using the smallest value $\nu_m(k_y, h)$, the value k_0 of k_y is determined for which $\nu_m(k_y, h)$ reaches its minimum value ν_0. Then, from (4.49) one obtains the value of ε/h for a given h, or, conversely, the value of h as a function ε [$= a^2/\xi^2(T)$]. Thus the nucleation field $H_{\parallel}$ is a function of the ratio $a/\xi(T)$.

The variation $h = h(\varepsilon)$ has been computed by Saint-James and de Gennes (1963) and is shown in Fig. 4.6. It is seen that for

$$\frac{a^2}{\xi^2(T)} \ll 1 \qquad \frac{H_{\parallel}}{H_{c2}} \text{ tends towards infinity}$$

† See, for example Erdelyi et al., *Higher Transcendental Functions* (McGraw-Hill), vol. II, p. 115 and ff.

Reversible Properties

while for

$$\frac{a^2}{\xi^2(T)} \gg 1 \qquad \frac{H_{//}}{H_{c2}} \text{ tends toward } 1\cdot 69 \text{ as expected}$$

since the behaviour approaches that of a semi-infinite medium.

In order to check this predicted variation one can either change the thickness of the slab, or, more easily, the temperature since $\xi(T)$ is a function of the temperature. This was proposed and done by Burger et al. (1965a) who obtained the experimental points shown in Fig. 4.6. The agreement between theory and experiment is very good and, at a given temperature T, the simultaneous measurement of H_{c2} [i.e. of $\xi(T)$] and of $H_{//}$ may be used to calculate a.

Having determined the value k_0 of k_y which corresponds to the minimum ε_0 for a given h, the ratio α_{v0}/β_{v0} can be obtained from the boundary conditions and

$$\frac{\alpha_{v0}}{\beta_{v0}} = \frac{D'_{v_0}[-\sqrt{(2h)}(1 - k_y/h)]}{D'_{v_0}[\sqrt{(2h)}(1 - k_y/h)]} = \frac{D'_{v_0}[\sqrt{(2h)}(1 + k_y/h)]}{D'_{v_0}[-\sqrt{(2h)}(1 + k_y/h)]}. \quad (4.51)$$

However, (4.50) is not modified when k_y is changed into $-k_y$. This means that to a given pair of h and v values there correspond two values k_y and $-k_y$. This is related to the fact that the equation for f is invariant when k_y is changed to $-k_y$ and X to $-X$. To the lowest eigenvalue, v_0, correspond $\pm k_0$ and two degenerate eigenfunctions, $f_{k_0}(t)$ and $f_{-k_0}(t)$.

For $k_y = k_0$ ($k_0 > 0$) the expression for $f(x)$ reads

$$f_{-k_0}(X) = \alpha_{v_0} D_{v_0}\left[\sqrt{(2h)}\left(X - \frac{k_0}{h}\right)\right] + \beta_{v_0} D_{v_0}\left[-\sqrt{(2h)}\left(X - \frac{k_0}{h}\right)\right], \quad (4.52)$$

where the ratio α_{v_0}/β_{v_0} is given by (4.51).

For $k_y = -k_0$, it is seen from (4.51) that the roles of α_{v_0} and β_{v_0} are exchanged so that

$$f_{k_0}(X) = \alpha_{v_0} D_{v_0}\left[-\sqrt{(2h)}\left(X + \frac{k_0}{h}\right)\right] + \beta_{v_0} D_{v_0}\left[\sqrt{(2h)}\left(X + \frac{k_0}{h}\right)\right], \quad (4.53)$$

and that

$$f_{-k_0}(X) = f_{k_0}(-X). \quad (4.54)$$

The most general solution of (4.40) and (4.41) is then

$$\psi(X, Y) = \alpha e^{ik_0 Y} f_{k_0}(X) + \beta e^{-ik_0 Y} f_{k_0}(-X), \quad (4.55)$$

Surface Superconductivity

where α and β are constants. It is clear that one can take for $f_{k_3}(X)$ a function of the form (4.53) which is real and normalized in the domain $-1 \leq X \leq 1$ without restricting the generality of (4.55). Actually, α and β are to be obtained by solving the non-linear Ginzburg–Landau equations supplemented by the boundary conditions (4.38).

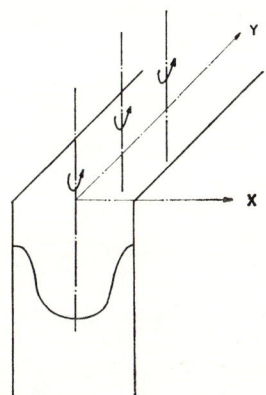

FIG. 4.7a. Schematic representation of the vortex structure for $h \gg h_0^c$

As the plane $X = 0$ is a symmetry plane, it is expected that the modulus of ψ will exhibit the symmetry†

$$|\psi(X, Y)| = |\psi(-X, Y)| \tag{4.56}$$

so that

$$|\alpha| = |\beta|. \tag{4.57}$$

The order parameter is then

$$\psi(X, Y) = \alpha_0 \left[e^{i(k_0 Y + \phi)} f_{k_0}(X) + e^{-i(k_0 Y + \phi)} f_{k_0}(-X) \right], \tag{4.58}$$

where α_0 is any complex constant and ϕ an arbitrary phase. (The value of α_0 and ϕ should be determined by solving the non-linearized equations.) It is seen that $\psi(X, Y)$ vanishes for

$$\cos(k_0 Y + \phi)[f_{k_0}(X) + f_{k_0}(-X)] = 0, \tag{4.59a}$$

$$\sin(k_0 Y + \phi)[f_{k_0}(X) - f_{k_0}(-X)] = 0. \tag{4.59b}$$

According to the form (4.53) of $f_{k_0}(X)$ and (4.51), the only solutions of these two equations are

$$X = 0, \tag{4.60a}$$

$$Y = -\frac{\phi}{k_0} + (2p + 1)\frac{\pi}{2k_0}. \tag{4.60b}$$

† This is in agreement with the symmetry properties of the non-linear Ginzburg–Landau equation.

Reversible Properties

The order parameter vanishes in the $X = 0$ plane, at the points given by (4.60b). Thus one expects a regular array of vortices with distance $L = \pi a/k_0$ between them (Fig. 4.7a). The variation of L with $a/\xi(T)$ will be studied in the next section.

(b) *The variation of the parallel nucleation field*

In order to determine the behaviour of h and k_y with ε it is interesting to study the behaviour of these quantities close to a certain value h_0 of h.

To this value h_0 there correspond a particular value v_0 of v and two values $\pm k_0$ of k_y which give the "ground state" of (4.41) and (4.43), i.e. the smallest value ε_0 of ε. We shall study the behaviour of ε, h and k_y close to ε_0, h_0, $+k_0$ by second order perturbation theory by writing (4.43) for f as

$$-\frac{d^2 f}{dX^2} + [(k_0 - h_0 X)^2 + (k_y - hX)^2 - (k_0 - h_0 X)^2] f = \varepsilon f, \quad (4.61)$$

where k_y and ε are the values corresponding to h. This equation must be supplemented with the boundary condition:

$$\frac{df}{dX} = 0 \quad \text{for} \quad X = \pm 1. \quad (4.62)$$

Equation (4.50) gives all the values of v corresponding to k_0 and h_0, i.e. all the "excited state" ε_v values corresponding to the "ground state" (v_0, k_0, h_0) (or ε_0). The corresponding eigenfunctions normalized to unity in the domain $-1 \leq X \leq 1$ will be denoted by $|v\rangle$. To second order, with $(k_y - hX)^2 - (k_0 - h_0 X)^2$ as the perturbation potential, ε is given by

$$\varepsilon = \varepsilon_0 + \langle 0|[k_y + k_0 - (h + h_0)X][k_y - k_0 - (h - h_0)X]|0\rangle$$
$$- \sum_{\substack{v \\ v \neq v_0}} \frac{|\langle v|[k_y + k_0 - (h + h_0)X][k_y - k_0 - (h - h_0)X]|0\rangle|^2}{\varepsilon_v - \varepsilon_0}.$$
$$(4.63)$$

Setting $h = h_0 + dh$, $k_y = k_0 + dk$, ε reads

$$\varepsilon = \varepsilon_0 + 2[k_0 - h_0 \langle 0|X|0\rangle] dk + 2[h_0 \langle 0|X^2|0\rangle - k_0 \langle 0|X|0\rangle] dh$$
$$+ \left[1 - \sum_{\substack{v \\ v \neq v_0}} \frac{4}{\varepsilon_v - \varepsilon_0} |\langle 0|[k_0 - h_0 X]|v\rangle|^2\right] dk^2 - 2\left[\langle 0|X|0\rangle\right.$$
$$\left. - \frac{1}{2} \sum_{\substack{v \\ v \neq v_0}} \frac{4}{\varepsilon_v - \varepsilon_0} \langle 0|[k_0 - h_0 X]|v\rangle\langle v|X(k_0 - h_0 X)|0\rangle + cc\right] dh\, dk$$
$$+ \left[\langle 0|X^2|0\rangle - \sum_{\substack{v \\ v \neq v_0}} \frac{4}{\varepsilon_v - \varepsilon_0} |\langle 0|X(k_0 - h_0 X)|v\rangle|^2\right] dh^2 \quad (4.64)$$

Surface Superconductivity

Comparing this result with the Taylor expansion it is seen that

$$\frac{\partial \varepsilon}{\partial k} = 2[k_0 - h_0 \langle 0|X|0\rangle], \tag{4.65a}$$

$$\frac{\partial \varepsilon}{\partial h} = 2[h_0 \langle 0|X^2|0\rangle - k_0 \langle 0|X|0\rangle]. \tag{4.65b}$$

$$\frac{\partial^2 \varepsilon}{\partial k^2} = 2\left[1 - \sum_{v \neq v_0} \frac{4}{\varepsilon_v - \varepsilon_0} |\langle 0|k_0 - h_0 X|v\rangle|^2\right], \tag{4.65c}$$

$$\frac{\partial^2 \varepsilon}{\partial h \, \partial k} = -2\left[\langle 0|X|0\rangle - \frac{1}{2}\sum_{v \neq v_0} \frac{4}{\varepsilon_v - \varepsilon_0} \langle 0|k_0 - h_0 X|v\rangle \right.$$
$$\left. \langle v|X(k_0 - h_0 X)|0\rangle + cc\right], \tag{4.65d}$$

$$\frac{\partial^2 \varepsilon}{\partial h^2} = 2\left[\langle 0|X^2|0\rangle - \sum_{v \neq v_0} \frac{4}{\varepsilon_v - \varepsilon_0} |\langle 0|X(k_0 - h_0 X)|v\rangle|^2\right]. \tag{4.65e}$$

The minimum of ε is obtained for

$$\frac{\partial \varepsilon}{\partial k} = 0 \quad \text{and} \quad \frac{\partial^2 \varepsilon}{\partial k^2} > 0. \tag{4.66}$$

The equality yields

$$\frac{k_0}{h_0} = \langle 0|X|0\rangle. \tag{4.67}$$

For $k_0 = 0$, the equation for f is even and the lowest eigenfunction is also even and has no nodes. It is seen that for this function $\langle 0|X|0\rangle = 0$ so that

$$\frac{\partial \varepsilon_0}{\partial k} = 0 \quad \text{for } k_0 = 0. \tag{4.68}$$

Thus, for a given h, $k_0 = 0$ always corresponds to an extremum of ε. This extremum will be a minimum, i.e. will correspond to the nucleation situation, as long as $\partial^2 \varepsilon / \partial k^2 \geq 0$, i.e. for

$$\sum_{v \neq v_0} \frac{4}{\varepsilon_v - \varepsilon_0} |\langle 0|h_0 X|0\rangle|^2 \leq 1. \tag{4.69}$$

For $h_0 = 0$ it is seen that the lowest solution is obtained for $k_0 = 0$, and this will be the case up to a field h_0^c for which $\partial^2 \varepsilon / \partial k^2$ vanishes and changes sign. This value h_0^c corresponds to a certain value ε_0^c.

Reversible Properties

For $h_0 < h_0^c$ (or $\varepsilon_0 < \varepsilon_0^c$), $k_0 = 0$ and the nucleation starts with its maximum located in the centre of the slab (Fig. 4.7b).

For $h_0 > h_0^c$ (or $\varepsilon_0 > \varepsilon_0^c$), k_0 is different from zero. There are two values $\pm k_0$ corresponding to the two functions $f_{k_0}(X)$ and $f_{k_0}(-X)$. The maximum of $f_{k_0}(X)$ is located between the axis $x = 0$ and the surface of the slab (Fig. 4.7a and b).

Thus the point $h_0 = h_0^c$ may be considered as a threshold value for which "surface superconductivity" begins to occur, or alternatively for which the vortices appear in the sample.

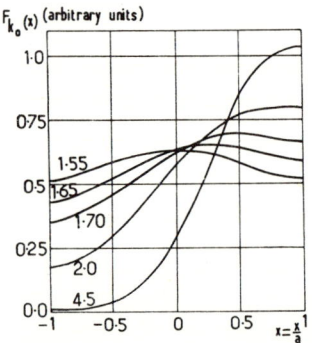

FIG. 4.7b. Variation of the function $f_{k_0}(X)$ for various values of h showing the progressive appearance of surface superconductivity

In order to investigate the appearance of these vortices it is necessary to study the behaviour of k_0 near h_0^c. Equation (4.67) yields the implicit relation between k_0 and h_0. Differentiating this expression yields

$$dk_0 = \langle 0|X|0\rangle dh_0 + h_0 d(\langle 0|X|0\rangle). \qquad (4.70)$$

The differential of $\langle 0|X|0\rangle$ is easily computed by a first order expansion of the wave function $|0\rangle$ near h_0 and k_0, and after some easy algebra one obtains

$$\frac{\partial^2 \varepsilon}{\partial k_0^2} dk_0 = \frac{\partial^2 \varepsilon}{\partial h_0 \, \partial k_0} dh_0. \qquad (4.71)$$

For $h_0 < h_0^c$, $k_0 = 0$ and $\partial^2 \varepsilon / \partial h_0 \partial k_0 \equiv 0$ as each eigenfunction $|\nu\rangle$ has a given parity. Thus $dk \equiv 0$.

For $h_0 > h_0^c$, $k_0 \neq 0$, $|\nu\rangle$ no longer has a definite parity and $\partial^2 \varepsilon / \partial h_0 \partial k_0$ is different from zero. The ratio dk_0/dh_0 is finite and positive for $k_0 > 0$ (negative for $k_0 < 0$).

For $h_0 = h_0^c$, $\partial^2 \varepsilon / \partial k_0^2$ and $\partial^2 \varepsilon / \partial h_0 \partial k_0$ vanish simultaneously (for $k_0 = 0$). However for $h_0 = h_0^c$ it can be seen that $\partial^2 \varepsilon / \partial k_0^2$ is proportional to k_0^2 while $\partial^2 \varepsilon / \partial h_0 \, \partial k_0$ is proportional to k_0.

This means that dk_0/dh_0 is infinite for $h_0 = h_0^c$. There is a discontinuity of the second kind in the behaviour of k_0 for $h_0 = h_0^c$ (Fig. 4.8) and consequently in the behaviour of $\partial\varepsilon/\partial h_0 = d\varepsilon/dh_0$ (Fig. 4.9). A numerical calculation has yielded $\varepsilon_0^c \sim 0.8$, $h_0^c \sim 1.61$ and the ratio $h_0^c/\varepsilon_0^c \sim 2$.

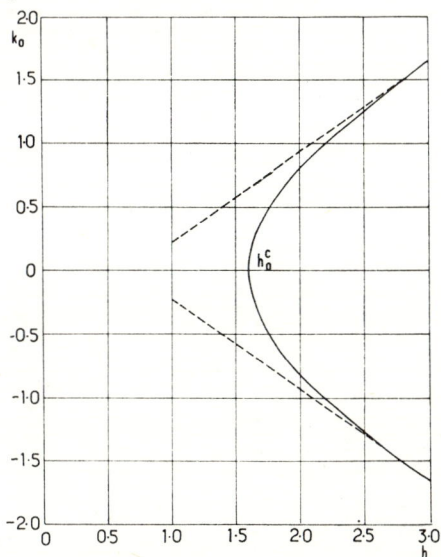

FIG. 4.8. Behaviour of the parameter k_0 as a function of h showing the value h_0^c at which surface superconductivity begins to occur. The dashed line represents the asymptotic formula for k_0 (see page 99)

Thus, for $h_0 = h_0^c$, vortices appear in the sample in the plane $x = 0$. The distance between these vortices is $L = a\pi/k_0$. Due to the particular variation of k_0 close to h_0^c it is seen that the number of vortices increases very quickly with h_0, while the distance between them decreases. This is a situation somewhat similar to the behaviour of vortices close to H_{c1} in an ideal reversible substance (cf. Chapter 3). It must be emphasized that when the thickness of the slab increases in comparison to $\xi(T)$, the amplitude of the order parameter at $x = 0$ decreases, so that the variation of H becomes very small (see Fig. 4.7a and b).

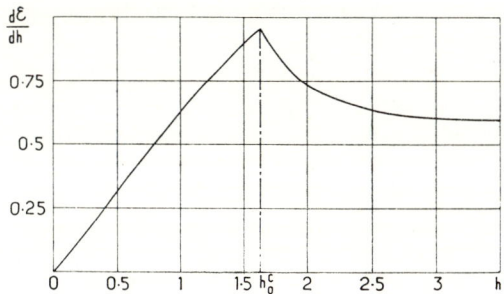

FIG. 4.9. Variation of $d\varepsilon/dh$ versus h showing the discontinuity at $h = h_0^c$

For h_0 very small, i.e. for a very thin sample, or very close to T_0, one can obtain the expression of ε_0 as a function of h_0 by a perturbation calculation. For $h_0 = 0$, the eigenfunctions and the eigenvalues are

$$\cos n\pi X \qquad \text{corresponding to} \quad \varepsilon = n^2\pi^2$$

$$\sin(2n+1)\frac{\pi}{2}X \qquad \text{corresponding to} \quad \varepsilon = (2n+1)^2\frac{\pi^2}{4}$$

and to second order in h_0^2 one obtains ($h_0^2 x^2$ is the perturbing potential):

$$\varepsilon_0 = \frac{h_0^2}{3} - \frac{8h_0^4}{3 \times 5 \times 7 \times 9} \tag{4.72}$$

If one retains only the term in h_0^2, it is seen that

$$h_0^2 = 3\varepsilon_0. \tag{4.73}$$

or

$$H_{//} = \frac{\sqrt{3}}{2\pi}\frac{\phi_0}{a\xi(T)} = \sqrt{6}\frac{\lambda(T)}{a}H_c \tag{4.74}$$

This formula was first obtained by Ginzburg (1952) by another technique.

For $h_0 \gg h_0^c$, i.e. for $h_0 \gtrsim 3$, the ratio ε_0/h_0 tends, as expected, towards the constant value μ^2, obtained in section (4.1). It can be shown that for high values of h_0, k_0 has the asymptotic form:

$$k_0 \simeq h_0 - \mu\sqrt{h_0}.$$

This means that there is a sheath on the two surfaces; thus for high values of ε_0, i.e. of $a/\xi(T)$, the perfect surface nucleation is obtained, and this occurs as soon as $a \gtrsim \sqrt{2}\,\xi(T)$.

4.3. Angular dependence of the nucleation field

It is interesting to study the behaviour of the nucleation field when the direction of the applied field varies from parallel to perpendicular to the surface of the sample. It might be expected that the nucleation field will decrease progressively from $H_{//}$ to H_{c2} and this has actually been observed by Tomasch and Joseph (1964), Hart and Swartz (1964) as well as by Burger et al. (1965a). However, the results obtained by these different experimentalists do not agree together completely, most likely because of the differences in the thickness of the samples used. It is thus convenient to consider a slab of thickness $2a$, where the direction of the applied field makes an angle θ with the sample surface (Fig. 4.10).

In the linearized Ginzburg–Landau equation the vector potential A may be chosen in the gauge

$$A_x = 0, \quad A_y = Hx\cos\theta - Hz\sin\theta, \quad A_z = 0, \tag{4.75}$$

so that (4.2) reads

$$-\frac{\hbar^2}{2m}\frac{\partial^2}{\partial x^2}\psi + \frac{1}{2m}\left[-i\hbar\frac{\partial}{\partial y} - \frac{2e}{c}Hx\cos\theta + \frac{2e}{c}Hz\sin\theta\right]^2\psi$$
$$-\frac{\hbar^2}{2m}\frac{\partial^2}{\partial z^2}\psi = -\alpha\psi. \qquad (4.76)$$

The boundary conditions are given by (4.1), i.e.

$$\frac{\partial\psi}{\partial x} = 0 \quad \text{for} \quad x = \pm a. \qquad (4.77)$$

Moreover, ψ must be bounded and normalizable.

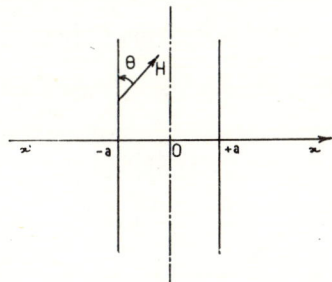

FIG. 4.10.

Using the dimensionless quantities (4.39), eqns. (4.76) and (4.77) become

$$-\frac{\partial^2\psi}{\partial X^2} - \frac{\partial^2\psi}{\partial Z^2} + \left[-i\frac{\partial}{\partial Y} - hX\cos\theta + hZ\sin\theta\right]^2\psi = \varepsilon\psi, \qquad (4.78)$$

$$\frac{\partial\psi}{\partial X} = 0 \quad \text{for} \quad X = \pm 1. \qquad (4.79)$$

The solution of this equation may be written as

$$\psi(X, Y, Z) = e^{ikY\cos\theta}\psi(X, Z,), \qquad (4.80)$$

where $\psi(X, Z)$ obeys the following differential equation:

$$-\frac{\partial^2\psi}{\partial X^2} - \frac{\partial^2\psi}{\partial Z^2} + [k\cos\theta - hX\cos\theta + hZ\sin\theta]^2\psi = \varepsilon\psi, \qquad (4.81)$$

$$\frac{\partial\psi}{\partial X} = 0 \quad \text{for } X = \pm 1. \qquad (4.82)$$

Reversible Properties

The eigenvalue of the above problem is a function $\varepsilon(k, h, \theta)$ and for a given value of h and θ, one should look for the value of k for which ε is a minimum. The solution of the two-dimensional problem (4.81) and (4.82), which involves a complicated coupling of the X and Z variables has not yet been achieved. What can be done, however, is to calculate the value of ε close to $\theta = 0$, i.e. to find the behaviour of the nucleation field close to $H_{//}$ (Saint-James, 1965). If θ is very small, the value of $\varepsilon(k, h, \theta)$ may be obtained by a perturbation method. The unperturbed equation is chosen to be

$$-\frac{\partial^2 \psi}{\partial X^2} - \frac{\partial^2 \psi}{\partial Z^2} + (k - hX)^2 \cos^2\theta \psi + h^2 Z^2 \sin^2\theta \psi = \varepsilon_0 \psi, \quad (4.83)$$

where ε_0 is the unperturbed eigenvalue and the boundary conditions are given by (4.82). The perturbation potential is

$$V = 2 h(k - hX) Z \cos\theta \sin\theta. \quad (4.84)$$

The solution of (4.83) can be written in the form

$$\psi(X, Z) = f(X) g(Z) \quad (4.85)$$

and $f(X)$ and $g(Z)$ obey the following equations:

$$-\frac{d^2 f}{dX^2} + (k - hX)^2 \cos^2\theta f = \varepsilon_0' f, \quad (4.86)$$

$$-\frac{d^2 g}{dZ^2} + h^2 Z^2 \sin^2\theta g = \varepsilon_0'' g, \quad (4.87)$$

$$\frac{df}{dX} = 0 \quad \text{for } X = \pm 1. \quad (4.88)$$

f and g are bounded and normalizable functions. The eigenvalue ε_0 of the unperturbed problem is obviously

$$\varepsilon_0 = \varepsilon_0' + \varepsilon_0''. \quad (4.89)$$

As g must be a bounded and normalizable function it is clear that (4.87) is the ordinary harmonic oscillator equation. The eigenvalues are

$$\varepsilon_0''(n) = (2n + 1) h \cos\theta,$$

while the corresponding eigenfunctions are

$$g_n(Z) = \left(\frac{1}{n!}\right)^{1/2} \frac{1}{\sqrt{2}} \left(\frac{h \sin\theta}{\pi}\right)^{1/4} D_n[Z\sqrt{(2h \sin\theta)}] \quad (4.90)$$

Surface Superconductivity

where n is an integer and D_n the Weber function associated with the harmonic oscillator.† On the other hand, the equation for f is identical to (4.43) where k_y is changed to $k \cos \theta$ and h to $h \cos \theta$. The eigenvalues are obtained from (4.50) (with the same modification of h and k_y) and are

$$\varepsilon_0'(v) = (2v + 1) h \cos \theta, \tag{4.91}$$

while the corresponding eigenfunctions are

$$f_v(X) = \alpha_v D_v \left[\sqrt{(2h \cos \theta)} \left(X - \frac{k}{h} \right) \right] + \beta_v D_v \left[-\sqrt{(2h \cos \theta)} \left(X + \frac{k}{h} \right) \right]. \tag{4.92}$$

α_v and β_v are related by (4.51) (with h changed into $h \cos \theta$ and k_y into $k \cos \theta$). It is clear that v is a function of k, h and $\cos \theta$.

α_v will be determined by normalizing $f_v(X)$ to unity in the domain $-1 \leq X \leq +1$. The normalized function $f_v(X)$ will be designated by $|v>$ and the normalized function $g_n(Z)$ by $|n>$.

Thus the unperturbed eigenvalues are

$$\varepsilon_0^{n,v} = (2v + 1) h \cos \theta + (2n + 1) h \sin \theta. \tag{4.93}$$

The perturbed eigenvalue $\varepsilon^{n,v}$ is obtained by a perturbation expansion, which introduces the matrix element of V, i.e.

$$\langle n v | V | n' v' \rangle. \tag{4.94}$$

Since $|n>$ and $|v>$ are actually functions of $Z \sqrt{(2h \sin \theta)}$ and $(X - k/h) \sqrt{(2h \cos \theta)}$ it is convenient to write V as

$$V = h(\cos \theta \sin \theta)^{1/2} \left(\frac{k}{h} - X \right) \sqrt{(2h \cos \theta)} Z \sqrt{(2h \sin \theta)} = \lambda^{1/2} \mathscr{V}, \tag{4.95}$$

$$\text{where } \lambda^{1/2} = h(\text{soc } \theta \sin \theta)^{1/2}. \tag{4.96}$$

It is seen that the perturbation expansion introduces a power series of $\lambda^{1/2}$:

$$\varepsilon^{n,v} = \varepsilon_0'(v) + \varepsilon_0''(n) + \sum_p \lambda^{p/2} F_{p/2}(\theta, h, k), \tag{4.97}$$

where $F_{p/2}(\theta, h, k)$ may be expressed as a function of the matrix elements of $\mathscr{V}$. The first order correction to $\varepsilon^{n,v}$ is

$$\lambda^{1/2} \langle v | \left(\frac{k}{h} - X \right) \sqrt{(2h \cos \theta)} | v \rangle \langle n | Z \sqrt{(2h \sin \theta)} | n \rangle. \tag{4.98}$$

† $D_n(x) = e^{-x^2/4} 2^{-n/2} H_n(\sqrt{2} x)$ when H_n is the Hermite polynomial of degree n. $D_n(x) = (-)^n D_n(-x)$ and $D_n(x)$ is the only function which yields the correct behaviour for g.

TS 8

Reversible Properties

This correction is clearly equal to zero since the eigenfunction $|n\rangle$ has a definite parity. This will also be the case for all the odd order corrections in (4.97) so that

$$\varepsilon^{n,v} = (2v+1)h\cos\theta + (2n+1)h\sin\theta + \sum_{p=1}^{\infty} h^{2p}(\cos\theta\sin\theta)^p [F_p(\theta, h, k)]. \tag{4.99}$$

In order to obtain the lowest eigenvalue, it is first necessary to set $n = 0$, and to select the lowest eigenvalue $v_0(k)$ of v for a given value of k. Then, as in the preceding section, one has to select the value of k for which ε^{0,v_0} is a minimum, so that

$$\frac{\partial \varepsilon^{0,v_0}}{\partial k} = 0, \tag{4.100}$$

i.e.
$$2\frac{\partial v_0}{\partial k} h\cos\theta + \sum_{p=1}^{\infty} h^{2p}(\cos\theta\sin\theta)^p \frac{\partial F_p}{\partial k} = 0. \tag{4.101}$$

In order to ensure that ε^{0,v_0} is the required minimum eigenvalue this condition must be supplemented by:

$$\frac{\partial^2 \varepsilon^{0,v_0}}{\partial k^2} > 0. \tag{4.102}$$

Thus the lowest eigenvalue will be

$$\varepsilon = (2v_0 + 1)h\cos\theta + h\sin\theta + \sum_{p=1}^{\infty} h^{2p}(\cos\theta\sin\theta)^p F_p(\theta, h, k). \tag{4.103}$$

Expression (4.103) gives an implicit relation between h, ε and θ, provided that k is obtained from (4.101), i.e. an implicit relation between the nucleation field $H(\theta)$, the ratio $a/\xi(T)$ and the angle θ.

In order to check this expression two kinds of experiment may be performed.

(a) At a given temperature T, a field is applied parallel to the sample surface and equal to the nucleation field $H_{//}(T)$. Then the sample is tilted by an angle θ.

In order to restore the nucleation situation one has to modify the temperature, i.e. $\xi(T)$ and consequently ε, if the magnitude of the applied field is maintained equal to $H_{//}$ (i.e. h constant). Differentiating (4.103) with respect to θ, h being constant, yields

$$\frac{d\varepsilon}{d\theta} = -(2v_0 + 1)h\sin\theta + 2\frac{\partial v_0}{\partial k}\frac{dk}{d\theta}h\cos\theta + 2\frac{\partial v_0}{\partial \theta}h\cos\theta + h\cos\theta$$

$$+ \sum_{p=1}^{\infty} h^{2p}\left\{\frac{2p}{2^p}\cos 2\theta(\sin 2\theta)^{p-1}F_p(\theta, h, k)\right\} + \sum_{p=1}^{\infty} h^{2p}(\sin\theta\cos\theta)^p$$

$$\left(\frac{\partial F_p}{\partial \theta} + \frac{\partial F_p}{\partial k}\frac{dk}{d\theta}\right) \tag{4.104}$$

Surface Superconductivity

The condition (4.101) shows that the coefficient of $dk/d\theta$ is equal to zero and

$$\frac{d\varepsilon}{d\theta} = -(2v_0+1)h\sin\theta + 2\frac{\partial v_0}{\partial\theta}h\cos\theta + h\cos\theta + \sum_{p=1}^{\infty} h^{2p}(\sin\theta\cos\theta)^{p-1}$$

$$\left\{p\cos 2\theta[F_p(\theta,h,k)] + \sin\theta\cos\theta\frac{\partial F_p}{\partial\theta}\right\}. \qquad (4.105)$$

For $\theta = 0$ this relation is particularly simple. v_0 being a function of $\cos\theta$, $\partial v_0/\partial\theta$ tends towards zero for $\theta = 0$ so that

$$\frac{1}{h}\left(\frac{d\varepsilon}{d\theta}\right)_{\theta=0} = 1 + hF_1(0), \qquad (4.106)$$

where $F_1(0)$ is the value of $F_1(\theta, h, k)$ for $\theta = 0$ and for k given by (4.102) where $\theta = 0$. It is seen that the corresponding value of k is the value k_0 determined in section 4.2 which gave the lowest eigenvalue of (4.43) [i.e. of eqn. (4.86) with $\theta = 0$].

It is clear that expression (4.106) is an exact relation, since in the expression for ε, all the terms in the expansion have been retained. This is due to the particular form of λ.

(b) The second kind of experiment has been performed by Burger et al. (1965b). At a given temperature T, they applied a field parallel to the sample surface and equal to the parallel nucleation field $H_{//}(T)$. The sample was then tilted by an angle θ, and the magnitude of the field altered to restore the nucleation situation, while the temperature was held constant. In this case ε is a constant, and differentiating (4.103) with respect to θ yields

$$\frac{d\varepsilon}{d\theta} = -(2v_0+1)h\sin\theta + 2\frac{\partial v_0}{\partial\theta}h\cos\theta + 2\frac{\partial v_0}{\partial k}\frac{dk}{d\theta}h\cos\theta$$

$$+ 2\frac{\partial v_0}{\partial h}\frac{dh}{d\theta}h\cos\theta + (2v_0+1)\frac{dh}{d\theta}\cos\theta + h\cos\theta + \frac{dh}{d\theta}\sin\theta$$

$$+ \sum_{p=1}^{\infty} 2ph^{2p-1}\frac{dh}{d\theta}(\cos\theta\sin\theta)^p F_p(\theta, h, k) + \sum_{p=1}^{\infty} h^{2p}(\sin\theta\cos\theta)^{p-1}$$

$$\left\{p\cos 2\theta[F_p(\theta,h,k)] + \sin\theta\cos\theta\left(\frac{\partial F_p}{\partial\theta} + \frac{\partial F_p}{\partial k}\frac{dk}{d\theta} + \frac{\partial F_p}{\partial h}\frac{dh}{d\theta}\right)\right\} = 0.$$

$$(4.107)$$

As in (4.104) the coefficient of $dk/d\theta$ is equal to zero. The simplest situation occurs for $\theta = 0$ and, since $\partial v_0/\partial\theta = 0$ for $\theta = 0$, the expression becomes

$$2h\frac{\partial v_0}{\partial h}\frac{dh}{d\theta} + (2v_0+1)\frac{dh}{d\theta} + h + h^2 F_1(0) = 0, \qquad (4.108)$$

Reversible Properties

where $F_1(0)$ is defined as in (4.106). Thus expression (4.108) leads to the logarithmic derivative of h close to $\theta = 0$, i.e.

$$\frac{1}{h_0}\left(\frac{dh}{d\theta}\right)_0 = -\frac{1 + h_0 F_1(0)}{\partial \varepsilon_0/\partial h_0}. \tag{4.109}$$

In this expression h_0 is the value of h corresponding to $H_{//}(T)$, and $\varepsilon_0 = (2v_0 + 1)h_0$ is the lowest eigenvalue of (4.43) corresponding to h_0; k is obviously equal to k_0.

Let us now investigate the relation between $h_0 F_1(0)$ and the properties of the function $\varepsilon_0 = \varepsilon_0(h_0)$. $F_1(0)$ arises from the second order term in the perturbation expansion (4.98), i.e.

$$h_0 F_1(0) = -\lim_{\theta \to 0} \sum_{v,n} \frac{h |\langle 0v_0 | (k/h - X)\sqrt{(2h\cos\theta)}Z\sqrt{(2h\sin\theta)} | nv \rangle|^2}{\varepsilon^{v,n} - \varepsilon^{0,v_0}}$$

$$= -\lim_{\theta \to 0} \sum_{v,n} \frac{h |\langle v_0 | (k/h - X)\sqrt{(2h\cos\theta)} | v \rangle|^2 |\langle 0 | Z\sqrt{(2h\sin\theta)} | n \rangle|^2}{2(v - v_0)h\cos\theta + 2nh\sin\theta}.$$

(4.110)

The series in n is readily computed since Z has matrix elements only between $|n\rangle$ and $|n+1\rangle$. Moreover from (4.90) it is seen that

$$\langle 0 | Z\sqrt{(2h\sin\theta)} | 1 \rangle = 1. \tag{4.111}$$

Using this result and setting $\theta = 0$ in (4.110) yields

$$h_0 F_1(0) = -2 \sum_v \frac{|\langle v_0 | k - hX | v \rangle|^2}{\varepsilon_v - \varepsilon_0}. \tag{4.112}$$

Thus

$$h_0 F_1(0) = -\frac{1}{4}\left(2 - \frac{\partial^2 \varepsilon_0}{\partial k^2}\right) \tag{4.113}$$

and

$$-\frac{1}{h_0}\left(\frac{dh}{d\theta}\right)_{\theta=0} = \frac{1}{4}\frac{2 + \partial^2\varepsilon_0/\partial k^2}{\partial \varepsilon_0/\partial h} \tag{4.114}$$

As the right-hand side of (4.114) is positive, the nucleation field decreases when θ increases. In section 4.2 it was shown that, for $h_0 = h_0^c$, $\partial^2\varepsilon/\partial k^2$ vanishes, while $\partial \varepsilon/\partial h$ has a second order discontinuity. This will result in a "knee" in the logarithmic derivative of h_0 at $h_0 = h_0^c$. This can be seen in Fig. 4.11 where the variation of $-(1/h_0)(dh/d\theta)_{\theta=0}$ versus h_0 is shown. An alternative representation of this variation, this time versus $H_{//}/H_\perp$, is given in Fig. 4.12 as well as the experimental points obtained by Burger *et al.* (using a tunnelling technique in Sn–In films). The agreement between theory and experiment is strikingly good, and Burger et al., were able to determine the threshold field h_0^c and obtained $H_{//}/H_\perp \sim 2$, i.e. $\varepsilon/h \sim 0.5$. This figure also shows the results of a measurement by Hart and Swartz (1964) and of a calculation by Tinkham (1964a), made with a similar method in the limit $a/\xi(T) \gg 1$.

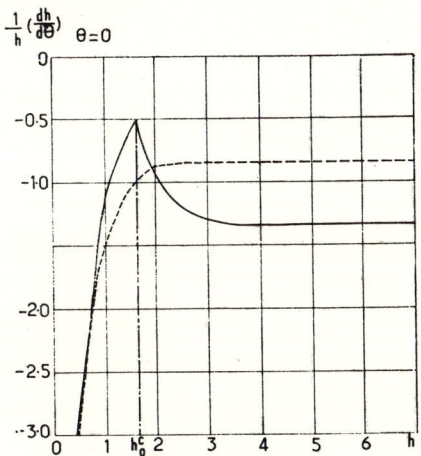

FIG. 4.11. Variation of the logarithmic derivative $(1/h)(dh/d\theta)_{\theta=0}$ as a function of h, as computed by Saint-James (1965), showing the discontinuity at $h = h_0^c$. The dashed curve represents the variation obtained from Tinkham's formula (4.122)

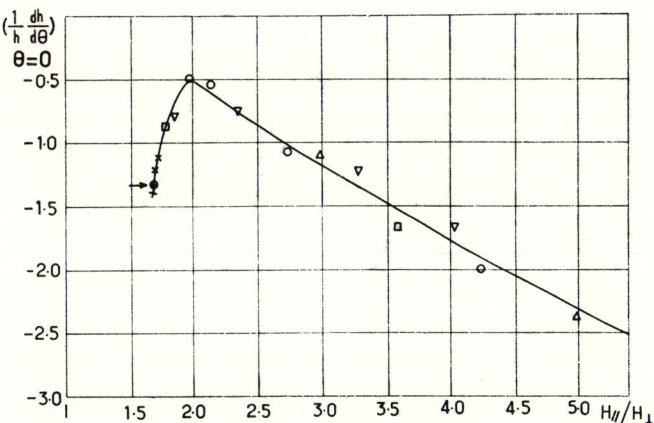

FIG. 4.12. Variation of $(1/h)(dh/d\theta)_{\theta=0}$ versus $H_{//}/H_\perp$. The theoretical curve is an alternative representation of the curve computed by Saint-James shown in Fig. 4.13. The experimental points are due to Burger et al. (1965) and refer to various samples. The arrow indicates the point calculated by Tinkham, while the circles indicate measurements made by Hart and Swartz

Reversible Properties

In very thin slabs, i.e. for $a/\xi(T) \ll 1$, one can obtain a very simple formula giving the angular dependence of the nucleation field over the whole of the θ range.

For small h, one can retain $k = 0$ in (4.81) and treat the potential $h^2 X^2 \cos^2 \theta - 2hXZ \cos \theta \sin \theta$ as a perturbation, the unperturbed equation being†

$$-\frac{\partial^2 \psi}{\partial X^2} - \frac{\partial^2 \psi}{\partial Z^2} + h^2 Z^2 \sin^2 \theta \psi = \varepsilon_1 \psi . \qquad (4.115)$$

The first order perturbation correction yields

$$\varepsilon = \varepsilon_1 + \langle h^2 X^2 \cos^2 \theta - 2h^2 XZ \sin \theta \cos \theta \rangle . \qquad (4.116)$$

ε is the lowest eigenvalue of (4.115) and is obviously

$$\varepsilon_1 = h \sin \theta , \qquad (4.117)$$

while the correction term is

$$\frac{1}{3} h^2 \cos^2 \theta , \qquad (4.118)$$

since the matrix elements of Z and X are equal to zero. Thus

$$\varepsilon = \frac{h^2}{3} \cos^2 \theta + h \sin \theta . \qquad (4.119)$$

For every given ε value there corresponds a nucleation field parallel to the sample such that

$$\varepsilon = \frac{h_{//}^2}{3} \qquad (4.120)$$

as shown at the end of section 4.2 [eqn. (4.73)], and (4.119) may be written as

$$\frac{h^2}{h_{//}^2} \cos^2 \theta + \frac{h}{\varepsilon} \sin \theta = 1 \qquad (4.121)$$

or alternatively

$$\left[\frac{H(\theta)}{H_{//}} \cos \theta \right]^2 + \frac{H(\theta)}{H_\perp} \sin \theta = 1 , \qquad (4.122)$$

where $H_{//}$, $H_\perp$ and $H(\theta)$ are the nucleation fields for $\theta = 0$, $\theta = \pi/2$ and $\theta = \theta$. This elegant formula was obtained by Tinkham (1964b) by a

† One must retain the term $h^2 Z^2 \sin^2 \theta$ in (4.115) since this potential is infinite for $Z = \pm \infty$.

different argument. It should be noted that this formula is valid for any value of θ provided that $a/\xi(T) \ll 1$. The corresponding behaviour of $-(1/h)(dh/d\theta)_{\theta=0}$ is shown in Fig. 4.11.

4.4. Persistence of the surface sheath below $H_{//}$

Little is known about the behaviour of the order parameter for an applied field parallel to the surface of the sample, below $H_{//}$. It has been observed experimentally (Tomash, 1964; Swartz and Hart, 1965) that the surface sheath persists in fields up to H_{c1}, i.e. that in bulk materials the presence of the surface still enhances superconductivity so that the order parameter is larger close to the surface than in the bulk. However, it is not known how the surface state modifies itself in the presence of a vortex structure which must exist below H_{c2}. In two cases several attempts have been made to compute the behaviour of the order parameter. These are briefly reviewed below.

(a) *Thin films*

For a thin film of thickness $2a \ll \xi(T)$ and for an applied field parallel to the surface (Fig. 4.5), one can compute the value of the order parameter simply by assuming that it is almost constant in space,† but depends on the applied field H. It has been shown [eqn. (4.74)] that the parallel nucleation field of such a film is

$$H_{//} = \frac{\sqrt{3}}{2\pi}\phi_{0/a\xi(T)} = \sqrt{6}\frac{\lambda(T)}{a}H_c. \qquad (4.123)$$

For an applied field H lower than $H_{//}$, the problem can be considered in terms of the Ginzburg–Landau theory by assuming that H and ψ depend only on x. Using the reduced variables (2.37), the one-dimensional Ginzburg–Landau equations take the form (2.52), i.e.

$$-\frac{1}{\kappa^2}\frac{d^2 f_0}{d\xi^2} - \frac{1}{f_0^3}\left(\frac{dh}{d\xi}\right)^2 + f_0 - f_0^3 = 0, \qquad (4.125)$$

$$\frac{d^2 h}{d\xi^2} - \frac{2}{f_0}\frac{dh}{d\xi}\frac{df_0}{d\xi} - f_0^2 h = 0. \qquad (4.126)$$

The boundary conditions are

$$\left.\begin{array}{r}h = h_0 \\[4pt] \dfrac{df}{dx} = 0\end{array}\right\} \quad \left[\text{for} \quad \xi = \pm\frac{a}{\lambda(T)}.\right] \qquad \begin{array}{l}(4.127)\\[10pt](4.128)\end{array}$$

† This avoids the difficulty that the term in $|\nabla\psi|^2$ would contribute a large amount to the free energy.

Reversible Properties

Since the field is directed along $0z$ and since its magnitude depends only on x, the current is directed along $0y$ and is

$$\left(\frac{4\pi}{c}j\right)_y = (\text{curl } H)_y = \frac{d}{dx}h_z \qquad (4.129a)$$

With the above reduced variables this equation reads:

$$j_r \simeq \frac{d}{d\xi}h \qquad (4.129b)$$

The condition $H = H_0$ on both sides of the sample ensures that the total current is zero

$$\int_{-a/\lambda(T)}^{a/\lambda(T)} j_r d\xi = \int_{-a/\lambda(T)}^{a/\lambda(T)} \frac{dh}{d\xi} d\xi = 0. \qquad (4.130)$$

In order to investigate the situation in more detail, let us take the order parameter as a constant in (4.126). This equation takes the form

$$\frac{d^2h}{d\xi^2} = f_0^2 h \qquad (4.131)$$

which is, as expected, the London equation.
Thus the convenient solution for h is

$$h = h_0 \frac{\cosh(\xi f_0)}{\cosh[(a/\lambda(T))f_0]}. \qquad (4.132)$$

The value of f_0 is obtained from the equation by integrating over ξ from $-a/\lambda(T)$ to $+a/\lambda(T)$. The term $d^2f/d\xi^2$ vanishes and

$$\int_{-a/\lambda(T)}^{a/\lambda(T)} \frac{1}{f_0^3}\left(\frac{dh}{d\xi}\right)^2 d\xi = \int_{-a/\lambda(T)}^{a/\lambda(T)} (f_0 - f_0^3) d\xi. \qquad (4.133)$$

In computing these integrals, it can be assumed that f_0 is constant in the domain $-a/\lambda(T), +a/\lambda(T)$ and the condition reads

$$\frac{1}{f_0^3}\int_{-a/\lambda(T)}^{a/\lambda(T)}\left(\frac{dh}{d\xi}\right)^2 d\xi = \frac{2a}{\lambda(T)}(f_0 - f_0^3). \qquad (4.134)$$

Or, by using expression (4.132),

$$h^2 = 2f_0^2(1 - f_0^2)\left[\frac{\text{sihn}(2a/\lambda(T))f_0}{(2a/\lambda(T))f_0} - 1\right]^{-1} \cosh^2\left(\frac{a}{\lambda(T)}f_0\right). \qquad (4.135)$$

i.e.

$$\left(\frac{H}{H_c}\right)^2 = 4f_0^2(1 - f_0^2)\left[\frac{\sinh(2a/\lambda(T))f_0}{(2a/\lambda(T))f_0} - 1\right]^{-1} \cosh^2\left(\frac{a}{\lambda(T)}f_0\right). \qquad (4.136)$$

f_0 is the magnitude $|\psi|/\psi_0$ of the order parameter, H is the applied field, and H_c the thermodynamic critical field.

It is clear that the value of f_0 deduced from expression (4.136) ensures that the free energy is a minimum since the Ginzburg–Landau equation was obtained by minimizing the free energy with respect to the order parameter.

For $f_0 \to 0$ one obtains the transition field for a second order transition†

$$\left(\frac{H}{H_c}\right)^2 = 6\frac{\lambda^2(T)}{a^2} \qquad (4.137)$$

which is the expression (4.123).

If $a/\lambda(T)$ is very small, one can expand the various functions to order f_0^2 and

$$\left(\frac{H}{H_c}\right)^2 = 6\frac{\lambda^2(T)}{a^2}\left[1 - f_0^2 + \frac{4a^2 f_0^2}{5\lambda^2(T)}\right] \qquad (4.138)$$

and using the expression for $H_{//}$,

$$\frac{H^2 - H_{//}^2}{H_{//}^2} = \left[\frac{4a^2}{5\lambda^2(T)} - 1\right]\left|\frac{\psi}{\psi_0}\right|^2. \qquad (4.139)$$

This gives the value of the order parameter as a function of $(H^2 - H_{//}^2)/H_{//}^2$. As expected, close to $H_{//}$, $|\psi|$ is proportional to $(H_{//} - H)^{1/2}$.

This expression for the order parameter has been used to interpret tunnelling experiments in the gapless region. This will be discussed in Chapter 6.

(b) *Bulk specimens*

The situation is not so clear in bulk specimens. As already stated, it has been observed that the surface sheath persists below H_{c3}. The theoretical investigation of this problem would involve the solution of the two Ginzburg–Landau equations including the boundary condition (4.1). The only simplification that arises here is that the microscopic field h may be assumed to be parallel to the applied field (i.e. to the surface). This means that the vector potential A can be written as

$$A = [0, A_y(x, y), 0], \qquad (4.140)$$

where $A_y(x, y)$ may be obtained from the solution of the Ginzburg–Landau equations.

---

† This calculation was first made by Ginzburg who concluded that the transition is of the first order if $a > (\sqrt{5/2})\lambda(T)$. However the computation is performed for $a \ll \xi(T)/2$. For type II superconductors $\lambda(T) \gg \xi(T)/2$, so that the transition is of the second order. For a discussion of the order of the transition as a function of κ and of the thickness of the slab see "Superconductivity" R. D. Parks Ed. M. Dekker New York (chapter XVIII) (To be published).

Reversible Properties

For a semi-infinite medium at the nucleation field the order parameter takes the form

$$\psi(x, y) = e^{iky}D(x), \qquad (4.141)$$

where $D(x)$ is a real function of (x) (see section 4.1 and Fig. 4.1) and it is tempting to assume that this will be the case below H_{c3}. The range of validity of this assumption is not known, and, indeed, is certainly incorrect for slabs of thickness of the order of $\xi(T)$ since, in this case, one must introduce the two vectors $+k$ and $-k$ in the expression for ψ (see section 4.2). It is also clear that the form (4.141) will not describe the vortex structure which appears below H_{c2}.† Fink and Kessinger (1965) have used (4.141) as the starting point for a calculation of the order parameter and of the field h below H_{c3}. If ψ has the form (4.141) the vector potential may be taken as dependent on x only, and A_y written as

$$A_y = H_0 x + \Delta A_y(x) + A_y(0), \qquad (4.142)$$

where H_0 is the applied field, $A_y(0)$ the value of A_y for $x = 0$, and $\Delta A_y(x)$ is a function to be determined. $\Delta A_y(x)$ vanishes for $x = 0$. ψ and $A_y(x)$ must obey the Ginzburg–Landau equations, namely,

$$\frac{1}{2m}\left[i\hbar\nabla + \frac{2e}{c}A\right]^2 \psi + \alpha\psi + \beta\psi|\psi|^2 = 0, \qquad (4.143)$$

$$-\frac{4\pi}{c}\mathbf{j} = -\operatorname{curl}\operatorname{curl}\mathbf{A} = \frac{16\pi e^2}{mc^2}|\psi|^2 A + \frac{4\pi ie\hbar}{mc}(\psi^*\nabla\psi - \psi\nabla\psi^*), \qquad (4.144)$$

with the boundary condition

$$\frac{1}{2m}\left[i\hbar\nabla + \frac{2e}{c}A\right]_x \psi = 0 \quad \text{for } x = 0. \qquad (4.145)$$

The parameter k is as yet undetermined and may be found by making the free energy, namely,

$$F_s(H_0) - F_s(0) = \int dv \left[\frac{H_c^2}{8\pi} + \alpha|\psi|^2 + \beta|\psi|^4 + \frac{(\operatorname{curl}A)^2}{8\pi}\right.$$
$$\left. + \frac{1}{2m}\left|\left[i\hbar\nabla + \frac{2e}{c}A\right]\psi\right|^2\right], \qquad (4.146)$$

a minimum.

† It is not known whether a vortex structure exists between H_{c3} and H_{c2} in bulk superconductors.

Surface Superconductivity

In order to simplify the above equations, let us make the following transformations:

$$\begin{aligned}
\zeta &= \frac{x}{\xi(T)}\left(\frac{H_0}{H_{c2}}\right)^{1/2}, \\
\mu^2 &= \frac{H_{c2}}{H_0}, \\
a(\zeta) &= \mu[\sqrt{2}H_c\lambda(T)]^{-1}\Delta A(\zeta), \\
C_0 &= \mu[\sqrt{2}H_c\lambda(T)]^{-1}A_y(0), \\
F(\zeta) &= \frac{D(\zeta)}{D_0}, \\
D_0^2 &= mc^2/[16\pi e^2\lambda^2(T)], \\
\zeta_0 &= k\xi(T)\left(\frac{H_{c2}}{H_0}\right)^{1/2}, \\
\Gamma &= \zeta_0 - C_0.
\end{aligned} \qquad (4.147)$$

With the above transformation the applied field H_0 is equal to 1.
Using the forms (4.142) for A and (4.141) for ψ the free energy and the Ginzburg–Landau equations become

$$\frac{4\pi}{H_c^2}\left[\frac{F_s(H_0) - F_s(0)}{l_y l_z}\right] = B + C\int_0^\infty d\zeta \left[\left(\frac{dF}{d\zeta}\right)^2 + [\zeta + a - \Gamma]^2 F^2 \right. \\
\left. - \mu^2 F^2(1 - F^2/2) + \left(\frac{\kappa}{\mu}\right)^2 \frac{da}{d\zeta}\left(2 + \frac{da}{d\zeta}\right)\right], \qquad (4.148)$$

$$\frac{d^2 F}{d\zeta^2} + [\mu^2(1 - F^2) - (\zeta + a - \Gamma)^2]F = 0, \qquad (4.149)$$

$$\left(\frac{\kappa}{\mu}\right)^2 \frac{d^2 a}{d\zeta^2} = [\zeta + a - \Gamma]F^2 = -\frac{4\pi}{c}\mu j(\zeta), \qquad (4.150)$$

where B and C are constants, and l_y and l_z the dimensions of the superconductor in the y and z directions.

It is seen that k appears in the above equation in the parameter Γ, i.e. in the form $\{k - (2e/c)A_y(0)\}$.

To these equations must be added boundary conditions. For $\zeta = 0$, i.e. on the surface of the specimen, they are

$$\begin{aligned}
H &= \frac{dA}{dX} = H_0 \quad \text{for } X = 0, \\
\frac{dF}{dX} &= 0 \quad \text{for } X = 0,
\end{aligned} \qquad (4.151)$$

Reversible Properties

or equivalently

$$\left.\begin{array}{l} \dfrac{da}{d\zeta} = 0 \quad \text{for } \zeta = 0, \\[8pt] \dfrac{dF}{d\zeta} = 0 \quad \text{for } \zeta = 0. \end{array}\right\} \qquad (4.152)$$

In order to determine Γ, one must minimize the free energy with respect to Γ. F and a are explicit functions of Γ. However, it can be seen from (4.149) and (4.150) that they are actually functions of $(\zeta - \Gamma)$, so that

$$\frac{\partial}{\partial \Gamma} = -\frac{\partial}{\partial \zeta}.$$

The minimizing condition reads

$$\int_0^\infty d\zeta \left[\frac{dF}{d\zeta}\frac{d^2F}{d\zeta^2} + [\zeta + a - \Gamma]^2 F^2 + [\zeta + a - \Gamma] F^2 \frac{da}{d\zeta} \right.$$
$$+ [\zeta + a - \Gamma]^2 F \frac{dF}{d\zeta} - \mu^2 F \frac{dF}{d\zeta} + \mu^2 F^3 \frac{dF}{d\zeta} + \left(\frac{\kappa}{\mu}\right)^2 \frac{d^2a}{d\zeta^2}$$
$$\left. + \frac{\kappa}{\mu}\frac{da}{d\zeta}\frac{d^2a}{d\zeta^2} \right] = 0. \qquad (4.153)$$

Or

$$\int_0^\infty d\zeta \left[\frac{d^2F}{d\zeta^2}\frac{dF}{d\zeta} + \left(\frac{\kappa}{\mu}\right)^2 \frac{d^2a}{d\zeta^2} + \frac{\kappa}{\mu}\frac{da}{d\zeta}\frac{d^2a}{d\zeta^2} \right] = 0, \qquad (4.154)$$

i.e.

$$\left| \left(\frac{dF}{d\zeta}\right)^2 + \left(\frac{\kappa}{\mu}\right)^2 \frac{da}{d\zeta}\left[2 + \frac{da}{d\zeta}\right] \right|_0^\infty = 0. \qquad (4.155)$$

In deriving this formula, use has been made of the Ginzburg–Landau equations. For $\zeta = 0$, the above expression is zero, and one obtains an extra boundary condition:

$$\left(\frac{dF}{d\zeta}\right)^2 + \left(\frac{\kappa}{\mu}\right)^2 \frac{da}{d\zeta}\left(2 + \frac{da}{d\zeta}\right) = 0 \qquad (4.156)$$

valid for $\zeta = \infty$.

This equation may be written as

$$\left(\frac{dF}{d\zeta}\right)^2 + \left(\frac{\kappa}{\mu}\right)^2 (h^2 - h_0^2) = 0, \qquad (4.157)$$

where $h_0 = 1$ is the reduced applied field.

Surface Superconductivity

In order to define the problem entirely an extra boundary condition is needed. It is clear that for $\zeta = \infty$, $dF/d\zeta$ cannot be finite, since F cannot be divergent for $\zeta \to \infty$. Two cases may then be possible as $\zeta \to \infty$:

(α) F and H oscillate on
(β) F and $dF/d\zeta$ go to zero.

In case (β) one obtains:

$$\frac{da}{d\zeta} = 0, \quad \text{i.e.} \quad h_\infty = h_0,$$

or

$$\frac{da}{d\zeta} = -2, \quad \text{i.e.} \quad h_\infty = -h_0.$$

The condition $h_\infty = -h_0$ probably yields a maximum of the free energy since it requires a huge magnetic energy. However, it is not known whether $h_\infty = h_0$ gives a lower free energy than the condition (α). It is possible that for $H > H_{c2}$ the situation $H_\infty = H_0$ is more favourable than the oscillating one and that for $H < H_{c2}$, the oscillating situation will become more stable, since for $H < H_{c2}$ the Abrikosov structure must appear far from the surface.

It is nevertheless to be noted that in the case of a slab the boundary conditions will be $dF/d\zeta = 0$ on the two surfaces of the slab. It is then natural to choose $dF/d\zeta = 0$ at infinity in a semi-infinite medium. According to (4.157) this leads to $h_\infty = \pm h_0$.

In any case, the calculation for $h_\infty = h_0$ is much simpler and this asumption was used by Fink and Kessinger (1965).

If $da/d\zeta$ vanishes for $\zeta = 0$ and $\zeta = \infty$, it is seen that the overall current vanishes i.e.†

$$\frac{4\pi}{c}\int_0^\infty j\,d\zeta = -\frac{\kappa^2}{\mu^3}\int_0^\infty \frac{d^2a}{d\zeta^2}\,d\zeta = 0. \tag{4.158}$$

The boundary condition can be written in another way. Setting $\xi = \zeta - \Gamma$ the free energy is of the form

$$B + C\int_{-\Gamma}^{\infty-\Gamma} d\xi \left[\left(\frac{dF}{d\xi}\right)^2 + [\xi + a]^2 F^2 - \mu^2 F^2(1 - F^2/2)\right.$$
$$\left. + \left(\frac{\kappa}{\mu}\right)^2 \frac{da}{d\xi}\left(2 + \frac{da}{d\xi}\right)\right] \tag{4.159}$$

and the minimum condition reads

$$\Gamma^2 = \mu^2\left(1 - \frac{1}{2}F_0^2\right). \tag{4.160}$$

† The current flows parallel to the y direction. There is a certain point ζ_c at which the current will reverse so that the total current is zero.

Reversible Properties

A condition which relates the parameter Γ (i.e. the phase of the order parameter and the vector potential for $\zeta = 0$) to the order parameter F_0 for $\zeta = 0$, or equivalently the slope of the magnetic field, i.e. the magnitude of the current at $\zeta = 0$ to the order parameter is

$$\frac{dh}{d\zeta} = \frac{d^2 a}{d\zeta^2} = \frac{1}{\kappa^2} F_0^2 \left(1 - \frac{1}{2} F_0^2\right). \qquad (4.161)$$

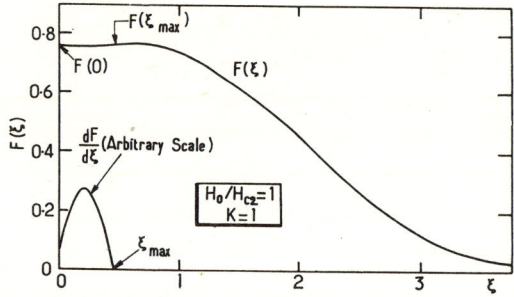

Fig. 4.13. Variation of the modulus of the order parameter close to the surface, for an applied field $H = H_{c2}$ and $\kappa = 1$ according to the calculation by Fink and Kessinger (1965)

Fink and Kessinger have computed F and a using an analogue computer. They have found that for $H_0 < H_{c3}$ the order parameter F increases up to a maximum F_M reached at $\zeta = \zeta_M$ and then decreases to zero, while the magnetic field decreases to a minimum H_m reached for $\zeta = \zeta_c$ ($\zeta_c > \zeta_M$), a point at which the current reverses sign. The value F_0 of the order para-

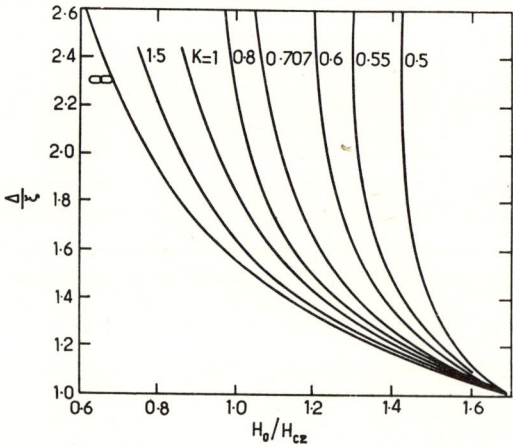

Fig. 4.14. Variation of the width $\varDelta$ of the sheath below H_{c3} (after Fink and Kessinger)

meter varies with κ and H_0/H_{c2}. The variation of the order parameter F, of the thickness of the sheath $\Delta/\xi(T)$, where Δ is given by

$$\Delta = \frac{1}{F_0^2} \int_0^\infty F^2(\zeta)dx,$$

and of F_0 for various values of κ and H_0/H_{c2} are shown on Figs 4.13 to 4.15.

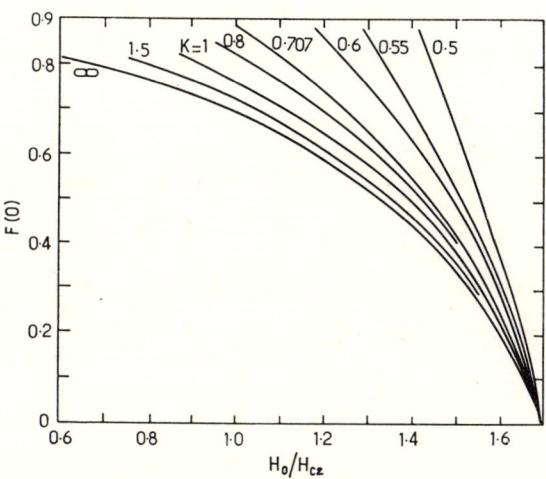

FIG. 4.15. Variation of F_0 the value of the modulus of the order parameter at the surface below H_{c3} (after Fink and Kessinger)

It is seen that:
(a) when the field is decreased, F_0 increases continuously as well as the width of the sheath;
(b) no discontinuous behaviour is found at $H = H_{c2}$;
(c) when H_c is approached from above in a type I superconductor with $0.417 < \kappa < 0.707$ the thickness of the sheath increases considerably;
(d) the same conclusion is obtained for $H < H_{c1}$ in type II superconductors;
(e) close to H_{c3} the magnitude F_0 of the order parameter is proportional to $[(H_{c3} - H_0)/H_{c3}]^{1/2}$.

4.5. Superconductor coated by a normal metal

If the superconductor is coated with any conducting material (for example, a normal metal) the boundary condition (4.1) is no longer valid. It is known from the microscopic theory (de Gennes, 1964) that the pair potential $\Delta(r)$ (to which ψ is proportional) will vary rapidly over a range ξ_0 close to the surface and will be different from zero in the normal material. Two cases may be encountered depending on whether the interaction between the electrons in the normal metal is attractive or repulsive. These two situations are roughly pictured in Fig. 4.16a and b.

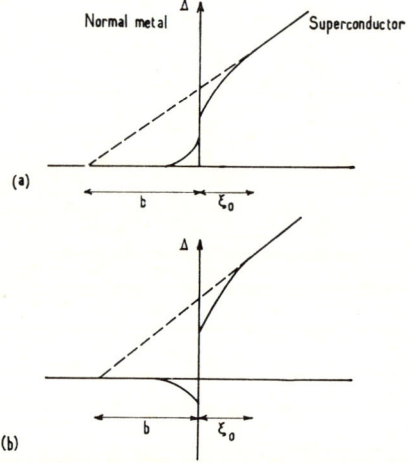

FIG. 4.16. Schematic variation of the pair potential $\Delta(r)$ (proportional to the order parameter) for a superconductor coated (a) by a normal metal with attractive interaction between the electrons, (b) by a normal metal with repulsive interaction between the electrons

In the Ginzburg–Landau scheme ψ varies in the region $\xi(T) \gg \xi_0$. The actual variation of $\Delta(r)$ may be replaced by the variation represented in Fig. 4.17, where ψ decreases from ψ_s, the value on the surface, to zero over a distance b. The boundary condition then reads

$$\frac{df}{dx} = \frac{1}{b} f. \qquad (4.162)$$

By comparing this expression with condition (2.22), it is seen that no current flows from the superconductor. b may be considered as a parameter

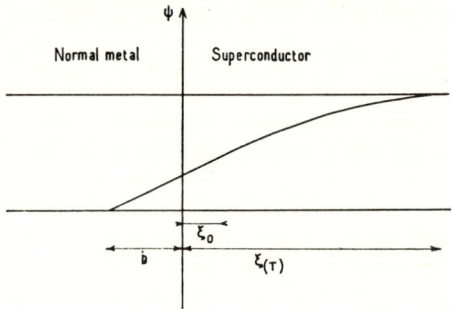

FIG. 4.17. Schematic variation of $\psi(r)$ for a superconductor coated with a normal metal in the Ginzburg–Landau approximation

depending on the structure of the junction, and on several other parameters such as the temperature, the density of states in the superconductor and the normal metal. If b is assumed to be field independent at a given temper-

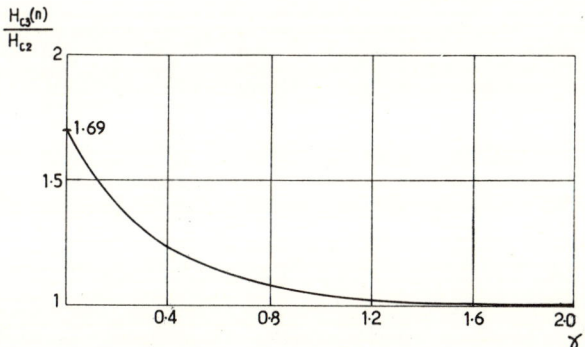

FIG. 4.18. Variation of $H_{c3}(n)/H_{c2}$ as a function of the parameter γ as defined in the text. When $\gamma \to \infty$, $H_{c3}(n)/H_{c2} \to 1$. The presence of a normal coating decreases the surface effect

ature, the nucleation field $H_{c3}(n)$ for a semi-infinite superconductor coated by a normal metal can be computed as a function of the parameter γ defined by

$$\gamma = \frac{\xi(T)}{b}\left[\frac{H_{c3}(n)}{H_{c2}}\right]^{1/2}.$$

The variation of $H_{c3}(n)/H_{c2}$ as a function of γ is shown in Fig. 4.18. It is seen that $H_{c3}(n)/H_{c2}$ tends towards 1 when γ increases. In particular if

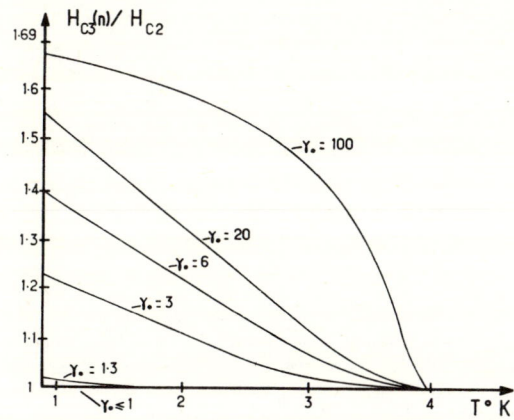

FIG. 4.19. Theoretical values of the H/H_{c2} ratio versus temperature for an In–Bi/Zn system for various values of the parameter γ_0 (the critical temperature T_{cs} of the superconductor is 4°K) as obtained by Hurault (1966). For $\gamma_0 = \sigma_s/\sigma_n < 1$ the surface nucleation disappears at all temperatures

Reversible Properties

$T \to T_0$, $\xi(T) \to \infty$. One therefore expects that, for a coated cylinder, the parallel nucleation field tends towards H_{c2} when T reaches T_0.

Due to technical difficulties this behaviour of $H_{c3}(n)$ has not been observed. At lower temperatures, no comparison has been made with the curve (4.18) because b has been found to depend on the applied field, and thus a more detailed analysis is necessary. Recently, Hurault (1966) has shown that b depends on the ratio $\gamma_0 = \sigma_s/\sigma_n$, the conductivities of the superconductor in the normal state, and of the coating metal. For $\gamma_0 \leq 1$ the surface superconductivity disappears and $H_{c3}(n) = H_{c2}$. For $\gamma_0 > 1$ surface superconductivity still exists but the ratio $H_{c3}(n)/H_{c2}$ depends on the temperature. Hurault has calculated this ratio for an In–Bi alloy coated with Zn for various values of γ_0. His results are reported on Fig. 4.19. Recent experiments by Burger et al. (1966) have revealed the impossibility of surface nucleation in the system for $\gamma_0 < 1$.

For a slab of thickness much greater than $\xi(T)$, covered on one side only by a normal metal, it is clear that the nucleation field H_{c3} of the free surface, which is larger than the field $H_{c3}(n)$, will be observed.

For a more complete discussion of these various effects, the reader is referred to the work of Deutscher (1966) and also chapter XIX. of "Superconductivity" R. D. Parks Ed, H. Dekker New York (To be published)

Bibliography of Chapter 4

BON MARDION, G., GOODMAN, B. B. and LACAZE, A. (1964), *Phys. Letters*, **8**, 15.
BURGER, J. P., DEUTSCHER, G., GUYON, E. and MARTINET, A. (1965a), *Phys. Rev.* **137 A**, 853.
BURGER, J. P., DEUTSCHER, G., GUYON, E. and MARTINET, A. (1965b), *Phys. Letters* **16**, 220.
BURGER, J. P., DEUTSCHER, G., GUYON, E. and MARTINET, A. (1966), To be published.
CARDONA, M. and ROSENBLUM, B. (1964), *Phys. Letters*, **8**, 308.
CAROLI, C., DE GENNES, P. G. and MATRICON, J. (1962), *J. Phys. Rad.* **23**, 707.
DEUTSCHER, G. (1966) *Thesis*
DRUYVESTEYN, W. F., VAN OIJEN, D. J. and BERBEN, T. J. (1964), *Rev. Mod. Phys.* **36**, 58.
FABER, T. E. (1954), *Proc. Roy. Soc.* A **223**, 174; (1957), *Proc. Roy. Soc.* A **241**, 531.
FINK, H. J. and KESSINGER, R. D. (1965), *Phys. Rev.* **140 A**, 1937.
DE GENNES, P. G. (1964a), *Physics Condensed Matter*, **3**, 79; (1964b), *Rev. Mod. Phys.* **36**, 225.
GINZBURG. V. L., (1952), *Doklady Akad. Nauk. SSSR* **83**, 385. see also: *Zh. Eksperim. i Teor. Fiz.* (1958), *34*, 113 [English translation: *Sov. Phys. J. E. T. P.* (1958), *7*, 78]
GOLDSTEIN, Y. (1964), *Phys. Letters*, **12**, 169.
GUYON, E., MATRICON, J., MARTINET, A. and PINCUS, P. (1965), *Phys. Rev.* **138 A**, 746.
HART, H. R. Jr. and SWARTZ, P. S. (1964), *Phys. Letters*, **10**, 40.
HURAULT, J. P. (1966), *Phys. Letters*, **20**, 587.
JOINER, W. C. H. and BLAUGHER, R. D. (1964), *Rev. Mod. Phys.* **36**, 67.
MAKI, K. (1964), *Physics*, **1**, 21.
SAINT-JAMES, D. (1965), *Phys. Letters*, **16**, 218.
SAINT-JAMES, D. and DE GENNES, P. G. (1963), *Phys. Letters* **7**, 306,
SERAPHIM, D. P. (1962), *Superconductors*, Tannenbaum et Wright Eds.
SERIN, B. (1964), *Contribution to type II superconductors conference*, Cleveland, (Ohio).
STRONGIN, M., PASKIN, A., SCHWEITZER, D. G., KAMMERER, O. F. and CRAIG, P. P. (1964), *Phys. Rev. Letters*, **12**, 442.
SWARTZ, P. S., HART, H. R. Jr. (1965), *Phys. Rev.* **137 A**, 818.
TINKHAM, M. (1964a), *Contribution to type II superconductors conference*, Cleveland, (Ohio).
TINKHAM, M. (1964b), *Phys. Letters*, **9**, 217.
TOMASCH, W. J. (1964), *Phys. Letters*, **9**, 104.
TOMASCH, W. J. and JOSEPH, A. S. (1964), *Phys. Rev. Letters*, **12**, 148.

CHAPTER 5

Microscopic Theory

5.1. Introduction

It is not the purpose of this book to develop a complete general microscopic theory of superconductivity; this has been the subject of several books and papers, some of which are listed in the bibliography. This chapter is confined to justifying the results obtained from the phenomenological treatment and will be devoted to obtaining the equilibrium equations in the vicinity of the transition from the microscopic theory, without entering into the details of the latter. This will allow a derivation of the Ginzburg–Landau equations [a result first obtained by Gorkov (1959)] and also lead to the various parameters introduced in the preceding chapters. Moreover the discussion will introduce a distinction between "clean" and "dirty" superconductors which is of a great help in the study of type II superconductors. The "equilibrium equations" to be obtained here are also useful in discussing the validity of the Ginzburg–Landau scheme and the generalization of this approach. It will be shown that, in some cases, one can define generalized Ginzburg–Landau parameters and calculate their variation with temperature. The influence of the magnetic field on electron spins (paramagnetic effect) can be investigated using a similar approach and will be considered in Chapter 6. The treatment will use several features of field theory but will avoid the use of Green's function techniques and will obtain the main results using conventional methods.

5.2. General features of the microscopic theory. The Cooper pairs

The fundamental feature of the microscopic theory is that the normal state is unstable in the presence of an attractive interaction between the electrons (Cooper, 1956). The effect of this instability may be seen by considering a gas of free electrons and the interaction of two electrons located at r_1 and r_2. If the centre of mass of these two electrons is at rest, the wave function $\phi(r_1, r_2)$ depends only on $r_1 - r_2$ so that

$$\phi(r_1 - r_2) = \sum_k g(k) e^{ik(r_1 - r_2)} \tag{5.1}$$

where $g(k)$ is the probability amplitude of finding an electron in the state $\hbar k$ and another electron in the state $-\hbar k$.
In the presence of an interaction $V(r_1 - r_2)$, the Schrödinger equation for the two electrons reads

$$-\frac{\hbar^2}{2m}[\nabla_1^2 + \nabla_2^2]\phi(r_1 - r_2) + V(r_1 - r_2)\phi(r_1 - r_2) = (E + 2E_F)\phi(r_1 - r_2) \tag{5.2}$$

The energy E of the pair is measured from the state where the two electrons are at the Fermi level. The equation for $g(k)$ is easily found to be

$$\frac{\hbar^2}{m} k^2 g(k) + \sum_{k'} g(k') V_{kk'} = (E + 2E_F)\, g(k), \tag{5.3a}$$

$$V_{kk'} = \frac{1}{\Omega} \int V(r) e^{i(k-k')r} d\,r = \langle k' | V | k \rangle, \tag{5.3b}$$

$$g(k) \equiv 0 \quad \text{for } k < k_F \tag{5.3c}$$

where Ω is the volume of the system.

Equation (5.3c) expresses the fact that the states $k < k_F$ are already occupied (Pauli exclusion principle). The set of equations (5.3) (Bethe–Goldstone equations) has a continuous spectrum of solutions for $E > 0$ which describes the collision of the two electrons $\pm \hbar k$. However, Cooper pointed out that bound states exist for $E < 0$ provided that V is attractive, whatever the magnitude of this interaction.

This is most easily seen on the simple model

$$V_{kk'} = -\frac{V}{\Omega} \quad \text{for} \quad \begin{cases} E_F < \dfrac{\hbar^2 k^2}{2m} < E_F + \hbar\omega_D \\[6pt] E_F < \dfrac{\hbar^2 k'^2}{2m} < E_F + \hbar\omega_D \end{cases} \tag{5.4}$$

$V_{kk'} \equiv 0$ everywhere else,
where ω_D is a cut-off frequency. The interaction $V_{kk'}$ is attractive ($V > 0$) and constant in a band $\hbar\omega_D$ above the Fermi level. Equation (5.3a) reads

$$\left(\frac{\hbar^2}{m}k^2 - E - 2E_F\right) g(k) = \frac{V}{\Omega} \sum_{k'} g(k'), \tag{5.5}$$

where the summation over k' is restricted to the band $\hbar\omega_D$ above E_F. Dividing each term of (5.5) by $[(\hbar^2/m)k^2 - E - 2E_F]$ and summing over k yields

$$1 = \frac{V}{\Omega} \sum_k \frac{1}{(\hbar^2/m)k^2 - E - 2E_F} \tag{5.6}$$

Reversible Properties

with

$$E_F < \frac{\hbar^2 k^2}{2m} < E_F + \hbar\omega_D.$$

For each allowed value of $k > k_F$ there exists a pole E of the right-hand side of (5.6) with $2\hbar\omega_D > E > 0$ which describes the scattering solution. However, if V is positive a solution exists with $E < 0$, i.e. a solution for which the energy of the pair is smaller than $2E_F$. If the potential were repulsive (i.e. $V < 0$) this special state would have existed for $E > 2\hbar\omega_D$.

In order to calculate the bound state one can introduce a new definition of the energy

$$\xi_k = \frac{\hbar^2}{2m} k^2 - E_F \tag{5.7}$$

and the density of states (per unit energy and unit volume)

$$N(\xi_k) = \frac{1}{(2\pi)^3} \frac{4\pi k^2 dk}{d\xi_k}. \tag{5.8}$$

The discrete summation may be replaced by an integral and

$$1 = V \int_0^{\hbar\omega_D} N(\xi) \frac{d\xi}{2\xi - E}. \tag{5.9}$$

If $\hbar\omega_D \ll E_F$, $N(\xi)$ may be replaced by its value $N(0)$ at the Fermi level and

$$1 = \frac{1}{2} N(0) V \log \frac{E - 2\hbar\omega_D}{E} \tag{5.10}$$

and if $N(0) V \ll 1$ (weak interaction)

$$E = -2\hbar\omega_D e^{-2/N(0)V}. \tag{5.11}$$

Thus an allowed energy state exists with $E < 0$. When the interaction V is applied to a gas of free electrons, the electrons will form Cooper pairs releasing energy. The normal state is unstable. Several remarks can be made:

(1) the instability exists even for a very weak interaction V, provided that V is attractive;

(2) the particular form of E cannot be expanded in powers of V when V tends towards zero. This explains why the microscopic theory of superconductivity could not be obtained by conventional perturbation techniques;

(3) in the preceding calculation the Pauli exclusion principle was taken into account between the electron (r_1) and the Fermi surface electrons and between the electrons (r_2) and the Fermi surface electrons. The wave function of the pair must be symmetrical with respect to the exchange (r_1, r_2). With the simplified interaction V, ϕ (r_1, r_2) is symmetrical [cf. (5.1)], and the spin wave function must be antisymmetrical.

If the interaction $V_{kk'}$ was dependent on the angle between k and k' two results could have been obtained:

(a) several bound states exist;
(b) the spatial part of the wave function would have been anisotropic and the spin part much more complicated.

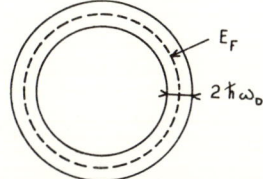

FIG. 5.1. Schematic diagram showing the band of width $2\hbar\omega_D$ around the Fermi level in which electrons are subject to an attractive potential (B. C. S. model). The width of the band is exaggerated in this diagram. In practice $E_F \gg 2\hbar\omega_D$

Actually, in superconductors, the major part of $V_{kk'}$ depends only on the modulus of k and k', so that the angular dependence can be neglected.

In the foregoing derivation nothing was said about the origin of the attractive interaction. A short account of the present position of this question is given below, but the reader is referred to more specialized works on this topic for further details.

In an electron gas, the interaction between the electrons is the Coulomb interaction which is repulsive. In order to get an attractive interaction, the electron gas must be coupled to another system of particles. In the non-transition metals, it is known that the mechanism responsible for superconductivity is due to the coupling of the electrons to the ionic lattice (Fröhlich, 1950). The interactions with the ions is pictured as an interaction of electrons via the phonons. With this model one can calculate the magnitude of the interaction term. It is found that $N(0)V$ is smaller than 0·5. Moreover, this mechanism explains the isotope effect (the transition temperature is proportional to M^{-a} where M is the isotopic mass and $a \sim 1/2$).

Using this picture Bardeen, Cooper and Schrieffer (1957) [B.C.S.] were able to give the first microscopic theory of superconductivity. They have solved the problem of a gas of electrons in the presence of an attractive interaction.

$$V = V \quad \text{for} \quad |\xi_k|, |\xi_{k'}| < \hbar\omega_D,$$
$$V = 0, \quad \text{elsewhere} \tag{5.12}$$

where ξ_k and $\xi_{k'}$ are the energies of the two electrons as measured from the Fermi level and $\hbar\omega_D$ is the Debye energy of the phonons. In this model the attractive interaction exists for electrons in a band $\pm \hbar\omega_D$ around the Fermi level E_F. This limitation to a band $\pm \hbar\omega_D$ is called the B.C.S. cut-off.

In the transition ions, the situation is not so clear, and there is no general agreement on the mechanism responsible for superconductivity in this case.

In the scope of this book we are not concerned with the actual form of the interaction, it is sufficient for us to assume, that an attractive interaction exists within a band of width $2\hbar\omega_D$ (Fig. 5.1).

5.3. Formulation of the microscopic theory. The self-consistent method†

(a) *The self-consistent hamiltonian*

The hamiltonian for a system of electrons, in the presence of an arbitrary potential $U_0(r)$ (for example, the periodic potential of the lattice), of a magnetic field $h(r) = \text{curl}\, A(r)$, and two body interactions may be written as:

$$\mathcal{H} = \mathcal{H}_0 + \mathcal{H}_1 + \mathcal{H}_2 - \mu N, \qquad (5.13)$$

where

$$\mathcal{H}_0 = \int dr \sum_\alpha \psi_\alpha^\dagger(r) \left[\frac{(p - (e/c)A)^2}{2m} + U_0(r) \right] \psi_\alpha(r), \qquad (5.14)$$

$$\mathcal{H}_1 = \frac{1}{2} \int dr\, dr' \sum_{\alpha\beta} \psi_\alpha^\dagger(r) \psi_\beta^\dagger(r') V(r - r') \psi_\beta(r') \psi_\alpha(r), \qquad (5.15)$$

$$\mathcal{H}_2 = \frac{\gamma}{2} \int dr [\psi_\uparrow^\dagger(r)\psi_\downarrow(r) - \psi_\downarrow^\dagger(r)\psi_\uparrow(r)] h(r). \qquad (5.16)$$

In the above expression $\psi_\alpha(r)$ is the second quantization field operator which annihilates an electron of spin α located at the point r. The operators $\psi_\alpha(r)$ obey the following anticommutation relations:

$$\psi_\alpha(r)\psi_\beta(r') + \psi_\beta(r')\psi_\alpha(r) = 0, \qquad (5.17a)$$

$$\psi_\alpha^\dagger(r)\psi_\beta^\dagger(r') + \psi_\beta^\dagger(r')\psi_\alpha^\dagger(r) = 0, \qquad (5.17b)$$

$$\psi_\alpha^\dagger(r)\psi_\beta(r') + \psi_\beta(r')\psi_\alpha^\dagger(r) = \delta_{\alpha\beta}\delta(r - r'). \qquad (5.17c)$$

With this notation the operator corresponding to the number of particles is

$$N = \sum_\alpha \int dr\, \psi_\alpha^\dagger(r) \psi_\alpha(r). \qquad (5.18)$$

† For further details of this method, see P. G. de Gennes, *Superconductivity of metals and alloys*, Benjamin, 1966.

Microscopic Theory

μ is the chemical potential for the grand canonical ensemble. $\mathcal{H}_0$ represents the kinetic energy of the electrons in the presence of the field h and of the potential $U_0(r)$. For the sake of simplicity $U_0(r)$ has been taken independent of the spins.

$\mathcal{H}_1$ is the part of the hamiltonian corresponding to the two-body interaction between the electrons.

$\mathcal{H}_2$ corresponds to the polarization of the electron spins due to the field h (paramagnetic effect), $\gamma = (e\hbar/2mc)g$ where g is the g factor of the electron ($g = 2$) and $e\hbar/2mc$ is the Bohr magneton μ_B. The self-consistent field method consists of replacing the interaction $V\psi^\dagger\psi^\dagger\psi\psi$ by a mean potential acting on one particle at a time (i.e. with two ψ operators). This defines an effective hamiltonian:

$$\mathcal{H}_{\text{eff}} = \sum_\alpha \int dr\, \psi_\alpha^\dagger(r) \left\{ \frac{[p - (e/c)A]^2}{2m} + U_0(r) \right\} \psi_\alpha(r) - \mu N$$

$$+ \sum_{\alpha\beta} \int dr dr' \left\{ U(r\alpha \mid r'\beta) \psi_\alpha^\dagger(r) \psi_\beta(r') \right.$$

$$+ \frac{1}{2} W(r\alpha \mid r'\beta) \psi_\alpha^\dagger(r) \psi_\beta^\dagger(r')$$

$$\left. + \frac{1}{2} W^*(r\alpha \mid r'\beta) \psi_\beta(r') \psi_\alpha(r) \right\} \tag{5.19}$$

The last two terms are unusual, but must be included in order to take into account the appearance of the order parameter in a self-consistent way. It will be shown that the order parameter in the condensed phase is proportional to the mean value $\langle\psi^\dagger\psi^\dagger\rangle$. In the term $U(r\alpha \mid r'\beta)$ one can separate a "local" and a "non-local" part, i.e.

$$U(r\alpha \mid r'\beta) = U_1(r)\delta_{\alpha\beta}\delta(r - r') + U_2(r\alpha \mid r'\beta).$$

$U_1(r)$ is the so called Hartree potential, while U_2 is the exchange potential. It is expected that these terms will not differ greatly between the superconducting state and the normal state, while only a few electrons, in a band $\hbar\omega_D$ around the Fermi level are influenced by the condensation process. They can be included in $\mathcal{H}_0$ which will read:

$$\mathcal{H}_0 = \int dr\, \psi_\alpha^\dagger(r) \left\{ \frac{[p - (eA/c)]^2}{2m} + U(r) \right\} \psi_\alpha(r). \tag{5.14a}$$

$U(r)$ is now the Hartree–Fock potential for the one electron hamiltonian. We shall now outline very briefly, the derivation of W.

First $\mathcal{H}_{\text{eff}}$ is diagonalized using a Bogoliubov transformation, i.e. by setting

$$\psi_\alpha(r) = \sum_n [u_{n\alpha}(r) \gamma_{n,\alpha} + v_{n\alpha}(r) \gamma_{n,\alpha}^\dagger], \tag{5.20}$$

Reversible Properties

where $\gamma_{n,\alpha}$ are fermion operators. $\mathcal{H}_{\text{eff}}$ will take the form

$$\mathcal{H}_{\text{eff}} = E_g + \sum_{n,\alpha} \varepsilon_n \gamma_{n,\alpha}^{\dagger} \gamma_{n,\alpha}, \tag{5.21}$$

where E_g is the ground state energy and ε_n is the energy of the excitation n. Actually, E_g and ε_n are functions of W. In order to determine W one uses the condition that

$$\langle \widetilde{\phi} | \mathcal{H} | \widetilde{\phi} \rangle \tag{5.22a}$$

is a minimum, where $\widetilde{\phi}$ is the state of no excitation, i.e.

$$\gamma_{n,\alpha} \widetilde{\phi} = 0 \tag{5.22b}$$

and $\mathcal{H}$ the hamiltonian (5.13).

It is found that

$$W(r\alpha | r'\beta) = V(r, r') \langle \widetilde{\phi} | \psi_\beta(r') \psi_\alpha(r) | \widetilde{\phi} \rangle, \tag{5.23a}$$

$$W^*(r\alpha | r'\beta) = - W(r'\beta | r\alpha). \tag{5.23b}$$

The matrix element $\langle \widetilde{\phi} | \psi_\beta(r') \psi_\alpha(r) | \widetilde{\phi} \rangle$ is different from zero if one considers an open system, so that the ground state wave function is a linear combination of functions corresponding to different numbers of particles.

Moreover, as usual, the chemical potential μ is equal to the Fermi energy E_F.

The effective hamiltonian reads

$$\mathcal{H}_{\text{eff}} = \mathcal{H}_0 - E_F N + \frac{1}{2} \int dr\, dr'\, V(r, r') \langle \psi_\beta(r') \psi_\alpha(r) \rangle \psi_\alpha^{\dagger}(r) \psi_\beta^{\dagger}(r')$$

$$+ \frac{1}{2} \int dr\, dr'\, V(r, r') \langle \psi_\alpha^{\dagger}(r) \psi_\beta^{\dagger}(r') \rangle \psi_\beta(r') \psi_\alpha(r). \tag{5.24}$$

The above calculation is valid at $T = 0$. If $T \neq 0$, one must consider the matrix elements $\langle \psi\psi \rangle$ as thermal averages, i.e.

$$\langle \psi\psi \rangle = \text{Trace}\,(e^{-\beta \mathcal{H}_{\text{eff}}} \psi \psi) / \text{Trace}\,(e^{-\beta \mathcal{H}_{\text{eff}}}),$$

where

$$\beta = 1/k_B T$$

and k_B is the Boltzmann constant.

As usual in the self-consistent method, the effective hamiltonian depends on the temperature through the unknown mean values $\langle \psi\psi \rangle$ and each hamiltonian $\mathcal{H}_{\text{eff}}(T)$ can only be used at temperature T.

Microscopic Theory

In order to derive the Ginzburg–Landau equations it is sufficient to assume that the interaction V is local and that $W(r\alpha \mid r'\beta)$ only connects antiparallel spin states, i.e.

$$V(r, r') = -V\delta(r - r'),$$

$$W(r\uparrow \mid r'\downarrow) = -V\langle\psi_\downarrow(r)\psi_\uparrow(r)\rangle = \Delta(r). \tag{5.25}$$

This introduces the quantity $\Delta(r)$ known as the pair potential. It will be shown later that $\Delta(r)$ is proportional to $\psi(r)$ the Ginzburg–Landau order parameter. The effective hamiltonian now reads:

$$\mathcal{H}_{\text{eff}} = \int dr\, \psi_\alpha^\dagger \left[\frac{1}{2m}\left(p - \frac{eA}{c}\right)^2 + U(r) - E_F\right]\psi_\alpha$$

$$+ \gamma \int dr\, h(r)\left[\psi_\uparrow^\dagger(r)\psi_\uparrow(r) - \psi_\downarrow^\dagger(r)\psi_\downarrow(r)\right]$$

$$+ \int dr\,[\Delta^*(r)\psi_\downarrow(r)\psi_\uparrow(r) + \Delta(r)\psi_\uparrow^\dagger(r)\psi_\downarrow^\dagger(r)] + \frac{1}{V}\int dr\,|\Delta(r)|^2. \tag{5.26}$$

The last term must be added in the self-consistent method because the two preceding terms introduce twice the interactions.

The free energy F_s will now be minimized with respect to the variations of $\Delta(r)$.

$$F_s = -k_B T \log Z, \tag{5.27a}$$

where

$$Z = \text{Trace}\, e^{-\beta \mathcal{H}_{\text{eff}}} \tag{5.27b}$$

and

$$\frac{\delta F}{\delta \Delta^*} = \int \left[\frac{\text{Trace}\, e^{-\mathcal{H}_{\text{eff}}}\psi_\downarrow(r)\psi_\uparrow(r)}{Z} + \frac{\Delta(r)}{V}\right]dr. \tag{5.28}$$

This variation is zero if $\Delta(r)$ obeys the self-consistent equation:

$$\Delta(r) = V(r)\langle\psi_\downarrow(r)\psi_\uparrow(r)\rangle = \frac{\text{Trace}\, e^{-\beta\mathcal{H}_{\text{eff}}}\psi_\downarrow(r)\psi_\uparrow(r)V(r)}{\text{Trace}\, e^{-\beta\mathcal{H}_{\text{eff}}}}. \tag{5.29}$$

The actual calculation of the free energy would be a terrible problem. It is greatly simplified when there is no magnetic field and, in this case, this leads to the B.C.S. theory. This will be outlined briefly below in order to compute the thermodynamic critical field through the relation already given in Chapter 1:

$$\frac{H_c^2(T)}{8\pi} = F_n - F_s.$$

Reversible Properties

(b) *The B.C.S. equation and the determination of the thermodynamic critical field*

The effective hamiltonian in the absence of an applied field is

$$\mathcal{H}_{\text{eff}} = \int d\mathbf{r}\, \psi_\alpha^\dagger \left(\frac{p^2}{2m} - E_F\right)\psi_\alpha + \int d\mathbf{r}\,[\Delta^*(\mathbf{r})\,\psi_\downarrow(\mathbf{r})\,\psi_\uparrow(\mathbf{r})$$

$$+ \Delta(\mathbf{r})\,\psi_\uparrow^\dagger(\mathbf{r})\,\psi_\downarrow^\dagger(\mathbf{r})] + \int \frac{|\Delta(\mathbf{r})|^2}{V}\,d\mathbf{r}. \quad (5.30)$$

ψ_α may now be expanded as

$$\psi_\alpha = \frac{1}{\sqrt{\Omega}} \sum_k e^{i\mathbf{k}\cdot\mathbf{r}}\, a_{k,\alpha},$$

where $a_{k,\alpha}$ is the annihilation operator of an electron of spin α and momentum $\hbar k$ and Ω is the volume of the system. The effective hamiltonian becomes

$$\mathcal{H}_{\text{eff}} = \sum_k \xi_k(a_{k\uparrow}^\dagger a_{k\uparrow} + a_{k\downarrow}^\dagger a_{k\downarrow}) + \sum_k \Delta(a_{-k\downarrow} a_{k\uparrow}$$

$$+ a_{k\uparrow}^\dagger a_{-k\downarrow}^\dagger) + \frac{\Delta^2}{V}. \quad (5.31)$$

In deriving this expression, it has been assumed that the interaction $V_{k,k'}$ is such that:

$$V = -V \quad \text{for } -\hbar\omega_D < \xi_k, \xi_{k'} < \hbar\omega_D$$

$$V = 0 \quad \text{everywhere else} \quad (5.32)$$

and that

$$\Delta_k = -V \sum_{|\xi_k|<\hbar\omega_D} \langle a_{-k\downarrow} a_{k\uparrow}\rangle = \Delta \quad \text{for } -\hbar\omega_D < \xi_k < \hbar\omega_D, |\xi_k| < \hbar\omega_D$$

$$\Delta_k = 0 \quad \text{everywhere else.} \quad (5.33)$$

ξ_k are the energies of the electrons in state k as measured from the Fermi level.

The hamiltonian is easily diagonalized using the transformation

$$\gamma_k = u_k a_{k\downarrow} - V_k a_{-k\uparrow}^\dagger$$

$$\gamma_{-k} = u_k a_{-k\uparrow} + V_k a_{k\downarrow}^\dagger \quad (5.34)$$

This yields

$$u_k^2 = \frac{1}{2}\left(1 + \frac{\xi_k}{E_k}\right); \quad V_k^2 = \frac{1}{2}\left(1 - \frac{\xi_k}{E_k}\right), \quad (5.35\text{a})$$

$$\mathcal{H}_{\text{eff}} = E_g + \sum_k E_k(\gamma_k^\dagger \gamma_k + \gamma_{-k}^\dagger \gamma_{-k}), \quad (5.35\text{b})$$

$$E_k = (\xi_k^2 + \Delta^2)^{1/2} \quad (5.35\text{c})$$

Microscopic Theory

The γ_k obey the commutation rule for the fermions and the Fermi statistics. It is seen that the spectrum of the excitations as defined by (5.33c) presents an energy gap of width Δ.

Δ must obey a self-consistent equation and this is readily obtained by calculating the thermal average of $a_{-k\uparrow} a_{k\downarrow}$ in the presence of the new fermions. This yields:

$$\Delta = V\Delta \sum_k \frac{1 - 2f(E_k)}{E_k}, \quad (5.36)$$

where $f(E_k)$ is the Fermi function. The summation over k is restricted to the band $-\hbar\omega_D < \xi_k < \hbar\omega_D$.

Clearly the above equation always has the solution $\Delta = 0$ which corresponds to the normal state. Solutions with a finite Δ also exist.

Replacing, as usual, the summation over k by an integral, $N(0)$ being the density of states per unit energy and unit volume at the Fermi level, the equation reads†

$$1 = N(0)V \int_{-\hbar\omega_D}^{\hbar\omega_D} \frac{\tanh\{(\beta/2)(\xi^2 + \Delta^2)^{1/2}\}}{2(\xi^2 + \Delta^2)^{1/2}} d\xi. \quad (5.37)$$

This is the well-known B.C.S. integral equation which gives the variation of Δ as a function of temperature. The transition temperature T_0 is the value at which (5.36) has a solution Δ different from zero. T_0 is obtained from (5.37) by setting $\Delta = 0$, i.e.

$$\frac{1}{N(0)V} = \int_{-\hbar\omega_D}^{\hbar\omega_D} \frac{\tanh \xi/2k_B T_0}{2\xi} d\xi. \quad (5.38)$$

Integrating by parts yields:

$$\frac{1}{N(0)V} = \log \frac{\hbar\omega_D}{k_B T_0} \tanh \frac{\hbar\omega_D}{2k_B T_0} - \int_0^{\hbar\omega_D/k_B T_0} dx \log x \frac{d}{dx} \tanh\left[\frac{x}{2}\right]. \quad (5.39)$$

Since $k_B T_0 \ll \hbar\omega_D$, $\tanh \frac{\hbar\omega_D}{2k_B T_0} \# 1$ and the integral may be extended to infinity. It is equal to $-\log \frac{2\gamma}{\pi}$ where $\gamma = e^C$ and $C = 0.55$ is the Euler constant. The transition temperature is obtained from

$$\frac{1}{N(0)V} = \log 1.14 \frac{\hbar\omega_D}{k_B T_0}. \quad (5.40)$$

For $T = 0$ the value of the gap Δ_0 is found to be

$$\Delta_0 = \frac{\pi}{\gamma} k_B T_0 \simeq 1.74 k_B T_0 \quad (5.41)$$

† In this equation, it is supposed that there is a spherical band. Since $\hbar\omega_D$ is very small in comparison to E_F, $N(\xi)$ can be taken as a constant and equal to the density of states at the Fermi energy.

131

Reversible Properties

while the variation of Δ with temperature is fairly well approximated by

$$\frac{\Delta(T)}{\Delta_0} = \tanh\left[\frac{T_0}{T}\frac{\Delta(T)}{\Delta(0)}\right] \quad (5.42)$$

except in the vicinity of T_0. Close to T_0 it is seen that

$$\Delta(T) = 3 \cdot 1\, k_B T_0 \left(1 - \frac{T_0}{T}\right)^{1/2}. \quad (5.43)$$

The variation of Δ with temperature is shown in Fig. 5.2.

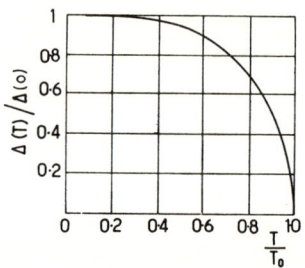

Fig. 5.2. Variation of the gap Δ as function of temperature according to the B. C. S. theory

The free energy is now easily obtained:

$$F_s(T) = -k_B T \log Z = -k_B T \log \text{Trace}\ e^{-\beta \mathcal{H}_{\text{eff}}} \quad (5.44)$$

Two parts appear in this expression, namely,

$$-k_B T\, \text{Trace}\ e^{-\beta E_g} \quad \text{and} \quad -k_B T\, \text{Trace}\ e^{-\beta \sum_k \xi_k (\gamma_k^\dagger \gamma_k + \gamma_{-k}^\dagger \gamma_{-k})} \quad (5.45)$$

The last term is the free energy of a set of independent fermions (quasi particles) and is equal to

$$2k_B T \sum_k \log[1 - f(E_k)]. \quad (5.46)$$

Using the value of E_g obtained from the Bogoliubov transformation, one obtains

$$F_s(T) = 2k_B T \sum_k \log[1 - f(E_k)] + \sum_k \left[\xi_k - \frac{\xi_k^2}{E_k} - \frac{\Delta^2}{2E_k} - \frac{\Delta^2 f(E_k)}{E_k}\right]. \quad (5.47)$$

In order to obtain the thermodynamic field, it is necessary to subtract

$F_s(T)$ from $F_n(T)$. This latter quantity is given by (5.47) with Δ set equal to zero. At zero temperature, it is then found that

$$\frac{H_c^2(0)}{8\pi} = \frac{N(0)\Delta_0^2}{2},\qquad(5.48)$$

where Δ_0 is the value of Δ for $T = 0$.

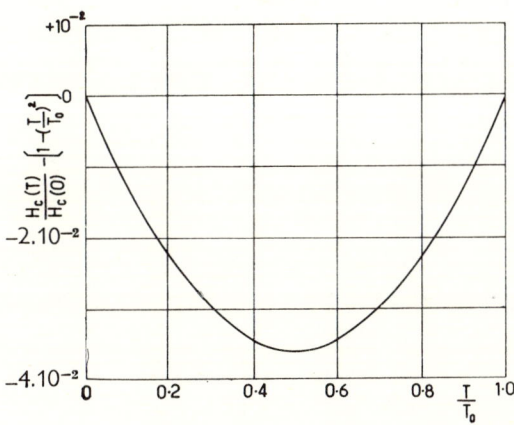

FIG. 5.3. Variation of $H_c(T)/H_c(0) - [1 - (T/T_0)^2]$ versus temperature in the B. C. S. theory according to Mühlschlegel (1959)

Close to T_0 where Δ is small $F_s(T)$ can be expanded in powers of Δ and

$$\frac{H_c(T)}{H_c(0)} = \gamma \left(\frac{8}{7\zeta(3)}\right)^{1/2} \left(\frac{T_0 - T}{T_0}\right) = 1\cdot74 \left(1 - \frac{T}{T_0}\right),\qquad(5.49)$$

where γ is the Euler constant and $\zeta(3)$ the Riemann Zeta function.
$H_c(T)$ vanishes at $T = T_0$ with a finite slope.
$H_c(T)/H_c(0)$ is slightly different from $1-(T/T_0)^2$ which is the empirical law suggested by early experiments. Mühlschlegel (1959) has computed the function

$$D\left(\frac{T}{T_0}\right) = \frac{H_c(T)}{H_c(0)} - \left[1 - \frac{T}{T_0}\right]^2.\qquad(5.50)$$

His results are given in Fig. 5.3.
Thus it is possible to compute the thermodynamic critical field. This will be used later.

Reversible Properties

5.4. Derivation of the Ginzburg–Landau equations. The distinction between clean and dirty superconductors

In order to simplify the writing, it is convenient to introduce the following notation.

$$T_\uparrow(r) = \frac{1}{2m}\left[p - \frac{eA}{c}\right]^2 + U(r) - E_F + \gamma h(r), \qquad (5.51a)$$

$$T_\downarrow(r) = \frac{1}{2m}\left[p - \frac{eA}{c}\right]^2 + U(r) - E_F - \gamma h(r), \qquad (5.51b)$$

$$\mathscr{A}(r) = \psi_\uparrow^\dagger(r)\psi_\downarrow^\dagger(r), \qquad (5.52a)$$

$$\mathscr{A}^\dagger(r) = \psi_\downarrow(r)\psi_\uparrow(r). \qquad (5.52b)$$

Then the effective hamiltonian reads

$$\mathscr{H}_{\text{eff}} = H_0 + H_1 + \frac{1}{V}\int dr\,|\Delta|^2, \qquad (5.53)$$

where

$$H_0 = \int dr\,\psi_\uparrow^\dagger(r)T_\uparrow(r)\psi_\uparrow(r) + \int dr\,\psi_\downarrow^\dagger(r)T_\downarrow(r)\psi_\downarrow(r), \qquad (5.54)$$

$$H_1 = \int dr\,[\Delta(r)\mathscr{A} + \Delta^*(r)\mathscr{A}^\dagger]. \qquad (5.55)$$

In the Ginzburg–Landau scheme $\psi(r)$ [i.e. $\Delta(r)$] is small, and H_1 can be treated as a perturbation of the hamiltonian H_0, which represents the normal electron in the presence of the field $h(r)$. The perturbation expansion of (5.29) will be written as

$$V(r)\langle\psi_\downarrow(r)\psi_\uparrow(r)\rangle = \frac{\text{Trace }e^{-\beta H_0}\left[1 - \int_0^\beta H_1(\beta_1)\,d\beta_1 + \int_0^\beta d\beta_1\int_0^{\beta_1} d\beta_2 H_1(\beta_1)H_1(\beta_2) + \ldots\right]V(r)\psi_\downarrow(r)\psi_\uparrow(r)}{\text{Trace }e^{-\beta H_0}\left[1 - \int_0^\beta H_1(\beta_1)\,d\beta_1 + \int_0^\beta d\beta_1\int_0^{\beta_1} d\beta_2 H_1(\beta_1)H_1(\beta_2) + \ldots\right]}. \qquad (5.56)$$

where

$$H_1(\beta_1) = \exp[\beta_1 H_0]\,H_1\,\exp[-\beta_1 H_0].$$

It is clear that (5.56) is equivalent to

$$-\Delta(r) = V(r)\langle\psi_\downarrow(r)\psi_\uparrow(r)\rangle =$$

$$\frac{\left\langle\left[1 - \int_0^\beta H_1(\beta_1)\,d\beta_1 + \int_0^\beta d\beta_1\int_0^{\beta_1} d\beta_2\,H_1(\beta_1)H_1(\beta_2) + \ldots\right]V(r)\psi_\downarrow(r)\psi_\uparrow(r)\right\rangle_0}{\left\langle 1 - \int_0^\beta H_1(\beta_1)\,d\beta_1 + \int_0^\beta d\beta_1\int_0^{\beta_1} d\beta_2\,H_1(\beta_1)H_1(\beta_2) + \ldots\right\rangle_0}.$$

$$(5.57)$$

where $\langle\ \rangle_0$ denotes the thermal average in the *normal state* (including the effect of the field on the spins).

The first Ginzburg–Landau equation can be obtained directly from an expansion of the right-hand side of (5.57) in powers of Δ, i.e.

$$-\Delta(r) = \int K_0(r, s)\,\Delta(s)\,ds + \int K_1(r, s_1, s_2, s_3)\,\Delta(s_1)\,\Delta^*(s_2)$$

$$\Delta(s_3)\,ds_1\,ds_2\,ds_3 + \ldots \qquad (5.58)$$

(a) *Derivation of the linear term in Δ*

In order to calculate the kernel $K_0(r, s)$, the following simplifying assumptions will be made:

(i) the paramagnetic effect will be neglected, as in the original Ginzburg–Landau equation. This means that γ is set equal to zero.

(ii) the interaction potential V will be taken to be a constant in the band $\pm\hbar\omega_D$ around the Fermi level (B.C.S. assumption) for electrons of opposite spins and opposite momenta, and to be zero outside this band. The linear term then reads

$$\Delta(r) = V\int ds\,\Delta(s)\int_0^\beta d\beta_1 \langle\mathcal{A}(s, \beta_1)\,\mathcal{A}^\dagger(r, 0)\rangle_0. \qquad (5.59)$$

Introducing the Fourier transform

$$\Delta(q) = \int \exp i q.r\,\Delta(r)\,dr, \qquad (5.60)$$

(5.59) changes into

$$\Delta(q) = V\iint e^{iq.s}\,\Delta(s)\,ds \int_0^\beta d\beta_1\,e^{iq.(r-s)}\langle\mathcal{A}(s, \beta_1)\,\mathcal{A}^\dagger(r, 0)\rangle_0\,dr. \qquad (5.61)$$

In a pure metal, for an infinite sample, it is clear that, due to the translational invariance, the mean value $\langle\mathcal{A}(s, \beta_1)\mathcal{A}^\dagger(r, 0)\rangle_0$, which is taken in the normal metal, depends only on $(r - s)$.

In an alloy, this translational invariance no longer exists. In order to restore it, an average must be taken over the impurity configuration. However, the right-hand side of (5.61) will contain:

$$\overline{\Delta(s)\langle\mathcal{A}(s, \beta_1)\,\mathcal{A}^\dagger(r, 0)\rangle_0}, \qquad (5.62)$$

where the bar denotes the average over the impurity configuration. A further approximation consists of replacing this by

$$\Delta(s)\overline{\langle\mathcal{A}(s, \beta_1)\,\mathcal{A}^\dagger(r, 0)\rangle_0} \qquad (5.63)$$

so that the mean value $\langle\mathcal{A}\mathcal{A}^\dagger\rangle_0$ depends only on $(r - s)$.

Reversible Properties

The approximation (5.63) is reasonable when the impurity potentials are weak (Caroli et al. 1962). It is essentially correct when the constituents of the alloy are similar from a chemical point of view.

One must therefore calculate the quantity

$$K(\boldsymbol{q}) = V \int e^{i\boldsymbol{q}\cdot(\boldsymbol{r}-\boldsymbol{s})} d(\boldsymbol{r}-\boldsymbol{s}) \int_0^\beta d\beta_1 \overline{\langle \mathcal{A}(\boldsymbol{s},\beta_1)\mathcal{A}^\dagger(\boldsymbol{r},0)\rangle_0}. \quad (5.64)$$

The field operator $\psi_\alpha(\boldsymbol{r})$ is first expanded as

$$\psi_\alpha(\boldsymbol{r}) = \sum_n w_{n\alpha}(\boldsymbol{r}) a_{n\alpha}, \quad (5.65)$$

where $a_{n\alpha}$ is the annihilation operator for an electron of spin α in the state n. $w_{n\alpha}(\boldsymbol{r})$ is the eigenfunction of the unperturbed hamiltonian H_0 such that

$$\left[\frac{1}{2m}\left[\boldsymbol{p} - \frac{e}{c}\boldsymbol{A}\right]^2 + U(\boldsymbol{r})\right] w_{n\alpha}(\boldsymbol{r}) = \xi_{n\alpha} w_{n\alpha}(\boldsymbol{r}), \quad (5.66)$$

where $\xi_{n\alpha}$ is the energy of the state $n\alpha$ as measured from the Fermi level. It is clear that $w_{n\alpha}(\boldsymbol{r})$ and $\xi_{n\alpha}$ do not depend of the spin α so that

$$\langle \mathcal{A}(\boldsymbol{s},\beta)\mathcal{A}^\dagger(\boldsymbol{r},0)\rangle_0 = \sum_{\substack{m,n \\ m',n'}} w^*_m(\boldsymbol{s}) w_n^*(\boldsymbol{s}) w_{m'}(\boldsymbol{r}) w_{n'}(\boldsymbol{r})$$

$$\langle a^\dagger_{m'\uparrow}(\beta_1) a^\dagger_{n\downarrow}(\beta_1) a_{m'\downarrow}(0) a_{n'\uparrow}(0) \rangle_0, \quad (5.67)$$

where

$$a_{m\alpha}(\beta) = a_{m\alpha} e^{\beta \xi_m}. \quad (5.68)$$

The calculation of the thermal average on the right-hand side is straightforward and yields

$$\langle \mathcal{A}(\boldsymbol{s},\beta_1)\mathcal{A}^\dagger(\boldsymbol{r},0)\rangle_0 = \sum_{m,n} w^*_m(\boldsymbol{s}) w_n^*(\boldsymbol{s}) w_m(\boldsymbol{r}) w_n(\boldsymbol{r}) e^{\beta_1(\xi_n+\xi_m)} f_m f_n, \quad (5.69)$$

where f_m and f_n are the Fermi function, i.e.

$$f_m = \frac{1}{e^{\beta \xi_m}+1}. \quad (5.70)$$

After a simple integration it is found that

$$K(\boldsymbol{q}) = \sum_{m,n} V \int e^{i\boldsymbol{q}\cdot(\boldsymbol{r}-\boldsymbol{s})} d(\boldsymbol{r}-\boldsymbol{s}) \left[f_m f_n\, w_m^*(\boldsymbol{s}) w_n^*(\boldsymbol{s}) w_m(\boldsymbol{r}) w_n(\boldsymbol{r}) \right.$$

$$\left. \frac{\exp\beta(\xi_m+\xi_n)-1}{\xi_m+\xi_n} \right]$$

$$= \sum_{m,n} \frac{V}{2} \int d(\boldsymbol{r}-\boldsymbol{s}) e^{i\boldsymbol{q}\cdot(\boldsymbol{r}-\boldsymbol{s})} \left[\tanh\frac{\beta\xi_m}{2} + \tanh\frac{\beta\xi_n}{2}\right]$$

$$\frac{w_m^*(\boldsymbol{s}) w_n^*(\boldsymbol{s}) w_m(\boldsymbol{r}) w_n(\boldsymbol{r})}{\xi_m+\xi_n}. \quad (5.71)$$

Microscopic Theory

If one expands the function $\tanh(\beta\xi_n)/2$ as a series of the form

$$\tanh\frac{\beta\xi_n}{2} = \frac{4}{\beta}\sum_{\nu=0}^{\infty}\frac{\xi_n}{[(2\nu+1)\pi/\beta]^2 + \xi_n^2} = \frac{2}{\beta}\sum_{\nu=-\infty}^{+\infty}\frac{1}{\xi_n + i\omega}$$

$$= \frac{2}{\beta}\sum_{\nu=-\infty}^{+\infty}\frac{1}{\xi_n - i\omega}, \qquad (5.72)$$

where

$$\omega = (2\nu+1)\frac{\pi}{\beta}$$

is the Matsubara frequency, $K(\boldsymbol{q})$ becomes, after some algebra

$$K(\boldsymbol{q}) = \sum_{m,n}\frac{V}{\beta}\int d(\boldsymbol{r}-\boldsymbol{s})\,e^{i\boldsymbol{q}\cdot(\boldsymbol{r}-\boldsymbol{s})}\,w_m^*(\boldsymbol{s})\,w_n^*(\boldsymbol{s})\,w_m(\boldsymbol{r})\,w_n(\boldsymbol{r})$$

$$\frac{1}{(\xi_m + i\omega)(\xi_n - i\omega)} \qquad (5.73)$$

It is convenient to rewrite $K(\boldsymbol{q})$ as

$$K(\boldsymbol{q}) = \sum_{\substack{m,n\\\omega}}\frac{V}{\beta\Omega}\int d\boldsymbol{r}\,d\boldsymbol{s}\,e^{i\boldsymbol{q}\cdot(\boldsymbol{r}-\boldsymbol{s})}\frac{w_m^*(\boldsymbol{s})\,w_n^*(\boldsymbol{s})\,w_m(\boldsymbol{r})\,w_n(\boldsymbol{r})}{(\xi_m + i\omega)(\xi_n + i\omega)} \qquad (5.74)$$

Ω is the volume of the sample, and in deriving (5.74) the translational invariance has been taken into account.

Let us write $w_n(\boldsymbol{r})$ in the form

$$w_n(\boldsymbol{r}) = e^{i(e/\hbar c)A(\boldsymbol{r})\cdot\boldsymbol{r}}\phi_n(\boldsymbol{r}). \qquad (5.75)$$

It is seen that $\phi_n(\boldsymbol{r})$ obeys the following equation:

$$\frac{1}{2m}\left\{\boldsymbol{p} + \frac{e}{c}\nabla[A(\boldsymbol{r})\cdot\boldsymbol{r}] - \frac{e}{c}A(\boldsymbol{r})\right\}^2\phi_n(\boldsymbol{r}) + U(\boldsymbol{r})\phi(\boldsymbol{r}) = \xi_n\phi_n(\boldsymbol{r}). \qquad (5.76)$$

If $A(\boldsymbol{r})$ is a slowly varying function of $\boldsymbol{r}$, its variation can be neglected and $\phi_n(\boldsymbol{r})$ is the solution of

$$H_{0N}\phi_n(\boldsymbol{r}) = \frac{p^2}{2m}\phi_n(\boldsymbol{r}) + U(\boldsymbol{r})\phi_n(\boldsymbol{r}) = \xi_n\phi_n(\boldsymbol{r}), \qquad (5.77)$$

where H_{0N} is the hamiltonian of the normal metal in the absence of magnetic field. Thus $\phi_n(\boldsymbol{r})$ and ξ_n are the eigenfunctions and eigenvalues for the electrons in the normal metal in the absence of the field $\boldsymbol{h}$. It is clear that the function $\phi_n(\boldsymbol{r})$ may be taken as real. This is the fundamental assumption for the derivation of the Ginzburg–Landau equations. Its domain of validity will be discussed in detail at the end of the section. Its main interest is to avoid solving the formidable problem linked with the Landau diamagnetism of the electron while calculating the Ginzburg–Landau parameters in terms of

Reversible Properties

simple properties of the normal metal. The equation for $\Delta(q)$ is now

$$\Delta(q) = \Delta(q) \int dr \left[\exp i \left(q + \frac{2e}{\hbar c} A \right) \cdot r \right] K(r) = K \left(q + \frac{2e}{\hbar c} A \right) \Delta(q) \tag{5.78}$$

where

$$K(r - s) = \frac{V}{\beta \Omega} \sum_{m,n} \phi_n(r) \phi_n(s) \phi_m(r) \phi_m(s) \frac{1}{(\xi_n + i\omega)(\xi_m - i\omega)}. \tag{5.79}$$

$\Delta(r)$ is obtained from the inverse Fourier transform, i.e.

$$\Delta(r) = \frac{1}{8\pi^3} \int e^{-iq \cdot r} K \left(q + \frac{2e}{\hbar c} A \right) \Delta(q) \, dq. \tag{5.80}$$

Since

$$K \left(i\nabla + \frac{2e}{\hbar c} A \right) e^{-iq \cdot r} = K \left(q + \frac{2eA}{\hbar c} \right) e^{-iq \cdot r}, \tag{5.81}$$

(5.80) becomes

$$\Delta(r) = K \left(- i\nabla - \frac{2e}{\hbar c} A \right) \Delta(r). \tag{5.82}$$

It is seen that it is sufficient to calculate $K(q)$ without taking into account the vector potential A. As expected, in the equation for $\Delta(r)$ [i.e. $\psi(r)$], the operator $[i\nabla - (2e/\hbar c)A]$ appears.

$K(q)$ has the simple form

$$K(q) = \frac{V}{\beta \Omega} \sum_{\substack{n,m \\ \omega}} \frac{\langle n | e^{iq \cdot r} | m \rangle \langle m | e^{-iq \cdot r} | n \rangle}{(\xi_n + i\omega)(\xi_m - i\omega)}. \tag{5.83}$$

It is now possible to introduce a correlation function for the electron in the normal metal since

$$\frac{1}{(\xi_n + i\omega)(\xi_m - i\omega)} = \frac{1}{(\xi_n - \xi_m + 2i\omega)} \left\{ \frac{1}{\xi_m - i\omega} - \frac{1}{\xi_n + i\omega} \right\}$$

$$= -\frac{i}{\hbar} \left\{ \frac{1}{\xi_m - i\omega} - \frac{1}{\xi_n + i\omega} \right\} \int_0^\infty dt \, \exp i(\xi_n - \xi_m + 2i\omega) \frac{t}{\hbar}. \tag{5.84}$$

$K(q)$ changes into:

$$K(q) = \operatorname{Im} \sum_{\omega_n} \frac{V}{\hbar \beta \omega} \int_0^\infty dt \, e^{-2|\omega|} \frac{t}{\hbar} \langle n | e^{iq \cdot r(t)} e^{-iq \cdot r(0)} | n \rangle$$

$$\left\{ \frac{1}{\xi_n + i\omega} - \frac{1}{\xi_n - i\omega} \right\}, \tag{5.85}$$

where

$$e^{i\mathbf{q}\cdot\mathbf{r}(t)} = e^{i\beta H_0 N} e^{i\mathbf{q}\cdot\mathbf{r}} e^{-i\beta H_0 N}. \quad (5.86)$$

The correlation function $\langle n | e^{i\mathbf{q}\cdot\mathbf{r}(t)} e^{-i\mathbf{q}\cdot\mathbf{r}(0)} | n \rangle$ can be deduced from the transport properties of the normal metal by simple physical considerations. Since only the electrons close to the Fermi level are involved in the process, it is sufficient to calculate this function just for these electrons. It is now convenient to introduce a distinction first proposed by Anderson (1959) between clean superconductors, in which the electron mean free path l is much larger that the coherence length ξ_0, and "dirty" superconductors in which $l \ll \xi_0$. In the class of "clean" superconductors one finds pure metals, and they are often referred to as "pure" superconductors, while in the "dirty" one, one finds alloys and metals containing impurities.

(α) *"Clean" superconductor.* In this case an electron at the Fermi surface moves with a velocity equal to v_F. If at $t = 0$ this electron is located at $\mathbf{r} = 0$ at time t, the quantity $\mathbf{q} \cdot \mathbf{r}(t)$ will be $qv_F\, t \cos\theta$ where θ is the angle between the velocity of the electron and the $\mathbf{q}$ vector.
Thus

$$\overline{\langle n | \exp[i\mathbf{q}\cdot\mathbf{r}(t)] \exp[-i\mathbf{q}\cdot\mathbf{r}(0)] | n \rangle} = \frac{1}{2} \int_0^\pi \sin\theta\, d\theta\, e^{iqv_F\cos\theta\, t}, \quad (5.87)$$

where the bar indicates an average over all states $| n \rangle$ of energy ξ_n, and

$$K(\mathbf{q}) = \mathrm{Im} \sum_n \frac{V}{\hbar\beta\Omega} \frac{1}{2} \int_0^\infty dt \int_0^\pi \sin\theta\, d\theta\, e^{iqv_F\cos\theta\, t}\, e^{-2|\omega|t/\hbar}$$

$$\left\{\frac{1}{\xi_n + i\omega} - \frac{1}{\xi_n - i\omega}\right\} = \sum_n \frac{V}{\hbar\beta\Omega} \frac{4\omega^2}{\xi_n^2 + \omega^2} \frac{1}{2|\omega|qv_F} \tan^{-1}\frac{\hbar qv_F}{2|\omega|}.$$

$$(5.88)$$

$K(\mathbf{q})$ depends only on the length of the vector $\mathbf{q}$.
The summation over n can be replaced by an integral. If $N(0)$ is the density of states per unit energy and unit volume at the Fermi level, then

$$K(\mathbf{q}) = \sum_\omega \frac{2\pi N(0) V}{\hbar\beta} \frac{1}{qv_F} \tan^{-1}\frac{\hbar qv_F}{2|\omega|}. \quad (5.89)$$

The summation over ω in this expression for $K(\mathbf{q})$ is divergent. This comes from the fact that the B.C.S. cut off has been neglected. In order to restore the convergence, it is conventional to proceed as follows. One writes $K(\mathbf{q})$ as

$$K(\mathbf{q}) = [K(\mathbf{q}) - K(0)] + K_t(0) \quad (5.90)$$

$K_t(0)$ is the value of $K(0)$ calculated in the presence of the B.C.S. cut off.

$$K(0) = \frac{2\pi N(0) V}{\beta} \sum_\omega \frac{1}{2|\omega|}, \quad (5.91)$$

Reversible Properties

while from (5.83)

$$K_t(0) = \frac{N(0)V}{\beta} \sum_\omega \int_{-\hbar\omega_D}^{\hbar\omega_D} \frac{d\xi}{\xi^2 + \omega^2} = N(0)V \int_{-\hbar\omega_D}^{\hbar\omega_D} \frac{\tanh \beta\xi/2}{2\xi} d\xi .$$

(5.92)

In the limit $\hbar\omega_D \gg k_B T$ this integral is equal to†

$$K_t(0) = N(0)V \log \frac{1 \cdot 14 \, \hbar\omega_D}{k_B T}$$

(5.93)

and the linear part of the equation reads

$$\Delta(\mathbf{q})[1 - K_t(0)] = [K(\mathbf{q}) - K(0)] \Delta(\mathbf{q}) .$$

(5.94)

(β) *"Dirty" superconductor.* In this case the motion of the electron is governed by a diffusion process. If the electron mean free path is small in comparison to $1/\mathbf{q}$, one has

$$\overline{\langle n | e^{i\mathbf{q} \cdot \mathbf{r}(t)} e^{-i\mathbf{q} \cdot \mathbf{r}(0)} | n \rangle} = e^{-\hbar D q^2 |t|} \quad (lq \ll 1),$$

(5.95)

where $D = v_F l/3$ is the diffusion coefficient. $K(\mathbf{q})$ becomes

$$K(\mathbf{q}) = \sum_{n \atop \omega} \frac{V}{\hbar\beta\Omega} \operatorname{Im} \int_0^\infty dt \exp\left[-\hbar D q^2 - 2\frac{|\omega|}{\hbar}\right] t \left[\frac{1}{\xi_n + i\omega} - \frac{1}{\xi_n - i\omega}\right].$$

(5.96)

As in the case of the pure metal, one integrates over the energies and over t and

$$K(\mathbf{q}) = \sum_\omega \frac{2\pi V N(0)}{\beta} \frac{1}{2|\omega| + \hbar D q^2} .$$

(5.97)

$K(\mathbf{q})$ depends only on the length of the vector $\mathbf{q}$.

This expression is, of course, divergent. It is necessary to use the same method as in the case of pure metals, and the linearized part of the equation reads:

$$\Delta(\mathbf{q})[1 - K_t(0)] = [K(\mathbf{q}) - K(0)] \Delta(\mathbf{q}) ,$$

(5.98)

where $K(0)$ and $K_t(0)$ respectively are given by (5.91) and (5.93).

† See (5.40).

(b) *Domain of validity of the above derivation*

(α) *Range of the kernel $K(R)$*. In order to estimate the range of the kernel $K(R)$, it is necessary to compute the inverse Fourier transform

$$K(R) = \frac{2\pi}{8\pi^3} \int_0^\infty q^2 dq \int_{-1}^1 \sin\theta d\theta \, e^{iqR\cos\theta} K(q). \qquad (5.99)$$

In pure metals one obtains, using (5.89),

$$K(R) = \frac{N(0)V}{2} \frac{k_B T}{\hbar v_F} \sum_\omega \frac{e^{-(2|\omega|/\hbar v_F)R}}{R^2}, \qquad (5.100a)$$

while in dirty superconductors, using (5.96)

$$K(R) = \frac{N(0)V}{2} \frac{k_B T}{\hbar D} \sum_\omega \frac{e^{-(2|\omega|/\hbar D)^{1/2}R}}{R}. \qquad (5.100b)$$

The range is of the order of

$$\frac{\hbar v_F}{2\omega_0} = \frac{\hbar v_F}{\pi k_B T} \qquad \text{for pure superconductors}, \qquad (5.101a)$$

$$\left(\frac{\hbar D}{2\omega_0}\right)^{1/2} = \left(\frac{\hbar v_F l}{\pi k_B T}\right)^{1/2} \qquad \text{for dirty superconductors}, \qquad (5.101b)$$

where ω_0 is the value of ω for $\nu = 0$.
The smallest range is obtained for $T = T_0$. It is seen from the definition (1.23) of ξ_0 that the range of $K(R)$ is of the order of

$$\xi_0 = 0.18 \frac{\hbar v_F}{k_B T_0} \qquad \text{in pure superconductors}, \qquad (5.102a)$$

and of

$$\sqrt{(l\xi_0)} \qquad \text{in dirty superconductors}. \qquad (5.102b)$$

The derivation for dirty superconductors was valid in the limit $ql \ll 1$, i.e. $R \gg l$. By comparing this result with the value obtained for the range it is seen that the concept of "dirty" superconductors is valid if

$$\sqrt{(l\xi_0)} \gg l, \quad \text{i.e.} \quad l \ll \xi_0. \qquad (5.103)$$

The concept of "dirty" superconductors, as defined by (5.103), has proved to be of a great help in the understanding of superconducting properties. It can be seen that it is easy to change a "pure" superconducting material into a "dirty" one. In pure aluminium, for example, $\xi_0 \sim 16{,}000$ Å but a very small amount of impurity ($\sim 0.1\%$) reduces the mean free path sufficiently to fulfil (5.103).

Reversible Properties

(β) Comparison with the variation of h and A. In order that the derivation of (5.89) and (5.97) should be correct the variations of A and h must be slow over a distance of the order of the range of $K(R)$. The first condition is satisfied if the penetration depth $\lambda(T)$ is much greater than ξ_0, the range of $K(R)$ in the pure metal:

$$\lambda(T) \gg \xi_0. \tag{5.104}$$

This is the condition already mentioned in Chapter 2.

The condition (5.104) ensures that h varies slowly, but not that the variation of A is sufficiently slow. Since $h = \text{curl } A$, the variation of A over a distance ξ_0 is of order $\xi_0 |h|$, if $|h|$ is assumed to be constant over this distance. This leads to a variation of the phase of the function $W_n(r)$ of the order of†

$$\frac{e}{\hbar c}|h|\xi_0^2 \simeq \frac{E_F}{(k_B T_0)^2} \hbar \frac{e|h|}{mc} = \frac{E_F \hbar \omega_c}{(k_B T_0)^2} \tag{5.105}$$

ω_c is the cyclotron frequency of the normal electrons in the field h. The variation of the phase will be negligible if

$$\hbar \omega_c \ll \frac{(k_B T_0)^2}{E_F}. \tag{5.106}$$

In a type I superconductor the greatest value of the field is obtained for $|h| = H_c(T)$, and condition (5.106) may be written as

$$\frac{e}{\hbar c} H_c(T)\xi_0^2 \ll 1,$$

i.e. using (3.14a) for $H_c(T)$,

$$\xi(T) \gg \xi_0/\sqrt{\kappa}. \tag{5.107}$$

Since $\kappa < 1/\sqrt{2}$ this condition is less restrictive than (5.104) which gives

$$\xi(T) \gg \xi_0/\kappa.$$

In type II superconductors, the greatest value of $|h|$ is attained for $|h| = H_{c2}(T)$ and the condition reads

$$\frac{e}{\hbar c} H_{c2}(T)\xi_0^2 \ll 1,$$

i.e.

$$\xi(T) \gg \xi_0. \tag{5.108}$$

These conditions have already been obtained in Chapter 2.

† See (5.75).

Microscopic Theory

(γ) *Influence of the Landau diamagnetism.* In order that the effect of the Landau diamagnetism can be neglected, the cyclotron radius $R_c = mcv_F/e|\mathbf{h}|$ must be much greater than the range of the kernel $K(r,s)$. This yields

$$\frac{mcv_F}{e|\mathbf{h}|} \gg \xi_0,$$

i.e.

$$\hbar\omega_c \ll k_B T_0 \qquad (5.109)$$

for pure metals. This condition is less restrictive than (5.106).

(δ) *Discussion of the preceding results.* In summary it has been shown that, in pure superconductors, the linear term in the equation for Δ is given by (5.94) provided that

$$\lambda(T) \gg \xi_0 \quad \text{in type I superconductors,}$$

$$\xi(T) \gg \xi_0 \quad \text{in type II superconductors.}$$

These are the conditions already mentioned in Chapter 2 and they both reduce to

$$\frac{T_0 - T}{T_0} \ll 1.$$

In dirty superconductors the range of the kernel $K(R)$ is $\sqrt{(\xi_0 l)}$, much smaller than ξ_0. It is therefore expected that the derivation of the linear term, will be valid in a more extended region that in pure superconductors. Indeed Maki (1964) and de Gennes (1964) have shown that the form for $K(q)$ is valid in the whole temperature range. The extra term is of order l/ξ_0 and is thus negligible.

This has important consequences because it allows the calculation of the critical field H_{c2} at any temperature as a function of $H_c(T)$.

In conclusion, the linear term takes the forms (5.94) and (5.98)

$$\Delta(q)[1 - K_t(0)] = [K(q) - K(0)]\Delta(q),$$

i.e.

$$\Delta(r)[1 - K_t(0)] = \left[K\left(-i\nabla - \frac{2e}{\hbar c}\mathbf{A}\right) - K(0)\right]\Delta(r), \qquad (5.110)$$

where $K(q)$ is given by (5.89) for pure superconductors provided that conditions (5.104) and (5.106) are fulfilled, and $K(q)$ is given by (5.97) for dirty superconductors in the whole temperature range.

For both pure and dirty superconductors, of type I and type II, close to T_0, in the absence of applied field, it is expected that $\Delta(r)$ will not depend on r. The linearized equation for Δ takes the form:

$$\Delta[1 - K_t(0)] = 0. \qquad (5.111)$$

Reversible Properties

This equation leads to
$$K_t(0) = 1, \quad (5.112a)$$
i.e.
$$\log \frac{1 \cdot 14 \hbar \omega_D}{T_0} = \frac{1}{N(0)V}. \quad (5.112b)$$

This is the B.C.S. result (5.40) and is expected since at $T = T_0$ the transition is of the second order and well described by the linearized equation. It is seen that

$$[1 - K_t(0)] \Delta(\mathbf{r}) = N(0)V \log \frac{T}{T_0} \Delta(\mathbf{r}). \quad (5.113)$$

(c) *The parameter $\kappa_1(T)$*

In the case of dirty superconductors it is possible to compute H_{c2} as a function of temperature. At H_{c2}, the transition is of the second order and it is sufficient to retain only the linear term in the expansion in powers of $\Delta(\mathbf{r})$ of the self-consistent equation. Moreover the vector potential $\mathbf{A}$ is taken as in Chapter 3 to be

$$A_x = 0, \quad A_y = Hx, \quad A_z = 0, \quad (5.114)$$

where H is the applied field, i.e. H_{c2}.

Due to the particular form of $K(\mathbf{q})$, $\Delta(\mathbf{r})$ may be taken to depend only on x and the equation for Δ at H_{c2} reads

$$\left\{ \log \frac{T}{T_0} + f_0 \left[\frac{\hbar D}{4\pi k_B T} \left(\frac{d^2}{dx^2} - \frac{4e^2}{\hbar^2 c^2} H^2 x^2 \right) \right] \right\} \Delta(x) = 0, \quad (5.115)$$

where $f_0(x) = \psi\left(\frac{1}{2} + x\right) - \psi\left(\frac{1}{2}\right)$

and $\psi(x) = \frac{d}{dx} \log \Gamma(x)$ is the digamma function.†

The upper critical field H_{c2} is determined by the highest eigenvalue H of the above equation, the boundary condition being that $\Delta(x)$ is bounded at infinity.

Since f_0 is a function of the operator $[(d^2/dx^2) - (4e/\hbar^2 c^2) H^2 x^2]$, it is easily seen that

$$\Delta(x) = e^{-(eH/\hbar c)x^2} \quad (5.116)$$

---

† For properties of this function and its relation with the series $\sum_{n=1}^{\infty} 1/(n + x)$ see Erdelyi *et al.*, *Higher Transcendental Functions*, vol. I (1953), McGraw-Hill.

and H_{c2} is related to T by

$$\log \frac{T}{T_0} + f_0\left(\frac{D}{2\pi k_B T} \frac{eH}{c}\right) = 0. \tag{5.117}$$

Using the asymptotic expansions for $\psi(x)$ it is found that

$$H_{c2} \simeq \frac{1}{2}\frac{c}{D}\frac{\Delta_0}{e}\left[1 - \frac{2}{3}\left(\frac{\pi k_B T}{\Delta_0}\right)^2\right] \text{ for } T \ll T_0 \tag{5.118a}$$

$$H_{c2} \simeq \frac{4k_B T_0}{\pi e}\frac{c}{D}\left(1 - \frac{T}{T_0}\right)\left\{1 - \left(\frac{1}{2} - \frac{28}{\pi^4}\zeta(3)\right)\theta\right\} \text{ for } T_0 - T \ll T_0, \tag{5.118b}$$

where $\theta = 1 - T/T_0$. $\Delta_0 = 1\cdot 74\, k_B T_0$ is the energy gap at $T = 0$ in zero field, and $\zeta(3)$ the Riemann zeta function.

If one defines a parameter $\kappa_1(T)$ such that

$$\sqrt{2}\,\kappa_1(T) = \frac{H_{c2}(T)}{H_c(T)}, \tag{5.119}$$

where $H_c(T)$ is obtained from (5.50) it is seen that

$$\kappa_1(T) = 1\cdot 20\,\kappa\left(1 - 1\cdot 06\frac{T^2}{T_0^2}\right) \text{ for } T \ll T_0, \tag{5.120a}$$

$$\kappa_1(T) = \kappa(1 + 0\cdot 13\,\theta) \text{ for } T_0 - T \ll T_0. \tag{5.120b}$$

The parameter κ is seen to be[†]

$$\kappa = \frac{3c}{2\pi^2\,ev_F l}\left[\frac{7\zeta(3)}{2\pi N(0)}\right]^{1/2}. \tag{5.121}$$

This defines the Ginzburg–Landau parameter as a function of the microscopic quantities. The above formula for κ was first obtained by Gorkov (1959) while formulae (5.115) to (5.120) are given by Maki (1964) and by de Gennes (1964). The variation of $\kappa_1(T)/\kappa$ is shown in Fig. 5.4. It is also clear that, if one imposes the boundary condition $d\Delta/dx = 0$ at the surface of the sample, the calculation of H_{c3} will remain valid. $\Delta(x)$ instead of being given by (5.116) will be a Weber function and the eigenvalue of the operator $(p - 2eA/c)^2$ will be $1\cdot 69\,H_{c2}$. This means that in the whole temperature range H_{c3}/H_{c2} is a constant.[‡]

[†] For the complete determination of κ at $T = T_0$ see below (section 5.4 f).
[‡] In the case of a slab of finite thickness the calculation of $H_\parallel$ is also valid in the whole temperature range provided that $\xi(T)$ is obtained from $H_{c2}(T)$ using (3.14b).

Reversible Properties

In the case of pure superconductors, the calculation of Maki and de Gennes is no longer valid since $l \gg \xi_0$. Far from T_0 one cannot use the kernel (5.88) and must return to the linearized equation in its integral form (5.59). This is discussed briefly in section 5.5.

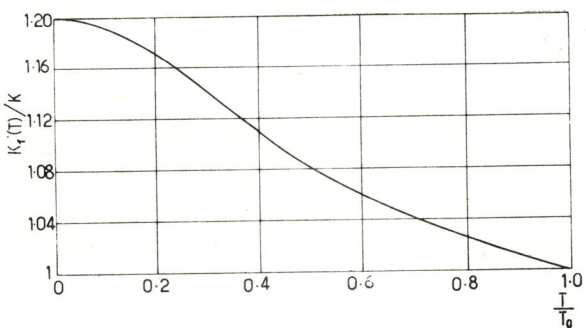

FIG. 5.4. Variation of the parameter $\kappa_1(T)$ versus temperature in a dirty superconductor (after Maki, 1964)

(d) *The non-linear terms. The parameter* $\kappa_2(T)$

It is in principle possible to compute the first non-linear term given by

$$\int K_1(r, s_1 s_2 s_3) \Delta(s_1) \Delta^*(s_2) \Delta(s_3) ds_1 ds_2 ds_3 . \tag{5.122}$$

This calculation has been done by Maki (1964) and by Caroli *et al.* (1966). It will not be developed here but the main features will be outlined. In a dirty superconductor, it can be shown that this term can be obtained over the whole temperature range. When this is done a treatment analogous to the Abrikosov treatment near H_{c2} can be performed which predicts for the magnetization that

$$\left(\frac{dM}{dH}\right)_{H=H_{cs}(T)} = \frac{1}{4\pi\beta_A[2\kappa_2^2(T) - 1]} . \tag{5.123}$$

β_A, which is a geometrical factor, is the same as that in Chapter 3. $\kappa_2(T)$ is a new generalized Ginzburg–Landau coefficient, which can be computed with the help of (5.123).

The calculation of $\kappa_2(T)$ was first made by Maki who predicted that $\kappa_2(T)/\kappa$ should decrease from 1 to 0·7 when T decreases from T_0 to zero. This predicts a strange behaviour for superconductors having a κ of the order of $1/\sqrt{2}$. However, the recent calculation by Caroli *et al.* (1966) has shown that Maki's result was in error.† The behaviour of $\kappa_2(T)$, in a

---

† Curiously enough the methods based on the Green function and on the correlation function techniques do not yield exactly the same result for the $|\Delta|^2 \Delta$ term but both methods yield the same result for $\kappa_2(T)$.

dirty superconductor is analogous to the behaviour of $\kappa_1(T)$. This is shown in Fig. 5.5.

The calculation of Caroli *et al.* predicts that $\kappa_2(T)$ will be equal to $\kappa_1(T)$ within 2%. The experimental determination of $\kappa_2(T)$ [which is more dif-

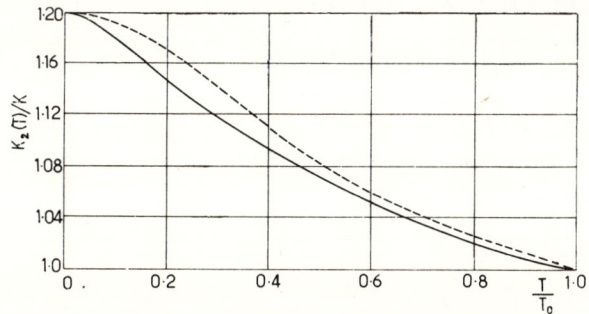

FIG. 5.5. Variation of $\kappa_2(T)$ versus temperature in a dirty superconductor (after Caroli *et al.*, 1966); the dotted line represents the variation of $\kappa_1(T)$

ficult than $\kappa_1(T)$] has always yielded $\kappa_2(T) \sim \kappa_1(T)$ within the experimental errors [Bon Mardion *et al.* (1965), McConville and Serin (1965), Orsay Group (1965)] (Fig. 5.6).†

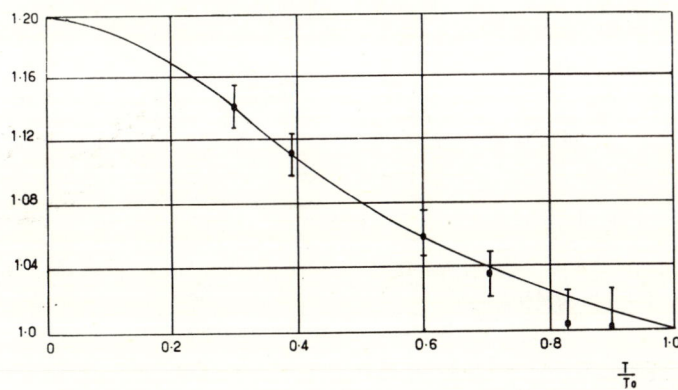

FIG. 5.6. Comparison of the experimental determination of $\kappa_1(T)/\kappa$ as a function of T/T_0 with the theoretical prediction for an In–2·5% Bi alloy (Orsay group, private communication). The comparison is difficult to make since there is great uncertainty in the value of $H_c(T)$, especially near T_0. This results in a poor determination of κ, the value of κ_1 at $T = T_0$. Usually, in the literature, one finds the variation of $H_{c2}(T)$. The above curve is given for the sake of illustration

† In films the variation of κ_2 as a function of T is not the same as in bulk superconductors. A complete study of this situation has been made by Guyon, Meunier and Thomson (1966.).

Reversible Properties

(e) *The first Ginzburg–Landau equation*

Close to T_0 one can deduce the first Ginzburg–Landau equation in pure and dirty superconductors. It is seen that the quantity $K(q) - K(0)$ as defined by (5.89) and (5.97) is negative for $q \neq 0$ and is therefore a maximum at $q = 0$. This quantity may be expanded in powers of q and close to T_0 it is sufficient to retain only the first term of the expansion which is proportional to q^2. In a pure superconductor

$$K(q) - K(0) \simeq -\frac{\pi N(0)V}{6\beta} \hbar^2(qv_F)^2 \sum_{\nu=0}^{\infty} \frac{\beta^3}{\pi^3(2\nu+1)^3}$$

$$\simeq -\frac{N(0)V}{6\pi^2} \left(\frac{\hbar v_F}{k_B T_0}\right)^2 \sum_{\nu=0}^{\infty} \frac{1}{(2\nu+1)^3} q^2$$

$$\simeq -\frac{N(0)V}{6\pi^2} \left(\frac{\hbar v_F}{k_B T_0}\right)^2 \frac{7}{8} \zeta(3) q^2, \qquad (5.124)$$

where $\zeta(3)$ is the Riemann zeta function.

In a dirty superconductor

$$K(q) - K(0) \simeq -\frac{\pi N(0)V}{\beta} \hbar D q^2 \sum_{\nu=0}^{\infty} \frac{\beta^2}{\pi^2(2\nu+1)^2}$$

$$\simeq -\frac{\pi N(0)V}{24} \frac{\hbar v_F l}{k_B T_0} q^2. \qquad (5.125)$$

In deriving these two expressions the temperature T has been replaced by T_0. Let us now examine the expression

$$1 - K_t(0)$$

which appears in the left-hand side of (5.94) and (5.98). Here

$$1 - K_t(0) = 1 - N(0)V \log \frac{1 \cdot 14 \hbar \omega_D}{T}$$

$$= N(0)V \log \frac{1 \cdot 14 \hbar \omega_D}{T_0} - N(0)V \log \frac{1 \cdot 14 \hbar \omega_D}{T} \simeq N(0)V \frac{T - T_0}{T_0}.$$

$$(5.126)$$

In deriving this expression, use has been made of (5.40) which gives the transition temperature in zero field, i.e. T_0.

In (5.82) it is seen that the linear part of the equation which obeys $\Delta(r)$ is

$$\frac{T - T_0}{T_0} \Delta(r) = -\frac{7}{8} \zeta(3) \frac{v_F^2}{6\pi^2(k_B T_0)^2} \left[-i\hbar \nabla - \frac{2eA}{c}\right]^2 \Delta(r)$$

$$(5.127)$$

for a pure superconductor and

$$\frac{T-T_0}{T_0}\Delta(r) = -\frac{\pi}{24}\frac{v_F l}{\hbar k_B T_0}\left[-i\hbar\nabla - \frac{2e}{c}A\right]^2 \Delta(r) \qquad (5.128)$$

for a dirty one. This is, of course, the linear part of the first Ginzburg–Landau equation. It is seen that $\Delta(r)$ is proportional to $\psi(r)$. The proportionality coefficient is, at this stage, left arbitrary. However, the vector potential appears as expected in the form $(2e/c)A$ which is characteristic of the existence of the Cooper pairs.

For the sake of completeness it is also necessary to compute the coefficient R of the term $|\Delta|^2\Delta$. Close to T_0, as expected from the Ginzburg–Landau theory, this coefficient does not depend on A. It is therefore sufficient to calculate it in the limit where the field is zero, i.e. by expanding the B.C.S. self-consistent equation (5.37) as a power series of Δ. It is readily seen that

$$R = -\frac{\pi}{\beta}N(0)V\sum_{\nu=0}^{\infty}\frac{\beta^3}{\pi^3(2\nu+1)^3} = -\frac{7N(0)V}{8\pi^2 k_B^2 T_0^2}\zeta(3) \qquad (5.129)$$

and the first Ginzburg–Landau equation reads

$$\frac{7}{8}\zeta(3)\frac{v_F^2}{6\pi^2(k_B T_0)^2}\left[-i\hbar\nabla - \frac{2eA}{c}\right]^2 \Delta(r) + \frac{T-T_0}{T_0}\Delta(r)$$

$$+\frac{7\zeta(3)}{8\pi^2 k_B^2 T_0^2}|\Delta(r)|^2\Delta(r) = 0 \quad \text{pure case}, \qquad (5.130)$$

$$\frac{\pi}{24}\frac{v_F l}{\hbar k_B T_0^2}\left[-i\hbar\nabla - \frac{2eA}{c}\right]^2 \Delta(r) + \frac{T-T_0}{T_0}\Delta(r)$$

$$+\frac{7\zeta(3)}{8\pi^2 k_B^2 T_0^2}|\Delta(r)|^2\Delta(r) = 0 \quad \text{dirty case}. \qquad (5.131)$$

The equation for the current can be determined using a similar method and it is not necessary to reproduce this calculation in full. The results are as follows:

$$\frac{j(r)}{c} = \left\{\frac{ie}{mc}(\Delta\nabla\Delta^* - \Delta^*\nabla\Delta) - \frac{4e^2}{mc^2}|\Delta|^2 A\right\}\frac{7}{8}\zeta(3)\frac{2mv_F N(0)}{6\pi^2(k_B T_0)^2} \qquad (5.132)$$

in the pure case, and

$$\frac{j(r)}{c} = \left\{\frac{ie}{mc}(\Delta\nabla\Delta^* - \Delta^*\nabla\Delta) - \frac{4e^2}{mc^2}|\Delta|^2 A\right\}\frac{\pi}{24}\frac{v_F l N(0)}{\hbar k_B T_0} \qquad (5.133)$$

in the dirty case.

These results were obtained by Gorkov (1959). It is seen that the Ginzburg–Landau scheme is justified by the microscopic theory. It remains now

Reversible Properties

to calculate the various parameters and characteristic lengths in terms of the parameters of the microscopic theory. This is done in the next section.

It must be noted that the above results are calculated for the isotropic case, for both pure and dirty superconductors. If this assumption is no longer valid the kernel $K(q)$ will depend on the direction of q. This has been examined by Caroli et al. (1963) and the reader is referred to their paper.

(f) *Evaluation of* $\xi(T)$, $\lambda(T)$, $\kappa(T)$.

The above equations take the familiar form if one sets

$$\psi(r) = \left[\frac{7\zeta(3)mv_F^2 N(0)}{24\pi^2 k_B^2 T_0^2}\right]^{1/2} \Delta(r) \quad \text{in the pure case} \quad (5.134)$$

and

$$\psi(r) = \left[\frac{\pi m v_F l N(0)}{12\hbar k_B T_0}\right]^{1/2} \Delta(r) \quad \text{in the dirty case}. \quad (5.135)$$

The Ginzburg–Landau coefficients are then

$$\alpha_p = \frac{\hbar^2}{2m} \frac{48\pi^2}{7\zeta(3)} \left(\frac{k_B T_0}{\hbar v_F}\right)^2 \frac{T-T_0}{T_0} = 1\cdot 83 \frac{\hbar^2}{2m} \frac{1}{\xi_0^2} \frac{T-T_0}{T_0}$$

$$\text{in the pure case}, \quad (5.136)$$

$$\alpha_d = \frac{\hbar^2}{2m} \frac{24}{\pi} \frac{k_B T_0}{\hbar v_F l} \frac{T-T_0}{T_0} = 1\cdot 36 \frac{\hbar^2}{2m} \frac{1}{\xi_0 l} \frac{T-T_0}{T_0} \quad \text{in the dirty case.} \quad (5.137)$$

As expected α is negative and vanishes at $T = T_0$.

$$\beta_p = \frac{288\pi^2}{7\zeta(3)} \frac{1}{N(0)} \left[\frac{\hbar^2}{2m}\left(\frac{k_B T_0}{\hbar v_F}\right)^2\right]^2 \frac{1}{k_B^2 T_0^2}$$

$$= 0\cdot 35 \frac{1}{N(0)} \left(\frac{\hbar^2}{2m} \frac{1}{\xi_0^2}\right)^2 \frac{1}{(k_B T_0)^2} \quad \text{in the pure case}, \quad (5.138)$$

$$\beta_d = \frac{504\zeta(3)}{\pi^4} \frac{1}{N(0)} \left(\frac{\hbar^2}{2m} \frac{1}{\hbar v_F l}\right)^2$$

$$= 0\cdot 2 \frac{1}{N(0)} \left(\frac{\hbar^2}{2m} \frac{1}{\xi_0 l}\right)^2 \frac{1}{(k_B T_0)^2} \quad \text{in the dirty case.} \quad (5.139)$$

The temperature dependent coherence length $\xi(T)$, given by $\xi^2(T) = \hbar^2/2m|\alpha|$, is thus:

$$\xi_p^2(T) = \frac{7\zeta(3)}{48\pi^2} \frac{\hbar^2 v_F^2}{k_B^2 T_0^2} \frac{T_0}{T_0-T} \quad \text{in the pure case} \quad (5.140)$$

$$\xi_d^2(T) = \frac{\pi}{24} \frac{\hbar v_F l}{k_B T_0} \frac{T_0}{T_0-T} \quad \text{in the dirty case} \quad (5.141)$$

or

$$\xi_p(T) = 0{\cdot}74\, \xi_0 \left(\frac{T_0}{T_0 - T}\right)^{1/2} \quad \text{in the pure case} \quad (5.142)$$

$$\xi_d(T) = 0{\cdot}85(\xi_0 l)^{1/2} \left(\frac{T_0}{T_0 - T}\right)^{1/2} \quad \text{in the dirty case.} \quad (5.143)$$

The penetration depth $\lambda(T)$ is given by $\lambda^2(T) = mc^2\beta/16\pi e^2 |\alpha|$ and

$$\lambda_p^2(T) = \frac{3c^2}{16\pi e^2 v_F^2 N(0)} \frac{T_0}{T_0 - T}, \quad (5.144)$$

$$\lambda_d^2(T) = \frac{3c^2}{8\pi e^2 v_F^2 N(0)} \frac{7\zeta(3)}{4\pi^3} \frac{1}{l} \frac{\hbar v_F}{k_B T_0} \frac{T_0}{T_0 - T}, \quad (5.145)$$

or

$$\lambda_p(T) = \frac{1}{\sqrt{2}} \lambda_L(0) \left(\frac{T_0}{T_0 - T}\right)^{1/2}, \quad (5.146)$$

$$\lambda_d(T) = 0{\cdot}615\, \lambda_L(0) \left(\frac{\xi_0}{l}\right)^{1/2} \left(\frac{T_0}{T_0 - T}\right)^{1/2}, \quad (5.147)$$

where $\lambda_L(0) = 3c^2/8\pi e^2 v_F^2 N(0)$ is the London penetration depth. $\xi(T)$ and $\lambda(T)$ diverge as $[T_0/(T_0 - T)]^{1/2}$ when $T \to T_0$. The Ginzburg–Landau parameter is seen to be

$$\kappa_p = \frac{3ck_B T_0}{e\hbar v_F^2} \left[\frac{\pi}{7\zeta(3)} \frac{1}{N(0)}\right]^{1/2} = 0{\cdot}96 \frac{\lambda_L(0)}{\xi_0}, \quad (5.148)$$

$$\kappa_d = \frac{3c}{2\pi^2 ev_F l} \left[\frac{7\zeta(3)}{\pi N(0)}\right]^{1/2} = 0{\cdot}725 \frac{\lambda_L(0)}{l}. \quad (5.149a)$$

The formulae for κ were first obtained by Gorkov (1959). κ has also been computed by Gorkov for intermediate values of l. Goodman (1962) has shown that, within a few per cent, κ is given by

$$\kappa = 0{\cdot}96 \left(\frac{1}{\xi_0} + \frac{1}{1{\cdot}32 l}\right). \quad (5.149b)$$

It is interesting to relate κ to the resistivity and the electronic specific heat, of the normal metal. In the presence of scattering centres and of an electric field $\boldsymbol{E}$ the current $\boldsymbol{j}$ will be

$$\boldsymbol{j} = -eD\nabla n + \boldsymbol{E}/\rho, \quad (5.150)$$

where ρ is the normal state resistivity, D is the diffusion coefficient and n is the number of electrons per unit volume. At equilibrium in the presence

Reversible Properties

of the field E, the Fermi surface is displaced and the number of electrons is

$$n = n_0 - 2N(0)\,eV, \qquad (5.151)$$

where V is the potential corresponding to E (i.e. $E = -\nabla V$). At equilibrium, $j = 0$ and one obtains

$$\rho = 2e^2 N(0)\, D = \frac{2}{3} e^2 N(0)\, v_F l. \qquad (5.152)$$

$N(0)$ can be obtained from the electronic specific heat C_e at low temperature. It is well known that

$$C_e = \gamma T, \qquad (5.153)$$

where $\gamma = (2\pi^3/3)\, N(0)\, k_B^2$
and this yields

$$\frac{\lambda_L(0)}{l} = \frac{1}{2\pi\sqrt{\pi}} \frac{C_e}{k_B} \rho\gamma^{1/2}. \qquad (5.154)$$

If ρ is expressed in ohm^{-1} and γ in erg/cm^3 deg^2 the formula for κ reads

$$\kappa = 7\cdot 5 \cdot 10^3\, \rho\gamma^{1/2} \quad (\xi_0 \gg l). \qquad (5.155)$$

It is seen that, in dirty alloys, the value of κ depends only on the transport properties of the metal in the normal state.

Formula (5.155) is due to Gorkov and has been tested in several compounds. In the alloy In Bi as studied by Kinsel et al. (1962–3), magnetic measurements give $\kappa = 1\cdot 79$ while the value calculated using (5.155) is $1\cdot 7$. Another interesting verification is provided by indium alloys [Seraphim (1962)]. Adding impurities decreases the mean free path l and if l is smaller than the critical length l_c such that

$$l_c = 1\cdot 06\, \lambda_L(0), \qquad (5.156)$$

the superconductor becomes type II as can be seen from (5.149). Seraphim has studied a number of alloys In Bi–Pb–Cu–Cd–Tl–Hg. The critical length is found to be independent of the nature of the impurity, and is of the order of 440 ± 100 Å.

The London penetration depth is 400 Å.†

5.5. Conclusion and discussion

In the preceding section it was shown that it is possible to derive the Ginzburg–Landau equations from the microscopic theory and therefore to justify the phenomenological approach. In this derivation it has been use-

† The critical concentration corresponding to l_c varies with the nature of the impurity.

Microscopic Theory

ful to introduce the important concept of clean and dirty superconductors. In the latter case, due to the fact that the electron mean free path l is much smaller than the coherence length ξ_0, it has been possible to obtain a local equation for the order parameter close to the nucleation field (H_{c2} or H_{c3}), valid in the whole temperature range and hence to deduce the values of $H_{c2}(T)$ and $H_{c3}(T)$. This has lead to the introduction of the two generalized Ginzburg–Landau parameters $\kappa_1(T)$ and $\kappa_2(T)$ and has shown that the ratio $H_{c3}(T)/H_{c2}(T)$ is constant over the whole temperature range. It is also possible, in high κ dirty superconductors, to introduce a third

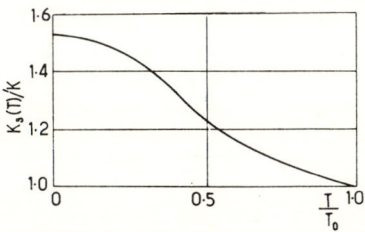

FIG. 5.7. Variation of $\kappa_3(T)$ versus temperature in a dirty superconductor with high κ values (after Maki, 1964)

parameter $\kappa_3(T)$ which relates the first penetration field $H_{c1}(T)$ to the field $H_c(T)$ by the equation

$$H_{c1}(T) = \frac{H_c(T)}{2\kappa_3(T)} \log \kappa_3(T).$$

The variation of $\kappa_3(T)$ with temperature has been computed by Maki (1964) and is shown in Fig. 5.7.

In a pure superconductor, one must return to the linearized equation for the pair potential, in its integral form (5.59), in order to calculate the critical field H_{c2} as a function of temperature. This was done by Gorkov (1959) using a variational procedure, starting from a Gaussian as a trial function for $\Delta(r)$. Helfand and Werthamer (1964, 1966) were able to show that the Gaussian is an exact solution of (5.59). Using this solution it is possible to define a parameter $\kappa_1(T)$ in the case of pure superconductors. The calculations of Gorkov and of Helfand and Werthamer predict a similar behaviour for $\kappa_1(T)$ in the pure and in the dirty case, $\kappa_1(0)$ is expected to be $1.25\ \kappa$ where κ is the value at $T = T_0$. The experimental results in pure niobium obtained by McConville and Serin (1965) are given in Fig. 5.8. The experimental values of $\kappa_1(T)$ are definitely higher than the theoretical one. At $T = 0$, $\kappa_1/\kappa = 1.6$ experimentally. More recent experiment by Finnemore et al. (1965) have given even higher values for this ratio.

Using the same procedure Helfand and Werthamer have also computed $\kappa_1(T)$ as a function of the mean free path. They conclude that $\kappa_1(T)$ varies smoothly from the value obtained in clean superconductors to the

Reversible Properties

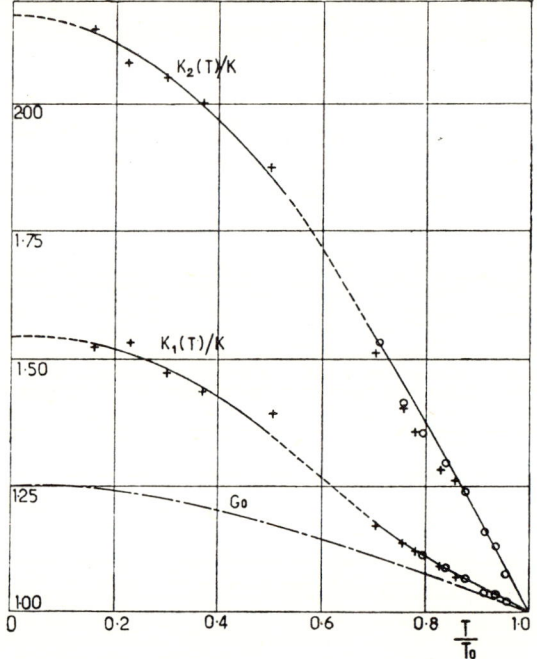

FIG. 5.8. Variation of the parameter $\kappa_1(T)$ in pure metals as computed by Gorkov (1959) (curve labelled G_0). This diagram also shows the experimental values of $\kappa_1(T)$ and $\kappa_2(T)$ obtained by McConville and Serin (1965) in pure niobium (the crosses and the dots refer to two different samples of high purity). The low temperature data have been obtained from magnetization measurements, the values in the vicinity of T_0 by specific heat measurement (see Section 7.1)

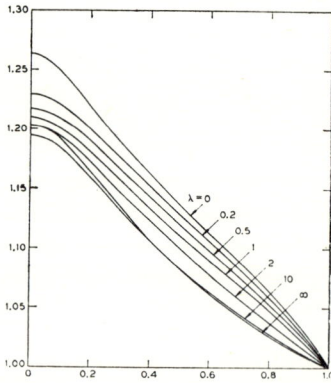

FIG. 5.9. Variation of $\kappa_1(T)$ versus temperature as a function of the normal electron mean free path l ($\lambda = 0.882\, \xi_0/l$) (Helfand and Werthamer, 1966). Note that the $l = \infty$ curve is exactly the curve computed by Gorkov and reproduced in Fig. 5.8.

value calculated in the dirty case when l varies from infinity to zero (Fig. 5.9). The discrepancy between experiment and theory in clean superconductors cannot be attributed to a mean free path effect. Helfand and Werthamer suggest that it can come from the inadequacy of the B.C.S. model in the case of strong-coupling, clean materials, in which certain consequences of the retarded electron phonon interaction and of anisotropy of the Fermi surface may be of importance. However, a recent experiment by Keesom and Radebaugh (1966) in pure vanadium (which is not a strong coupling superconductor) has shown that the variation of $\kappa_1(T)/\kappa$ is almost identical to that of niobium. The question of the validity of the Werthamer–Helfand's theory is still open.

McConville and Serin have also measured the parameter $\kappa_2(T)$ related to the magnetization slope at $H = H_{c2}(T)$. The variation of $\kappa_2(T)$ is more rapid than the variation of $\kappa_1(T)$ and at $T = 0$ $\kappa_2(0)/\kappa$ in pure niobium is of the order of 2·2 (Fig. 5.8). A recent calculation by Maki and Tsusuki (1965) predicts that $\kappa_2(T) > \kappa_1(T)$ and diverges for $T = 0$ in very pure superconductors ($l = \infty$) in rough agreement with the observed behaviour.

No calculation exists for the field H_{c3}, but it is clear that there is no *a priori* reason for the ratio $H_{c3}(T)/H_{c2}(T)$ in pure superconductors to remain constant over the whole temperature range.†

At the present time, the behaviour of pure superconductors at low temperatures is not as well understood as the behaviour of dirty ones.

Several other attempts have been made to extend the Ginzburg–Landau equations to all temperatures for low fields (see Tewordt, 1963, 1965; Werthamer, 1963, 1964).

---

† It is also clear that in the pure case all the calculations of $H_{//}$ performed in Chapter 4 are only valid in the vicinity of T_0.

Bibliography for Chapter 5

ANDERSON, P. W. (1959), *J. Phys. Chem. Solids*, **11**, 26.
BARDEEN, J., COOPER, L. N. and SCHRIEFFER, J. R. (1957), *Phys. Rev.* **108**, 1175.
BON MARDION, G., GOODMAN, B. B., LACAZE, A. (1965), *J. Phys. Chem. Solids*, **26**, 1143.
CAROLI, C., DE GENNES, P. G. and MATRICON, J. (1962), *J. Phys. Rad.* **23**, 707; (1963), *Phys. Cond. Mat.* **1**, 176.
CAROLI, C., CYROT, M. and DE GENNES, P. G. (1966), *Solid State Comm.* **4**, 17.
COOPER, L. N. (1956), *Phys. Rev.* **104**, 1189.
FRÖHLICH, H. (1950), *Phys. Rev.* **79**, 845.
DE GENNES, P. G. (1964), *Phys. Cond. Mat.* **3**, 79.
FINNEMORE, D, K., STROMBERG, T. F. and SWENSON, C, A., (1966), *Phys. Rev.* **149**, 231.
GOODMAN, B. B. (1962), *I. B. M. J. Res. Develop.* **6**, 63.
GORKOV, L. P. (1959), *Zh. Eksperim. i. Teor. Fiz.* **36**, 1918, and **37**, 833 and 1407 [English translation: *Soviet Phys. JETP*, **9**, 1364 (1959), and **10**, 593 and 998 (1960)].
GUYON, E., MEUNIER, F. and THOMPSON, R. S. (1966), To be published in *Phys. Rev.*
HELFAND, E. and WERTHAMER, N. R. (1964), *Phys. Rev. Letters*, **13**, 686.
HELFAND, E, and WERTHAMER, N. R. (1966), *Phys. Rev.*, **147**, 288.
KEESOM, P. H. and RADEBAUGH, R. (1966), LT X Conference.
KINSEL, T., LYNTON, E. A. and SERIN, B. (1962), *Phys. Letters*, **3**, 30; (1963), *Bull. Am. Phys. Soc.* **8**, 294.
MCCONVILLE, T. and SERIN, B. (1965), *Phys. Rev.* **140**, A, 1169.
MAKI, K. (1964), *Physics*, **1**, 21, 127 and 201.
MAKI, K. and TSUZUKI, T. (1965), *Phys. Rev.* **139** A, 868.
MUHLSCHLEGEL, B. (1959), *Z. Phys.* **155**, 313.
ORSAY GROUP (1965) Communication to the Symposium on Quantum Fluids, Brighton, England.
SERAPHIM, D. P. (1962), *Superconductors* (Tannenbaum and Wright, Eds.) Interscience, p. 25.
TEWORDT, L. (1963), *Phys. Rev.* **132** A, 595; (1965) *Phys. Rev.* **137** A, 1745.
WERTHAMER, N. R. (1963), *Phys. Rev.* **132**, A, 663; (1964), *Rev. Mod. Phys.* **36**, 292.
Several books, in which the microscopic theory of superconductivity is developed, are listed below. The given list does not pretend to be exhaustive.
ABRIKOSOV, A. A., GORKOV, L. P. and DZYALOSHINSKI, I. E., *Methods of Quantum Field Theory in Statistical Physics*, Prentice Hall, Englewood Cliffs, New Jersey, 1963. *Quantum Field Theoretical Methods in Statistical Physics*, Pergamon Press, London—Paris, 1963.
BLATT, J. M., *Theory of Superconductivity*, Academic Press, New York 1964.
BOGOLIUBOV, N. N., TOLMACHEV, V. V. and SHIRKOV, D. V., *A New Method in the Theory of Superconductivity*, Consultant Bureau, New York, 1959.
DE GENNES, P. G., *Superconductivity of Metals and Alloys*, Benjamin, New York, 1966.
RICKAYSEN, G., *Theory of Superconductivity*, J. Wiley, New York, 1965.
SCHRIEFFER, J., R., *The Theory of Superconductivity*, Benjamin, New York.

CHAPTER 6

Miscellaneous Properties of Type II Superconductors in High Fields

6.1. THE PARAMAGNETIC EFFECT

6.1.1. Introduction

In the preceding chapter the Ginzburg–Landau equations were deduced neglecting the influence of the magnetic field on the electron spins. Various properties of clean and dirty superconductors can be explained by calculating the corresponding parameters from the microscopic theory. This procedure is justified as long as the magnetic fields of interest are small. However, in type II superconductors, with high κ values, the upper critical field H_{c2} (or H_{c3}) can be very high and in this case the contribution of the electron spin susceptibility (Pauli paramagnetism) cannot be neglected in the thermodynamic balance.

In the normal state, the spin susceptibility is small but finite and is

$$\chi_n = \frac{1}{2} (g\mu_B)^2 N(0) \tag{6.1}$$

where g is the spectroscopic splitting factor of the electron ($g = 2$), μ_B the Bohr magneton and $N(0)$ the density of states at the Fermi level.

In the superconducting phase, due to the formation of Cooper pairs, the susceptibility is much reduced. Indeed, in order to polarize the superconductor one must break the pairs, i.e. one must apply a field H so that $\mu_B H \sim \Delta$, the energy of formation of the pair. In particular at $T = 0$, the spin susceptibility of the superconductor is expected to be equal to zero, i.e.

$$\chi_s = 0 \quad (T = 0). \tag{6.2}$$

Thus the normal state is more polarizable than the superconducting state but in high fields it is possible for the polarized normal state to have a lower free energy than the superconducting phase.

Reversible Properties

Qualitatively, the transition from the superconducting state to the polarized normal state will be reached as soon as the polarization energy is equal to the condensation energy $H_c^2(T)/8\pi$, i.e. for

$$\frac{1}{2}(\chi_n - \chi_s)H^2 = \frac{H_c^2(T)}{8\pi}. \tag{6.3}$$

Expression (6.3) defines a limiting field beyond which superconductivity will not exist. This is the so-called paramagnetic limit $H_p(T)$ given by

$$H_p(T) = \frac{H_c(T)}{[4\pi(\chi_n - \chi_s)]^{1/2}}. \tag{6.4}$$

This formula was first given by Clogston (1962) and Chandrasekhar (1962). At $T = 0$, the limiting $H_p(0)$ is

$$H_p(0) = \frac{H_c(0)}{\sqrt{[2\pi N(0)]}\, g\mu_B} \tag{6.5a}$$

or

$$H_p(0) = \frac{\sqrt{2}\Delta_0}{g\mu_B}. \tag{6.5b}$$

In this last expression, use has been made of the relation (5.48) between the gap at $T = 0$ and the critical field $H_c(0)$. Equation (6.5b) expresses the fact that the polarization energy is equal to the formation energy of the pair.

It is interesting to evaluate the order of magnitude of $H_p(0)$. Δ_0 is generally of the order of 20 cm^{-1}, so this yields with $g = 2$,

$$H_p(0) \simeq 300 \text{ kOe}.$$

In a type I superconductor with perfect Meissner effect, it can be shown that the first order transition between the superconducting and the normal phase would take place at a field

$$H_c^* = \left(\frac{1}{H_c^2} + \frac{1}{H_p^2}\right)^{-1/2} \tag{6.6}$$

Since H_p is expected to be of the order of 300 kOe and H_c of the order of 1 kOe at most, the paramagnetic effect is quite negligible in type I superconductors. On the contrary, in type II superconductors, as has been shown in the preceding chapters, the magnetic field can penetrate the sample. The transition takes place at the field H_{c2} and is of the second order. In high κ materials, where H_{c2} is expected to be high, the calculation of the actual transition field must include the spin paramagnetism and orbital

Type II Superconductors in High Fields

effects. If the transition is of the second order, one can use the method of Chapter 5, i.e. expand the self-consistent equation for $\Delta(r)$ as a power series of Δ. However, when the spin paramagnetism is especially strong [or if H_{c2} is particularly large in comparison to $H_p(0)$] the transition to the normal state may be of the first order. This was first pointed out by Sarma (1963) who has computed the transition field in the special case of

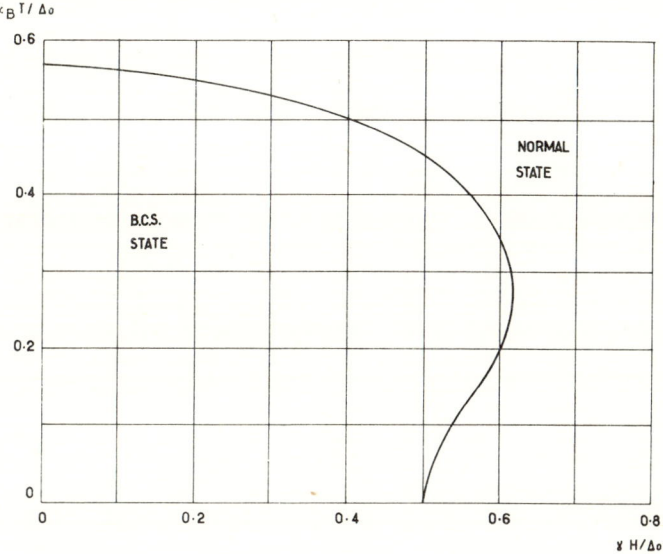

Fig. 6.1. Transition curve from the B.C.S. state to the normal state in the presence of a uniform field acting on the electron spins. The transition is assumed to be of the second order

a uniform field acting on the spins only (exchange field). This yields a curve $T(H)$ which represents the limit between the B.C.S. state and the polarized normal state (Figs 6.1 and 6.2). A triple point $T_c(H_0)$ exists at which the transition switches to the first order [for $T < T_c(H_0)$, $H > H_0$]. However, Fulde and Ferrell (1964) have suggested that the system might go into a "depaired" superconducting state in which all the Cooper pairs have a single non-vanishing center of mass momentum. The transition between the superconducting state and the depaired state would be of the first order while the transition between this last state and the normal state would still be of the second order. This situation has been studied by Sarma and Saint-James (1964). They concluded that the Fulde and Ferrell state might be possible in the clean limit but in the dirty limit, its formation is never favourable so that the superconducting-normal transition can be of the first order as predicted by Sarma.†

† This was also investigated by Gruenberg and Gunther (1966) who arrived at similar conclusions. See section 6. I. 3.

Reversible Properties

It is to be emphasized that the above considerations are valid in a model which takes into account only the action of the field on the electron spins. Maki (1964) has calculated the field H_{c2} as a function of temperature by including orbital effects. His results have been deduced in the case of dirty superconductors, and are valid if the transition is of the second order. Maki has discussed the order of the transition and concluded that for sufficiently

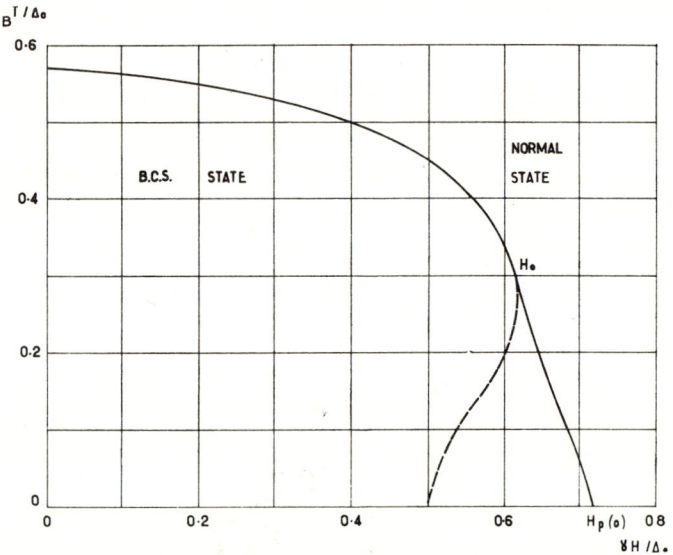

FIG. 6.2. Transition curve from the B.C.S. state to the normal state in the same conditions as in Fig. 6.1 as calculated by Sarma (1963). The curve between H_0 and $H_p(0)$ ($\gamma H_p^{(0)}/\Delta_0 = 1/\sqrt{2}$) corresponds to a first order transition

high ratios of $H_{c2}(0)/H_p(0)$ the transition becomes of the first order at low temperatures.

Theoretically, the spin susceptibility in the superconductor should be zero at $T = 0$. This is not in good agreement with the measurement of the Knight shift in small particles where this shift was found to be 75% of the shift in the normal phase. This led Ferrell (1959), Anderson (1959), Abrikosov and Gorkov (1962), Gorkov (1965) to investigate the influence of spin-orbit scattering. Due to spin-orbit interaction, the scattered electron can change its spin direction. This leads to a destruction of the correlation between a pair of electrons and consequently the superconductor spin susceptibility no longer vanishes at $T = 0$.

This effect has been taken into account by Werthamer et al. (1966) and independently by Maki (1966). They have included both spin paramagnetism and spin-orbit scattering and conclude that the effect of spin paramagnetism is reduced. In particular they have calculated the field H_{c2} in the dirty limit and show that this field is higher than the corresponding field calculated by Maki in the absence of spin-orbit effects.

Type II Superconductors in High Fields

All the above points will be now discussed on a more quantitative basis using the formalism of Chapter 5 although the order of the transition will not be discussed in detail. Only the main results will be quoted here and the reader is referred to the original papers for further details.

6.1.2. Calculation of the various kernels

We shall first calculate the various kernels neglecting the spin-orbit scattering. This will naturally lead to a discussion of the Fulde and Ferrell depaired state. In order to obtain the kernel $K(q)$ in the presence of the paramagnetic effect, it is necessary to get back to the microscopic formulation of Chapter 5 and to take into account the Pauli paramagnetism, namely,

$$\gamma \int h(r) [\psi_\uparrow^\dagger(r) \psi_\downarrow(r) - \psi_\downarrow^\dagger(r) \psi_\uparrow(r)] dr, \qquad (6.7)$$

where

$$\gamma = \frac{1}{2} g\mu_B = \frac{e\hbar}{2mc} \quad (g = 2).$$

The influence of this term is very difficult to evaluate in the most general situation. However, if one can assume that the microscopic field $h(r)$ is a constant, independent of r and equal to the applied field throughout the sample, the calculation can be performed. This assumption is clearly more restrictive than the assumption that the vector potential $A(r)$ is slowly varying. If the transition is of the second order, in the vicinity of H_{c2} where the order parameter [and therefore $\Delta(r)$] is small, this assumption will be correct at least in dirty materials. As in Chapter 5 the results to be obtained will be valid in the whole temperature range for dirty materials, while in the clean limit only general indications can be given. Moreover, a complication will occur in the case of dirty superconductors, since the transition will become of the first order when the field exceeds a certain value so that the above assumptions are no longer valid.

If the electron field operator ψ is expanded as

$$\psi_\alpha(r) = \sum_n \phi_{n\alpha}(r) a_{n\alpha}, \qquad (6.8)$$

where α is the spin index, $\phi_{n\alpha}(r)$ will obey the following equations:

$$\frac{1}{2m} \left\{ \left[p - \frac{e}{c} A \right]^2 + U + \gamma h \right\} \phi_{n\uparrow}(r) = \xi_{n\uparrow} \phi_{n\uparrow}(r), \qquad (6.9a)$$

$$\frac{1}{2m} \left\{ \left[p - \frac{e}{c} A \right]^2 + U - \gamma h \right\} \phi_{n\downarrow}(r) = \xi_{n\downarrow} \phi_{n\downarrow}(r). \qquad (6.9b)$$

h is the modulus of $h(r)$ and the z direction has been taken along h. If h is taken to be a constant and equal to the applied field H and if A is supposed

Reversible Properties

to be a slowly varying function of r, $\phi_{n\alpha}(r)$ can be written as

$$\phi_{n\alpha}(r) = e^{i(e/c)A \cdot r} w_{n\alpha}(r) \tag{6.10}$$

$w_{n\alpha}(r)$ satisfies the following equations:

$$\frac{1}{2m}[p^2 + U + \gamma H] w_{n\uparrow}(r) = \xi_{n\uparrow} w_{n\uparrow}(r), \tag{6.11a}$$

$$\frac{1}{2m}[p^2 + U - \gamma H] w_{n\downarrow}(r) = \xi_{n\downarrow} w_{n\downarrow}(r). \tag{6.11b}$$

It is clear that if H is constant the $w_{n\alpha}(r)$ may be taken as real and independent of the spin, while

$$\xi_{n\uparrow} = \xi_n + \gamma H, \tag{6.12a}$$

$$\xi_{n\downarrow} = \xi_n - \gamma H, \tag{6.12b}$$

where ξ_n is the one electron energy in the absence of the field. The calculation follows exactly the procedure of Chapter 5. It is seen that one obtains the same formulae, but ω must be replaced by $\omega - i\gamma H$. The kernel $K(q)$ will be a function $K(q, H, T)$ of q, H and T and is given by

$$K(q, H, T) = -\operatorname{Im} \sum_{n,\omega} \frac{VT}{\Omega} \int_0^\infty dt\, e^{-2|\omega|t} e^{2i\gamma Ht} \langle n|e^{i q \cdot r(t)} e^{-i q \cdot r(0)}|n\rangle$$

$$\left[\frac{1}{\xi_n + i\omega + \gamma H} - \frac{1}{\xi_n - i\omega - \gamma H}\right]. \tag{6.13}$$

The correlation function which appears in the right-hand side of (6.13) has to be calculated in the normal metal in the absence of field effects and has already been obtained in Chapter 5. Therefore the above expression for $K(q, H, T)$ will be valid in the same limits as in Chapter 5. However there is an extra limitation, since $K(q, H, T)$ is restricted to near H_{c2}. In the dirty as well as in the clean limit, $K(q, H, T)$ will be obtained from the corresponding kernels derived in Chapter 5 by replacing ω by $\omega - i\gamma H$, $K(q)$ is, as usual, a divergent series in ω. In order to restore the convergence the linear part of the equation for Δ will be written as

$$[K(q, H, T) - K(0, 0, T)]\Delta(q) = [1 - K_t(0, 0, T)]\Delta(q), \tag{6.14}$$

where the B.C.S. cut off has been taken into account in $K_t(0, 0, T)$. Clearly, $K_t(0, 0, T)$ is the kernel already calculated in Chapter 5 so that

$$K_t(0, 0, T) = 1 \tag{6.15}$$

Type II Superconductors in High Fields

defines the transition temperature T_0 in zero field and zero paramagnetic effect. The right-hand side of (6.14) takes the form

$$1 - K_t(0,0,T) = N(0)V \log \frac{T}{T_0}.$$

One can write down the linearized equation for the pair potential in the form

$$\left[K\left(\mathbf{p} - \frac{2e}{c}\mathbf{A}, H, T\right) - K(0,0,T)\right]\Delta(\mathbf{r}) = [1 - K_t(0,0,T)]\Delta(\mathbf{r}). \quad (6.16)$$

This equation can then be used to calculate the critical fields exactly as in Chapter 5. We postpone this calculation to section 6.1.4. since it is first necessary to discuss some special features of the kernel $K(\mathbf{q})$, which allow the investigation for the possible existence of the Fulde and Ferrell state.

6.I.3. The Fulde and Ferrell state

The expression (6.16) is a rather complicated equation for $\Delta(\mathbf{r})$. It is therefore interesting to study a somewhat idealized problem in which the B.C.S. state is compared to the polarized normal state, i.e. one retains only the action of the constant field on the spins and neglects the contribution of the vector potential $\mathbf{A}$. The discussion is easier if one rewrites (6.16) in the form

$$[K(\mathbf{q}, H, T) - K(0, H, T)]\Delta(\mathbf{q}) = [1 - K_t(0, H, T)]\Delta(\mathbf{q}). \quad (6.17)$$

This defines a new kernel $K_t(0, H, T)$ in which the B.C.S. cut off has been taken into account. If $\Delta(\mathbf{r})$ is taken to be a constant one obtains the B.C.S. state. In this case $\Delta(\mathbf{q})$ is different from zero only for $\mathbf{q} = 0$. It is seen that the equation

$$K_t(0, H, T) = 1 \quad (6.18)$$

defines a transition temperature $T_c(H)$ which corresponds to the transition between the B.C.S. state and the polarized normal state. This will give the variation of the paramagnetic limit as a function of temperature, i.e. will yield explicitly the form of (6.4.).

The corresponding variation of $T_c(H)$ as a function of H was first given by Baltensperger (1958) and has the form indicated in Fig. 6.1. It is seen that the behaviour of H_p versus T is rather peculiar. This is due to the fact that in deriving (6.18), it is assumed that the transition between the B.C.S. state and the normal state is of the second order. In 1963 Sarma pointed out that a critical point $[H_0, T_c(H_0)]$ exists for which the transition becomes of the first order. This was obtained by Sarma by considering the free energies of the B.C.S. state, of the normal state and of a polarized B.C.S. state in which the gap is field dependent. This will be recast here in the

Reversible Properties

framework of the Landau theory of phase transitions. The third term in the power series of Δ in the above simplified situation is readily obtained from the coefficient R given by (5.129) by replacing ω by $\omega - i\gamma H$.

In order that the transition is of the second order the coefficient R must be positive. If R vanishes, and changes sign, the second order transition changes to a first order transition and the expansion as a power series of Δ will not be justified in general.†

Therefore the critical point $[H_0, T_c(H_0)]$ is given by the simultaneous solution of the two equations:

$$\left. \begin{array}{l} K_t(0, H, T) = 1, \\ R(H, T) = 0. \end{array} \right\} \qquad (6.19)$$

From (5.129) it is seen that

$$R = -\frac{\pi k_B T}{2} \, Re \sum_\omega \frac{1}{(|\omega| + i\gamma H)^3} \qquad (6.20)$$

which vanishes for

$$\frac{\gamma H_0}{k_B T_c(H_0)} = 1 \cdot 911. \qquad (6.21)$$

The solution of (6.19) has been calculated by Sarma and is

$$\left. \begin{array}{l} \dfrac{\gamma H_0}{\Delta_0} = 0 \cdot 61, \\[1em] \dfrac{k_B T_c(H_0)}{\Delta_0} = 0 \cdot 31, \end{array} \right\} \qquad (6.22)$$

where $\Delta_0 = 1 \cdot 74 \, k_B T_0$ is the B.C.S. energy gap at $T = 0$. For temperatures $T < T_c(H_0)$ the transition between the superconducting state and the normal state becomes of the first order. Close to $T_c(H_0)$, the transition field and temperature can be calculated using higher terms in the expansion of the power series of Δ. The corresponding transition curve was calculated by Sarma using a different technique and is shown in Fig. 6.2. At $T = 0$, the transition field $H_p(0)$ is given by

$$\frac{\gamma H_p(0)}{\Delta_0} = \frac{1}{\sqrt{2}}, \qquad (6.23a)$$

i.e.

$$H_p(0) = \frac{\sqrt{2}\Delta_0}{g\mu_B}, \qquad (6.23b)$$

which is, as expected, the Chandrasekhar-Clogston limit.

† See Landau and Lifshitz, *Statistical physics*, Pergamon Press, 1958, p. 452 and ff.

Type II Superconductors in High Fields

All the above considerations were valid in the case of a constant Δ. However, as was first pointed out by Fulde and Ferrell, (6.16) (in the absence of A) gives solutions of the form

$$\Delta(\mathbf{r}) = \Delta(\mathbf{q}_0) e^{i\mathbf{q}_0 \cdot \mathbf{r}}, \tag{6.24}$$

where $\mathbf{q}_0$ is a vector and $\Delta(\mathbf{q}_0)$ a constant. The corresponding equation for $\Delta(\mathbf{q}_0)$ takes the form (6.17).

If $\mathbf{q}_0$ is different from zero, the pair potential $\Delta(\mathbf{r})$ corresponds to a "depaired" superconducting state in which Cooper pairs have a single non-vanishing centre of mass momentum.† If the transition between this depaired state and the polarized normal state is of the second order, from (6.17), it is seen that for each temperature T there will exist a limiting field H_{FF}, or conversely for each field H a transition temperature T_{FF} is obtained by solving the equation

$$K(\mathbf{q}_0, H, T) - K(0, H, T) = 1 - K_t(0, H, T). \tag{6.25a}$$

Or, equivalently,

$$K(\mathbf{q}_0, H, T) - K(0, 0, T) = 1 - K_t(0, 0, T) = N(0) V \log \frac{T}{T_0}. \tag{6.25b}$$

It is clear from (6.25b) that the Fulde and Ferrell state will correspond to the maximum of $K(\mathbf{q}_0, H, T) - K(0, 0, T)$ [or, equivalently, to the maximum of $K(\mathbf{q}_0, H, T) - K(0, H, T)$] as a function of $\mathbf{q}_0$. The transition curve $H = f(T)$ is thus given by the simultaneous solution of (6.25b) and,‡

$$\frac{dK(\mathbf{q}, H, T)}{d\mathbf{q}} = 0. \tag{6.26}$$

Before studying this equation in detail, it may be remarked, that if the depaired state is to exist, the transition between the B.C.S. state (which is obviously the correct solution in low fields) and the depaired state will be of the first order. It can then be predicted that $K(\mathbf{q}, H, T) - K(0, H, T)$ will not exhibit a maximum if $H < H_0$ where H_0 is given by (6.21) and (6.22).

As in Chapter 5 it is now necessary to introduce the concept of dirty and pure superconductors. In dirty superconductors the kernel $K(\mathbf{q})$ is readily obtained from the corresponding kernel of Chapter 5, by changing ω in $\omega - i\gamma H$ so that

$$K(\mathbf{q}, H, T) - K(0, H, T) = 2\pi k_B T N(0) V \operatorname{Re} \sum_{\omega} \left[\frac{1}{2|\omega| + Dq^2 - i\gamma H} - \frac{1}{2|\omega| - 2i\gamma H} \right], \tag{6.27}$$

† According to equation (5.133) for the current, the form (6.24) for $\Delta(\mathbf{r})$ gives a permanent current. This current is compensated, however, by an opposite current due to unpaired electrons [see Fuelde and Ferrell (1964)].

‡ Note that $K(\mathbf{q}, H, T)$ depends only on the modulus of $\mathbf{q}$.

Reversible Properties

while in pure superconductors the same transformation yields

$$K(q, HT) - K(0, H, T) = \frac{\pi V N(0) k_B T}{q v_F} \operatorname{Re} \sum_\omega \left\{ \tan^{-1} \frac{q v_F - 2\gamma H}{2|\omega|} \right.$$

$$\left. + \tan^{-1} \frac{q v_F + 2\gamma H}{2|\omega|} \right\} - \pi V N(0) k_B T \operatorname{Re} \sum_\omega \frac{1}{|\omega| + i\gamma H}.$$

(6.28)

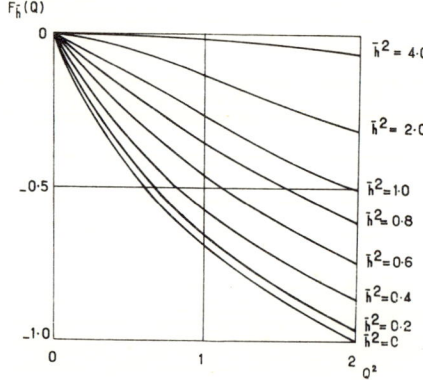

FIG. 6.3. Dirty superconductors. Variation of $F_{\bar{h}}(Q) =$

$$\sum_{n=0}^{\infty} \left\{ \frac{2n + 1 + Q^2}{(2n + 1 + Q^2)^2 + \bar{h}^2} - \frac{2n + 1}{(2n + 1)^2 + \bar{h}^2} \right\}$$

versus $Q^2 = Dq^2/2\pi k_B T$ for various values of $\bar{h} = \gamma H/\pi k_B T$: showing that this function is maximum for $Q = 0$.

These functions were computed by Sarma and Saint-James (1964) for various values of the ratio H/T, as a function of q. Their variations are shown in Fig. 6.3 and 6.4.

It is seen that in the dirty case $K(q, H, T) - K(0, H, T)$ exhibits no maximum and is always negative so that no depaired state is expected. This is in agreement with the fact that the diffusion process involved in the dirty case is unlikely to allow the sinusoidal variation of the pair potential. In the pure case, however, it is seen that a maximum exists for $q_0 \neq 0$ as soon as $H/T > > H_L/T$. Since

$$\left(\frac{dK(q, H, T)}{dq} \right)_{q=0}$$

is negative for $H/T < H_L/T$ and positive for $H/T > H_L/T$.

The limiting field H_L is given by

$$\left(\frac{d^2 K(q, H, T)}{dq^2} \right)_{q=0} = 0.$$

(6.29)

As expected, this equation is exactly $R = 0$ where R is given by (6.20) so that the limiting point is

$$H = H_0, \quad T = T_c(H_0).$$

It is then easy to compute the transition curve between the depaired state and the normal state by solving (6.25b) and (6.26). This curve is drawn on Fig. 6.5.

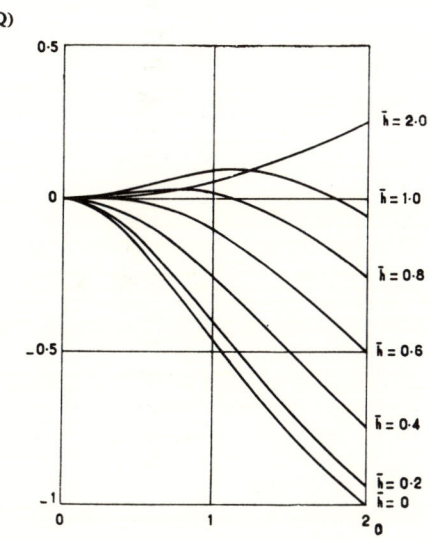

FIG. 6.4. Pure superconductors. Variation of $F_{\bar{h}}(Q) =$

$$\sum_{n=0}^{\infty} \left\{ \frac{1}{Q} \tan^{-1} \frac{Q-\bar{h}}{2n+1} + \frac{1}{Q} \tan^{-1} \frac{Q+\bar{h}}{2n+1} - \frac{2(2n+1)}{(2n+1)^2 + \bar{h}^2} \right\}$$

versus $Q = \hbar q v_F / 2k_B T$ for various values of $\bar{h} = \gamma H / \pi k_B T$ showing that $F_{\bar{h}}(Q)$ presents an extremum for $Q \neq 0$ as soon as $H > H_0$

For the sake of completeness, it should now be necessary to calculate the first order transition curve between the B.C.S. state and the depaired state but this has not been done for all temperatures. In their original paper Fulde and Ferrell studied the possible existence of the depaired state at $T = 0$ in a pure superconductor. They concluded that a general condition for the existence of this state is $H > H_0$, but the first order transition between B.C.S. state and the depaired state occurs for $\gamma H / \Delta_0 = 1/\sqrt{2}$ (i.e. the Chandrasekhar–Clogston limit) while the transition between the depaired and the normal state is of the second order and occurs at $\gamma H / \Delta_0 = 0.76$, which is the value obtained from (6.25b) and (6.26).†

† It can be shown that at $T = 0$, (6.25b) and (6.26) yield exactly the equation obtained by Fulde and Ferrell by a completely different approach.

Reversible Properties

The first order transition curve between the B.C.S. state and the depaired state will thus start from $[H_0, T_c(H_0)]$ and end at

$$\frac{\gamma H_0}{\Delta_0} = \frac{1}{\sqrt{2}}, \quad T = 0.$$

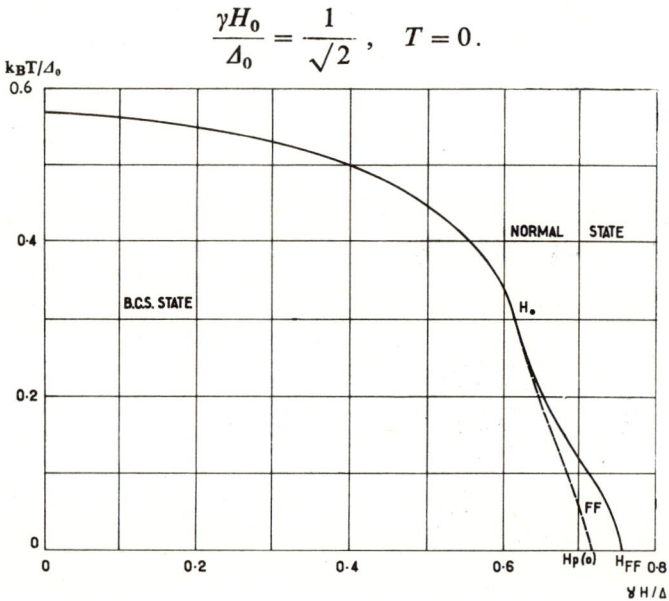

FIG. 6.5. Transition curve from the B.C.S. state and from the Fulde and Ferrell depaired state to the normal state (full curve). The domain of existence of the Fulde and Ferrell state is labelled F.F. The dashed line shows an approximate boundary between the B.C.S. and the F.F. state

In conclusion, it is seen from this simple model that the Fulde and Ferrell depaired state is not likely to occur in the dirty case. Since the kernel $K(q,H,T)$ is valid in the whole temperature range and since it presents no maximum this conclusion should hold over the whole temperature range, even in the presence of orbital effects. In the case of pure superconductors, the above model does not allow a definite conclusion since the kernel $K(q,H,T)$ is strictly valid only in the vicinity of T_0, the transition temperature in zero field.

In a recent paper Gruenberg and Gunther (1966), using a method similar to the technique of Helfand and Werthamer which was briefly outlined in section 5.5, have studied the possible existence of the Fulde and Ferrell state in the presence of orbital effects.

Their results are as follows:

(a) in pure superconductors:

(i) for each value of $\alpha = \sqrt{2} H_{c2}(0)/H_p(0) > 1\cdot 8$, where $H_{c2}(0)$ is the upper critical field in the absence of paramagnetic effect at $T = 0$, a temperature $T_{0,\alpha}$ exists below which the Fulde and Ferrell state is stable.

For $\alpha \to \infty$ one finds, as expected,†

$$T_{0,\alpha} = T_c(H_0) = 0.55 T_0,$$

where $T_c(H_0)$ is given by (6.22).

(ii) The transition from the mixed state to the Fulde and Ferrell state is of the first order while the transition from the Fulde and Ferrell state to the normal state is of the second order.

(b) In dirty superconductors:

For $\alpha \to \infty$, the optimum solution for Δ is of the form $e^{iq_0 z}\Delta(x,y)$, which seems to be in contradiction with the conclusion of the above section.‡ However by studying the first nonlinear correction to the pair potential equation Gunther and Gruenberg have shown that this solution is not a stable one, so that the transition field calculated appears as a supercooling field. No definite conclusion has been drawn by these authors on the possible observation of the Fulde and Ferrell state in the dirty case.

In summary, the Fulde and Ferrell state probably exists in pure superconductors. It must be pointed out that this state is important for high fields superconductors, i.e. for high κ materials but very few pure superconductors belong to this category. However, Gruenberg and Gunther suggest that this new state may be realized in V_3Ga at sufficiently low temperature and high fields.†††

6.1.4. The transition field H_{c2} in the dirty case

In the dirty case, one does not expect a depaired state and the kernel $K(q)$ is valid at all temperatures. The transition field H_{c2} can be calculated easily from the corresponding equations (5.115) and (5.117). The change of ω into $\omega - i\gamma H$ yields

$$\log \frac{T}{T_0} + f_0 \left[\frac{D}{2\pi k_B T} \frac{eH}{c} + \frac{i\gamma H}{2\pi k_B T} \right] = 0, \qquad (6.30)$$

where f_0 is now

$$f_0(Z) = Re \left\{ \psi\left(\frac{1}{2} + Z\right) - \psi\left(\frac{1}{2}\right) \right\}.$$

---

† It can be shown that for $\alpha \to \infty$ the orbital contribution may be neglected.

‡ According to Gruenberg (private communication) the discrepancy arises from the fact that the diffusion approximation is not sufficient. We have nevertheless used it for the sake of simplicity.

††† It would be interesting to study the behaviour of the Fulde and Ferrell state as a function of the electronic mean free path l. Qualitatively one can assert that when l is lowered the limiting curve between the depaired state and the normal state (corresponding to a second order transition) will depart from the curve represented in figure (6.5). Still starting from the field H_0 it will reach lower values of the field. For a certain value of l it will go below the first-order transition curve computed by Sarma (fig. 6.2). The depaired state will then be unstable and appear as a supercooling state. Eventually for $l = 0$, the curve will reach the Baltensperger curve shown in figure (6.1).

Reversible Properties

The solution of (6.30) has been calculated by Maki (1964) and gives the critical field H_{c2}^* in the presence of the paramagnetic effect as a function of the reduced temperature. It is convenient to introduce the parameter

$$\alpha = \frac{\gamma c}{eD}. \qquad (6.31a)$$

This parameter is related directly to the Chandrasekhar–Clogston limit at $T = 0$. According to (5.118a), the field $H_{c2}(0)$ at $T = 0$ in the absence of paramagnetic effect is

$$H_{c2}(0) = \frac{\Delta_0}{2D}\frac{c}{e} = 0.87\frac{k_B T_0}{D}\frac{c}{e} \qquad (6.32)$$

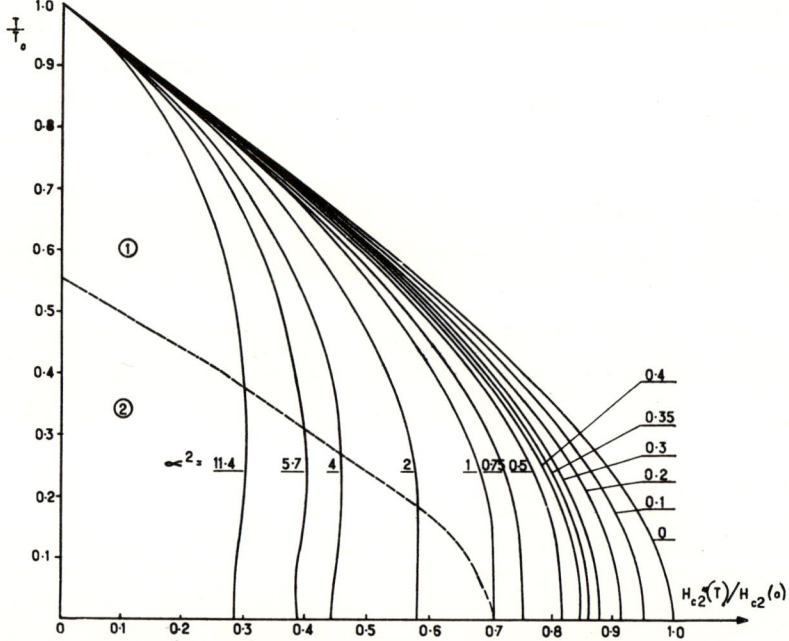

FIG. 6.6. Variation of $H_{c2}^*(T)/H_{c2}(0)$ versus T/T_0 for various values of the parameter α^2, supposing that the transition at H_{c2}^* is of the second order. The dashed curve represents the boundary between the second and the first order transitions domain (in the large κ limit). In domain ② where the transition should be of the first order no calculation of the transition curves exists

while $H_p(0)$ is given by (6.5b). α is thus

$$\alpha = \sqrt{2}\,\frac{H_{c2}(0)}{H_p(0)}. \qquad (6.31b)$$

and (6.30) reads

$$\log\frac{T}{T_0} + f_0\left\{\frac{0.87}{2\pi}\frac{T_0}{T}\frac{H_{c2}^*(T)}{H_{c2}(0)}(1+i\alpha)\right\} = 0. \qquad (6.33)$$

The variation of $H_{c2}^*(T)/H_{c2}(0)$ versus T/T_0 is given in Fig. 6.6 for various values of α. Using the asymptotic form of the ψ function it is seen that

$$H_{c2}^* = \frac{\Delta_0}{2D}\frac{c}{e}(1+\alpha^2)^{-1/2}\left[1 - \frac{2}{3}\left(\frac{\pi k_B T}{\Delta_0}\right)^2 \frac{1-\alpha^2}{1+\alpha^2}\right] \text{ for } T \ll T_0 \quad (6.34\text{a})$$

$$H_{c2}^* = \frac{4k_B T_0}{D}\frac{c}{e}\left[1 - \frac{T}{T_0}\right]\left\{1 - \left(1-\frac{T}{T_0}\right)\left[\frac{1}{2} - \frac{28\zeta(3)}{\pi^4}(1-\alpha^2)\right]\right\} \quad (6.34\text{b})$$

for $|T - T_0| \ll T_0$.

In particular, at $T = 0$ the value of $H_{c2}^*(0)$ is given by

$$H_{c2}^*(0) = \frac{H_{c2}(0)\, H_p(0)}{[2H_{c2}^2(0) + H_p^2(0)]^{1/2}}. \quad (6.34\text{c})$$

The calculation of H_{c3}^* (i.e. H_{c3} in the presence of the paramagnetic effect) proceeds in the same way. The operator $(p - 2e\mathbf{A}/c)^2$ which appears in (6.16) should be replaced by $0.59\, eH/\hbar c$ so that the critical field H_{c3}^* is given by

$$\log\frac{T}{T_0} + f_0\left(\frac{0.59\, D}{2\pi k_B T}\frac{e}{c}H_{c3} + \frac{i\gamma H_{c3}}{2\pi k_B T}\right) = 0. \quad (6.35)$$

It is seen that the variation of H_{c3}^* with temperature is obtained from the variation of H_{c2}^* simply by changing α into $1.69\,\alpha$, i.e.

$$H_{c3}^* = 1.69\, H_{c2}^*(1.69\,\alpha). \quad (6.36)$$

The corresponding variation of H_{c3}^*/H_{c2}^* has been computed by Saint-James (1966) and is shown in Fig. 6.7. It is seen that the ratio H_{c3}^*/H_{c2}^* starts from the value 1.69 at $T = T_0$ and decreases as the temperature is decreased. One can obtain the corresponding asymptotic formula from (6.34a) and (6.34b). In particular at $T = 0$

$$H_{c3}^*(0) = \frac{H_{c3}(0)\, H_p(0)}{[2H_{c3}^2(0) + H_p^2(0)]^{1/2}}, \quad (6.37\text{a})$$

where $H_{c3}(0) = 1.69\, H_{c2}(0)$. The ratio $H_{c3}^*(0)/H_{c2}^*(0)$ is thus

$$\frac{H_{c3}^*(0)}{H_{c2}^*(0)} = \frac{1.69\sqrt{(\alpha^2+1)}}{\sqrt{(1.69^2\,\alpha^2+1)}}. \quad (6.37\text{b})$$

The simultaneous measurement of H_{c3}^* and H_{c2}^* should yield a direct determination of α.

All the above considerations are valid provided that the transition from the superconducting to the normal state is of the second order. If the transition becomes of the first order, H_{c2}^* and H_{c3}^* will appear only as supercooling fields. In order to determine the critical points at which the transition switches to first order, one should calculate the coefficient R of $|\Delta|^2\Delta$

Reversible Properties

and solve the equation $R = 0$, and (6.33) for H_{c2} [or (6.35) for H_{c3}]. This was first done by Maki (1964b). However, his result is wrong since he deduces the coefficient R from the incorrect coefficient calculated in the absence of paramagnetic effect. The correct equation is readily obtained from the R coefficient given by Caroli *et al.* (1966) by changing ω into $\omega - i\gamma H$. It is readily found that the critical point for large values of κ, neglecting

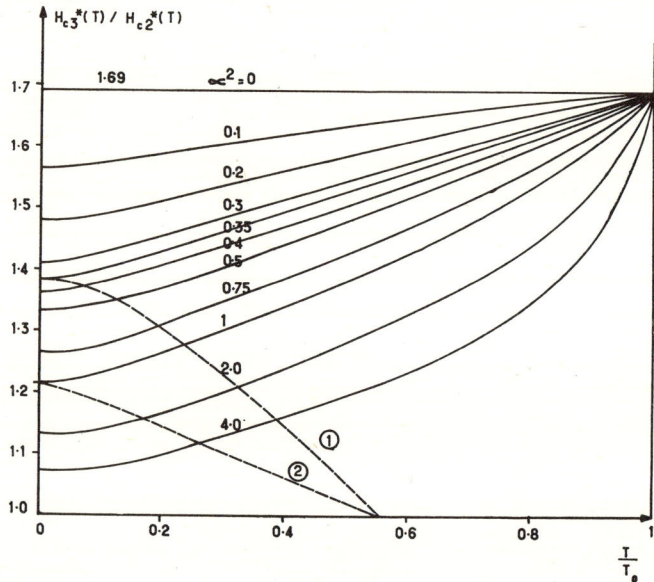

FIG. 6.7. Variation of the ratio H_{c3}^*/H_{c2}^* versus T/T_0 for various values of the parameter α^2 supposing that the transitions at H_{c3}^* and H_{c2}^* are of the second order. Below the curve labelled ① the transition becomes of the first order at $H = H_{c3}^*$: and below the curve labelled ② the transition switches to first order at $H = H_{c2}^*$ (these conclusions are valid in the large κ limit)

a term of order $1/\kappa^2$, is given by,

$$\operatorname{Re} \sum_n \frac{1}{[n + 1/2 + \bar{h}(1 + i\alpha)]^3} = 0, \qquad (6.38)$$

where

$$\bar{h} = \frac{0.87}{2\pi} \frac{T_0}{T} \frac{H_{c2}^*(T)}{H_{c2}(0)}.$$

Solving (6.33) and (6.38) yields the limiting curve between the second order and the first order transitions for the critical field H_{c2}^*. This curve is shown in Fig. 6.6. It is found that expression (6.38) cannot vanish for $\alpha^2 < 1$ and that for $\alpha^2 > 1$ the transition becomes of the first order at low temperatures. This is in agreement with the fact that $H_{c2}^*(T)$ becomes a two valued function for $\alpha^2 > 1$. It is to be noted that, whatever the value of α, the transition at H_{c2}^* is of the second order up to a temperature $T/T_0 = 0.557$.

It is also to be noted that the error in the coefficient R leads to an error in the magnetization slope at $H = H_{c2}$. Thus, the Maki coefficient κ_2^* (κ_2 when the paramagnetic effect is present) must be corrected. The corrected variation of κ_2^* is reported in Fig. 6.8. It is seen that for sufficiently high values of α the ratio κ_2^*/κ should decrease with temperature. This has been observed by Cape (1966) for a Ti–Mo alloy.

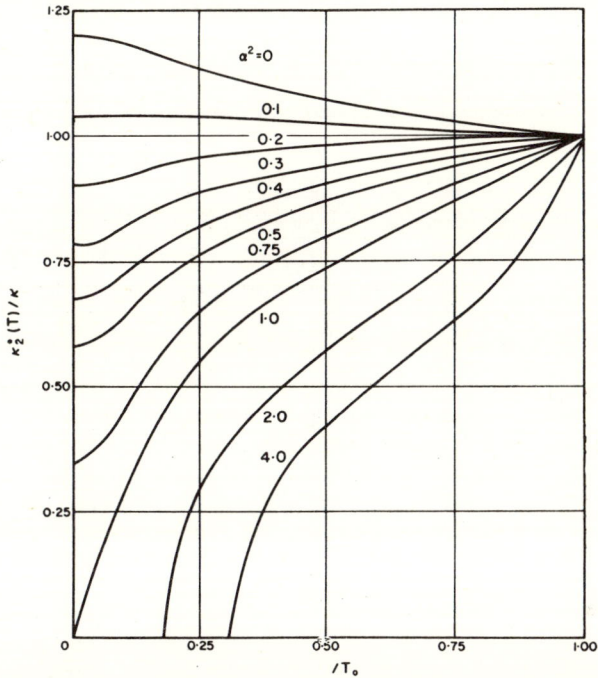

FIG. 6.8. Variation of Maki's parameter $\kappa_2^*(T)/\kappa$ for various values of α^2. When the transition becomes of the first order $(2\kappa_2^{*2} - 1)$ vanishes

In the case of H_{c3}^* similar conclusions can be drawn. Equation (6.38) is still valid but with α replaced by $1\cdot69\,\alpha$. It is seen that the transition becomes of the first order for $\alpha > 1/1\cdot69 = 0\cdot59$. H_{c3}^* is then a supercooling field. In Fig. 6.7 the boundaries between first order and second order transition at $H = H_{c3}^*$, or $H = H_{c2}^*$, are represented. Care must be taken in the experimental determination of the perpendicular and parallel nucleation fields to ensure that one actually measures the fields of interests. If $T/T_0 > 0\cdot56$ the transition in both situations will be of the second order whatever the value of α. No calculation exists at present for the parallel and the perpendicular nucleation fields when the transition is of the first order.†

† The transition at the nucleation field can still be of the second order in the perpendicular field and of the first order in the parallel situation. This could lead to an overestimate of H_{c3}^* and consequently of α.

Reversible Properties

All the above conclusions were valid in the limit of high κ values. If this is not the case the transition will become of the first order for lower values of α. The magnetization slope at H_{c2}^* is proportional to $-(2\kappa_2^{*2}-1)^{-1}$. Since this slope cannot be positive, this means that the transition will become of the first order if κ_2^* is smaller than $1/\sqrt{2}$. Since for $\alpha^2 > 0.13$, $\kappa_2^*(T)/\kappa$ is a decreasing function of T, close to T_0 one can have a superconductor in which $\kappa = \kappa_2(T_0) > 1/\sqrt{2}$ while below a certain temperature T_1, κ_2 is smaller than $1/\sqrt{2}$. For $T < T_1$ the transition in perpendicular field will be of the first order, and H_{c2}^* as given by (6.33) will appear as a supercooling field. In practice, however, the approximation $2\kappa_2^{*2} \gg 1$ is valid since the paramagnetic effect is of importance only in high κ materials.

6.1.5. Experimental situation. The spin-orbit scattering effect

Various experiments have been performed in alloys with high κ values especially in Ti–V, Ti–Nb and Ti–Mo compounds [Berlincourt and Hake (1963), Strnad and Kim (1965), Shapira and Neuringer (1966), Cape (1966)]. Unfortunately, no simultaneous measurements of H_{c3}^* and H_{c2}^* have been reported so that a direct determination of α by this method is not yet possible.† α, however, is not an adjustable parameter of the theory but may, in principle, be deduced from measurements performed in the normal phase. The diffusion coefficient which appears in (6.32) is related to the transport properties of the normal electrons, through the Fermi velocity v_F and the mean free path l. This yields for α the following expression:

$$\alpha = \frac{3e^2 \hbar \gamma_n \rho_n}{2m\pi^2 k_B^2}, \qquad (6.39)$$

where γ_n is the normal state electronic specific heat and ρ_n the normal state d.c. resistivity. This expression could also have been deduced from the ratio $H_{c2}(0)/H_p(0)$ which can be calculated easily from the Gorkov expression (5.155) for κ. Indeed

$$H_{c2}(0) = \frac{3\Delta_0 c}{2elv_F} = 3.09 \times 10^{-2} \rho_n \gamma_n T_0, \qquad (6.40)$$

where γ_n and ρ_n are expressed in C.G.S. units.

In the Ti alloys studied by the different authors quoted above, α is of the order of 1·5. Generally speaking, the measured values of $H_{c2}^*(T)$ are, as expected, smaller than $H_{c2}(T)$ and are in qualitative agreement with the predictions made by Maki. The results obtained for the perpendicular

---

† In a paper by Kim *et al.* (1965), H_{c3} and H_{c2} are given, but with no usable temperature data. These authors have successfully fitted their results to (6.37b) for low values of α. Recenty Hake (1967) has given values of H_{c3}^{**} and H_{c2}^{**}. No comparison can be made since H_{c3}^{**} du eto surface imperfection, vanishes at a higher temperature than H_{c2}^{**}.

critical field however are always larger than the predicted values obtained with the calculated value of α.

As already mentioned in the introduction, this effect can be attributed to the spin-orbit scattering which brings the spin susceptibility of the superconducting state closer to that of the normal state. We shall not enter into the details of the corresponding theory which was proposed simultaneously by Werthamer et al. (1966) and by Maki (1966). The formula for H_{c2}^{**} (the value of H_{c2} in the presence of the paramagnetic effect and the spin-orbit scattering) will be given here and several comments added. This formula is only valid in the dirty case and moreover in the limit:

$$l \ll l_{so},$$

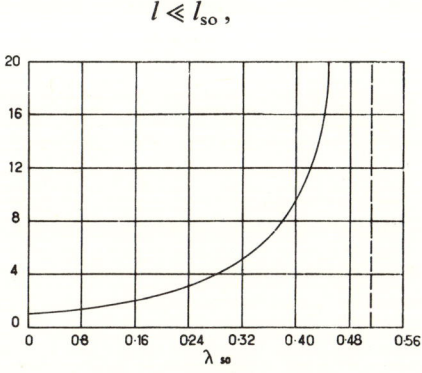

FIG. 6.9 The values of α_c at which the function $T(H_{c2})$ becomes a two valued function, as a function of λ_{so}. The dashed line represents the asymptote for $\lambda_{so} = \lambda_{so}^c$. [after Werthamer et al. (1966)]

where l is the spin-orbit scattering independent mean free path of the electrons in the normal metal and $l_{so} = v_F \tau_{so}$ is the spin-orbit scattering mean free path. For this assumption to be valid l must be very short and also the spin-flip scattering should be infrequent in comparison to non spin-flip scattering.† Following Werthamer et al., it is convenient to introduce the dimensionless variables

$$t = \frac{T}{T_0}, \qquad \bar{h} = \frac{\Delta_0}{4\pi k_B T_0} \frac{H_{c2}^{**}}{H_{c2}(0)}$$

$$\lambda_{so} = \frac{\hbar}{3\pi k_B T_0 \tau_{so}}$$

and the equation for H_{c2}^{**} reads

$$\log t = \sum_{n=-\infty}^{+\infty} \left\{ \frac{1}{|2n+1|} - \left[|2n+1| + \frac{\bar{h}}{t} + \frac{(\alpha \bar{h}/t)^2}{|2n+1| + (\bar{h} + \lambda_{so})/t} \right]^{-1} \right\}$$

(6.41)

† With the possible exception of V_3Si, all the type II superconductors with a field H_{c2} sufficiently high to present a paramagnetic effect are of the dirty type.

For $\lambda_{so} = 0$, this equation reduces to (6.30). It can be seen from (6.41) that if $\lambda_{so} > 0$, the spin paramagnetism effect is reduced, that is to say $H_{c2}^{**} > H_{c2}^{*}$. Werthamer *et al.* could also show that the value α_c at which t becomes a double valued function of $\bar{h}$ increases with λ_{so}. For $\alpha > \alpha_c$ the transition at low temperature will be of the first order.†

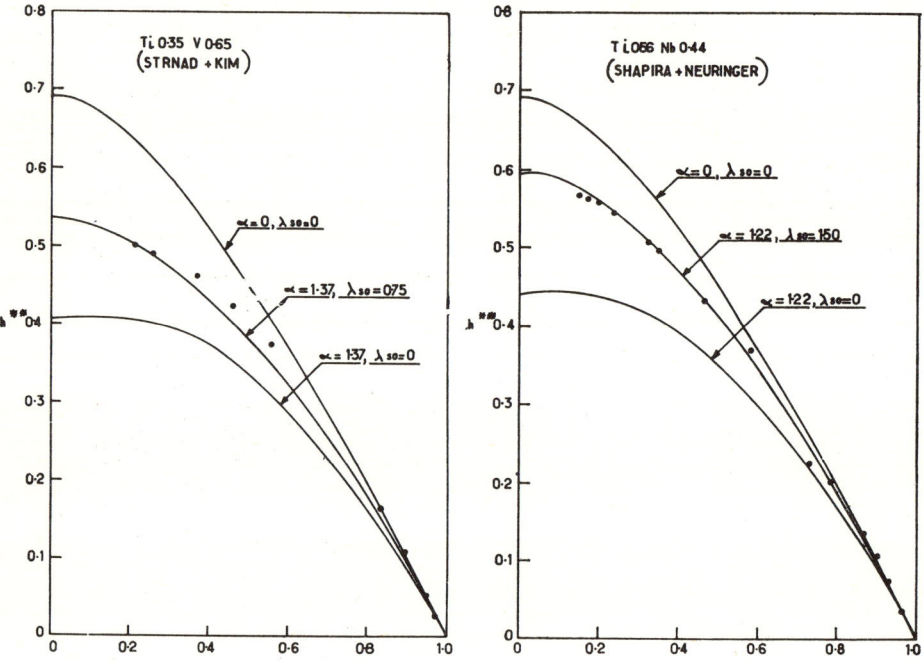

FIG. 6.10. Comparison of theory and experiment for the upper critical field H_{c2}^{**} in the presence of paramagnetic and spin-orbit effects after Werthamer *et al.* (1966). The figures represent the variation of $h^{**} = H_{c2}^{**}(t)/(- dH_{c2}^{**}) t = 1$ versus $t = T/T_0$ in

(a) a $Ti_{0.35} - V_{0.65}$ alloy studied by Strnad and Kim (1965),

(b) a $Ti_{0.56} - Nb_{0.44}$ alloy studied by Shapira and Neuringer (1966)

For $\lambda_{so} = 0$, $\alpha_c = 1$, a result already obtained in section 6.14. The variation of α_c with λ_{so} is given in Fig. 6.9. A critical value $\lambda_{so}^c = 0.5139$, also exists above which t never becomes a double valued function of h. For $\lambda_{so} > \lambda_{so}^c$ it is expected that the transition will remain of the second order over the whole temperature range.

† The boundary curve between the superconducting and the normal state cannot be a double valued function. If this were the case, in the low temperature region, the transition from the superconducting state (ordered phase) to the normal state (disordered phase) could be obtained by lowering the temperature and at the same moment by increasing the entropy. This is in contradiction with the fact that entropy is an increasing function of temperature.

Type II Superconductors in High Fields

The agreement between theory and experimental data obtained by Strnad and Kim (1966), Shapira and Neuringer (1966), and Cape (1966) is rather good. α is first determined using (6.39) and λ_{so} is considered as an adjustable parameter. Figure 6.10a, b and Fig. 6.11 compare the various experimental results with the theoretical predictions.†

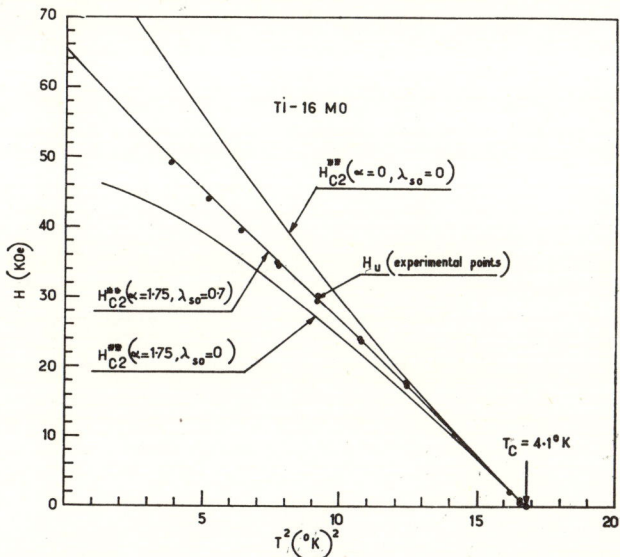

FIG. 6.11. Comparison of theory and experiment for the upper critical field H_{c2}^{**} in the presence of paramagnetic and spin-orbit effects [after Cape (1966)] for a $Ti_{0.84} - Mo_{0.16}$ alloy

It can also be shown that (6.41) is valid for H_{c3}^{**} if one replaces H_{c2} by H_{c3} and α by $1 \cdot 69\alpha$, i.e.

$$H_{c3}^{**} = 1 \cdot 69 \, H_{c2}^{**} \, (1 \cdot 69 \, \alpha, \lambda_{so}). \qquad (6.42)$$

No measurements of H_{c3}^{**} have been reported so that no comparison can be made with (6.42).

6.II. GAPLESS SUPERCONDUCTIVITY

6.II.1. Introduction

It has been shown in the preceding chapters that the superconducting state exhibits long range order. The order parameter $\psi(r)$ is proportional to the pair potential:

$$\Delta(r) = -V \langle \psi_\downarrow(r) \, \psi_\uparrow(r) \rangle \qquad (6.43)$$

† See also the work of Hake (1967) who compares experimental determinations with the values of κ_2^* shown in fig. 6.8.

Reversible Properties

which was introduced in Chapter 5. It should be emphasized again that the existence of the thermal average appearing in (6.43) is a quantum effect with no classical analogue, and that superconductivity displays quantum effects on a macroscopic scale.

In the simplest case of a uniform electron gas, in the absence of magnetic fields, $\Delta(r)$ reduces to a constant Δ. There is an energy gap 2Δ in the quasi-particle excitation spectrum (B.C.S.). However, $\Delta(r)$ is not always related to an energy gap in the excitation spectrum. $\Delta(r)$ can actually be different from zero for all values of r even though there is no energy gap in the spectrum. In this case, the spectrum in the superconducting phase is not *qualitatively* different from the normal state spectrum.

Abrikosov and Gorkov (1960) were the first to point out that this is the case for a superconductor containing magnetic impurities. The most striking property is that, in the absence of an applied field, Δ and the transition temperature T_0 vary with impurity concentration. For a given concentration, the gap in the excitation spectrum 2Δ can vanish while T_0 is still finite.

It has recently been recognized that this is also the case in all type II superconductors (dirty or clean) when the applied field is only slightly smaller than the upper critical field. This is the property which will be investigated here.

6.II.2. Order parameter and excitation spectrum

In Chapter 5 the main conclusions of the microscopic theory were obtained in the frame of the self-consistent method. The hamiltonian (5.13) which describes a system of electrons, in the presence of an arbitrary potential $U_0(r)$ and two-body interactions was replaced by the self-consistent hamiltonian (5.53), which in the absence of Pauli spin-paramagnetism, reads

$$\mathcal{H}_{\text{eff}} = \int dr\, \psi_\alpha^\dagger \left[\frac{1}{2m} \left(p - \frac{eA}{c} \right)^2 + U(r) - E_F \right] \psi_\alpha$$

$$+ \int dr\, [\Delta^*(r)\, \psi_\downarrow(r)\, \psi_\uparrow(r) + \Delta(r)\, \psi_\uparrow^\dagger(r)\, \psi_\downarrow^\dagger(r)] + \frac{1}{V} \int |\Delta(r)|^2\, dr.$$

(6.44)

In this equation $U(r)$ contains the effects of impurities, boundaries, etc. In order to diagonalize the hamiltonian (6.44), we can use a Bogoliubov transformation, i.e. set

$$\psi_\alpha(r) = \sum_n \{ \gamma_{n\alpha}\, u_{n\alpha}(r) + \gamma_{n\alpha}^+\, v_{n\alpha}(r) \}, \qquad (6.45)$$

where the γ_n are annihilation operators for fermions which satisfy the anticommutation relations. The γ_n correspond to the quasiparticles.

Type II Superconductors in High Fields

u and v will be the coefficients of a unitary transformation if

$$\sum_\alpha \int dr \left[u_{n\alpha}(r) u^*_{m\alpha}(r) + v_{n\alpha}(r) v^*_{m\alpha}(r) \right] = \delta_{n,m}, \quad (6.46a)$$

$$\sum_\alpha \int dr \left[u_{n\alpha}(r) v_{m\alpha}(r) + v_{n\alpha}(r) u_{m\alpha}(r) \right] = 0, \quad (6.46b)$$

$$\sum_n \{ u_{n\alpha}(r) u^*_{n\beta}(r') + v^*_{n\alpha}(r) v_{n\beta}(r') \} = \delta_{\alpha\beta} \, \delta(r - r'), \quad (6.46c)$$

$$\sum_n \{ u_{n\alpha}(r) v^*_{n\beta}(r') + v^*_{n\alpha}(r) u_{n\beta}(r') \} = 0. \quad (6.46d)$$

After the diagonalization $\mathscr{H}_{\text{eff}}$ will take the form

$$\mathscr{H}_{\text{eff}} = E_g + \sum_n \varepsilon_n \gamma^\dagger_{n\alpha} \gamma_{n\alpha} \quad (6.47)$$

and the commutator of $\mathscr{H}_{\text{eff}}$ with $\gamma^\dagger_{n\alpha}$ is easily found to be

$$[\mathscr{H}_{\text{eff}}, \gamma^\dagger_{n\alpha}] = \varepsilon_n \gamma_{n\alpha}. \quad (6.48)$$

In (6.47), E_g represents the energy of the ground state of $\mathscr{H}_{\text{eff}}$ while ε_n is the energy of the excitation.

Using expressions (6.44), (6.45) and (6.48), it is seen that the function $u_{n\alpha}(r)$ and $v_{n\alpha}(r)$ obey the following set of linear equations (Bogoliubov equations):

$$\varepsilon_n u_{n\uparrow}(r) = T u_{n\uparrow}(r) + \Delta(r) v_{n\downarrow}(r), \quad (6.49a)$$

$$\varepsilon_n v_{n\downarrow}(r) = -T^* v_{n\downarrow}(r) + \Delta^*(r) u_{n\uparrow}(r), \quad (6.49b)$$

$$\varepsilon_n u_{n\downarrow}(r) = T u_{n\downarrow}(r) - \Delta(r) v_{n\uparrow}(r), \quad (6.49c)$$

$$\varepsilon_n v_{n\uparrow}(r) = -T^* v_{n\uparrow}(r) - \Delta^*(r) u_{n\downarrow}(r), \quad (6.49d)$$

where $\quad T = \dfrac{1}{2m} \left[p - \dfrac{eA}{c} \right]^2 + U(r) - E_F. \quad (6.50)$

Equations (6.49) define a wave function with two components $\begin{pmatrix} u_{n\alpha}(r) \\ v_{n\alpha}(r) \end{pmatrix}$ which describes an excitation. u is called the "electron amplitude" while v is the "hole amplitude". T is an hermitian operator and the different solutions $\begin{pmatrix} u \\ v \end{pmatrix}$ are orthogonal so that (6.46a) is satisfied. Since $\begin{pmatrix} u^* \\ v^* \end{pmatrix}$ corresponds to $-\varepsilon$, (6.46b) is also satisfied.

Here ε_n must correspond to an excitation energy so that it is necessary for $\varepsilon_n > 0$. We must therefore retain in (6.49) the solution which corresponds to this condition. Hence the self-consistent equation for the pair potential (6.44) should be added to (6.49a). This can be written in term of the u_n and v_n and will be done here in the case where one can neglect the spin effect [i.e. the unperturbed hamiltonian T has the form (6.50)].

In this case one can take u and v as real and write

$$u_{n\uparrow}(r) = u_{n\downarrow}(r) = u_n(r), \qquad (6.51a)$$

$$v_{n\downarrow}(r) = -v_{n\uparrow}(r) = v_n(r), \qquad (6.51b)$$

so that (6.49) takes the simple form

$$\varepsilon_n u_n(r) = T u_n(r) + \Delta(r) v_n(r), \qquad (6.52a)$$

$$\varepsilon_n v_n(r) = -T^* v_n(r) + \Delta(r) u_n(r), \qquad (6.52b)$$

while the self-consistent equation for $\Delta(r)$ reads:

$$\Delta(r) = \sum_n u_n(r) v_n(r) [1 - 2f(\varepsilon_n)]. \qquad (6.53)$$

The solutions of the three equations (6.52) and (6.53) should yield the complete solution of the problem.†

(a) *The B.C.S. energy gap*

In general the solution of the three equations (6.52) and (6.53) is impossible to obtain. The problem is greatly simplified if one already knows the pair potential $\Delta(r)$ however. This is the case in a uniform electron gas in the absence of a magnetic field ($A \equiv 0$), the situation first treated by B.C.S. If $\phi_n(r)$ is the eigenfunction of T with eigenvalue ξ_n, one can take‡

$$u_n(r) = u_n \phi_n(r), \qquad (6.54a)$$

$$v_n(r) = v_n \phi_n(r) \qquad (6.54b)$$

and the equation reads

$$\varepsilon_n u_n = \xi_n u_n + \Delta v_n, \qquad (6.55a)$$

$$\varepsilon_n v_n = -\xi_n v_n + \Delta u_n. \qquad (6.55b)$$

It is seen that this system possesses a non-zero solution if

$$\varepsilon_n^2 = \xi_n^2 + \Delta^2 \qquad (6.58a)$$

or

$$\varepsilon_n = (\xi_n^2 + \Delta^2)^{1/2} \qquad (6.58b)$$

and that a gap exists in the quasi-particle spectrum, i.e. $\varepsilon \geq \Delta$. The existence

† For further details on the self-consistent method see De Gennes (1966a).
‡ The energies ξ_n are measured from the Fermi energy.

of this energy gap has several important consequences. At finite temperatures the population of the quasi-particle states is given by the Fermi function of ε_n. The specific heat behaves as $e^{-\Delta/k_B T}$ and this gives a method of measuring Δ. Other methods used consist of creating quasi-particles in the system. Two main classes of experiments can be distinguished:

(i) One can create single excitations. This is the case in tunnelling experiments in which electrons are injected into the superconductor through a

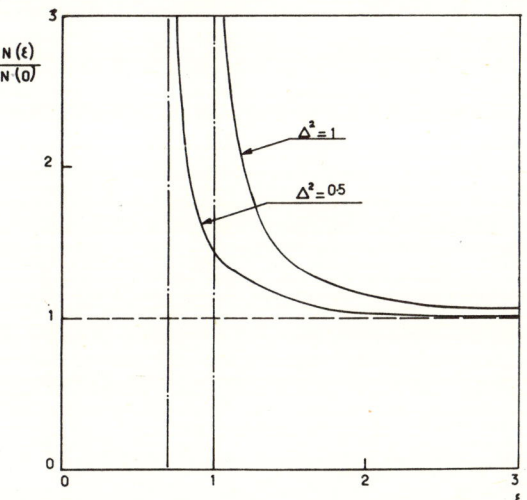

FIG. 6.12. The density of states in the B.C.S. model for two values of the energy gap

thin oxide layer, the tunnelling current being different from zero when the applied voltage is greater than Δ. This technique, which has been very widely used to measure Δ, was first proposed by Giaever (1960). In this case, one can measure the density of states in the superconductor. Using (6.58) for ε_n this density of states is readily seen to be

$$\frac{N(\varepsilon)}{N(0)} = \frac{\varepsilon}{(\varepsilon^2 - \Delta^2)^{1/2}} \quad \text{for } \varepsilon > \Delta \quad (6.59)$$
$$= 0 \quad \text{for } \varepsilon < \Delta,$$

where $N(0)$ is the density of state at the Fermi level in the normal state. The behaviour of $N(\varepsilon)$ as a function of ε in the B.C.S. model is shown in Fig. 6.12.

(ii) When the number of particles is conserved, excitations are created by "breaking pairs" and this needs an energy of at least 2Δ. This is, for example, the case in infrared absorption where a threshold value of 2Δ is observed. For more details of these and other methods see Lynton (1964), Chapter X. Values of the gap in several superconductors have been given in Table 1.1.

Reversible Properties

(b) *Calculation of the excitation spectrum in the general case. Gapless superconductivity*

The above calculation was easy because in the B.C.S. model $\Delta(\mathbf{r})$ is constant. In the general case, however, the pair potential will be dependent on $\mathbf{r}$ and one has to solve the two Bogoliubov equations (6.52) and the self-consistent equation (6.53). Close to T_0, one can use the pair potential $\Delta(\mathbf{r})$ as obtained from the solution of the Ginzburg–Landau equation, since this ensures that the self-consistent equation is satisfied. This was used by Caroli (1966) to compute the excitation spectrum of a type II superconductor in the presence of vortex lines close to H_{c1}.

One can also obtain the energies ε_n by considering the pair potential $\Delta(\mathbf{r})$ as a perturbation. It is to be expected that if there is a gap in the excitation spectrum this expansion should not converge close to the Fermi level. This is the basic principle of a calculation by de Gennes (1964 and 1966b), that will be outlined here.

Let us suppose that ϕ_n is the eigenfunction of the unperturbed hamiltonian T corresponding to the one electron excitation energy ξ_n. Actually, the functions ϕ_n are very complicated since they introduce all the multiple scattering effects of the normal phase. However, as in Chapter 5 in the case of dirty superconductors, the final result introduces only particular combinations of the values of ϕ_ns which have a simple physical meaning.

A perturbation technique may be used to solve (6.52). The functions u and v are expanded as

$$u_n(\mathbf{r}) = u_n^0(\mathbf{r}) + u_n^1(\mathbf{r}) + \ldots$$
$$v_n(\mathbf{r}) = v_n^0(\mathbf{r}) + v_n^1(\mathbf{r}) + \ldots \qquad (6.60)$$

where $u_n^i(\mathbf{r})$ is the correction of order $|\Delta|^i$ to u_n and $v_n^i(\mathbf{r})$ to v_n.

To order zero in Δ

$$u_n^0(\mathbf{r}) = \phi_n(\mathbf{r}) \quad v_n^0(\mathbf{r}) = 0 \quad \text{for } \xi_n > 0, \qquad (6.61\text{a})$$

$$u_n^0(\mathbf{r}) = 0, \quad v_n^0(\mathbf{r}) = \phi_n^*(\mathbf{r}) \quad \text{for } \xi_n < 0. \qquad (6.61\text{b})$$

The excitation energy corresponds to electrons for $\xi_n > 0$ and to holes for $\xi_n < 0$ and is

$$\varepsilon_n = |\xi_n|. \qquad (6.62)$$

To first order in Δ, the function $u_n^1(\mathbf{r})$ and $v_n^1(\mathbf{r})$ are of the form

$$u_n^1(\mathbf{r}) = \sum_{m \neq n} a_{mn} \phi_n, \qquad (6.63\text{a})$$

$$v_n^1(\mathbf{r}) = \sum_{m \neq n} b_{mn} \phi_n^*. \qquad (6.63\text{b})$$

Type II Superconductors in High Fields

It is readily seen that

$$a_{mn} = 0; \quad b_{mn} = -\frac{1}{\xi_n + \xi_m} \int d\mathbf{r}\, \phi_m(\mathbf{r})\, \Delta(\mathbf{r})\, \phi_n(\mathbf{r}) \quad \text{for } \xi_n < 0, \tag{6.64a}$$

$$b_{mn} = 0; \quad a_{mn} = -\frac{1}{\xi_n + \xi_m} \int d\mathbf{r}\, \phi_m^*(\mathbf{r})\, \Delta(\mathbf{r})\, \phi_n^*(\mathbf{r}) \quad \text{for } \xi_n > 0. \tag{6.64b}$$

To second order in Δ, the eigenvalue is thus

$$\varepsilon_n = \xi_n + \sum_{m \neq n} \frac{|\int d\mathbf{r}\, \phi_n^*(\mathbf{r})\, \Delta(\mathbf{r})\, \phi_m^*(\mathbf{r})|^2}{\xi_n + \xi_m}. \tag{6.65}$$

This result is somewhat similar to the result obtained from the conventional perturbation theory applied to T with $\Delta(\mathbf{r})$ as a perturbing potential. One should note the differences which are the star on the function ϕ_m^* and the $+$ sign in the denominator. These arise from the fact that the perturbation in the total hamiltonian has the form $\Delta(\mathbf{r})\, \psi\dagger(\mathbf{r})\, \psi\dagger(\mathbf{r})$.

In order to obtain a matrix element in (6.65), let us introduce the operator K which transforms ϕ_n into ϕ_n^*. Then†

$$\int \phi_n^*(\mathbf{r})\, \Delta(\mathbf{r})\, \phi_m^*(\mathbf{r})\, d\mathbf{r} = \int \phi_n^*(\mathbf{r})\, \Delta(\mathbf{r})\, K\phi_m(\mathbf{r})\, d\mathbf{r} = \langle n | \Delta(\mathbf{r})\, K | m \rangle \tag{6.66}$$

and (6.65) becomes

$$\varepsilon_n = \xi_n + \sum_{m \neq n} \frac{|\langle n | \Delta(\mathbf{r})\, K | m \rangle|^2}{\xi_n + \xi_m}. \tag{6.67}$$

One is led to investigate the properties of the matrix element (6.66) to determine the validity of the expansion (6.67).

Two cases which are qualitatively different may be investigated.

(a) There is no magnetic contribution in the normal metal hamiltonian ($\mathbf{A} \equiv 0$). In this case K will commute with T, i.e.

$$\left[\frac{p^2}{2m} + U\right] K = K \left[\frac{p^2}{2m} + U\right].$$

The eigenfunctions $\phi_n(\mathbf{r})$ of this hamiltonian are very complicated, but as first remarked by Anderson (1959)‡ it is possible to build up a "Cooper pair" state by considering the two functions $\phi_n(\mathbf{r})$ and $K\phi_n(\mathbf{r})$ which corre-

---

† The operator K changes the function $\phi_{n\alpha}$ into $\phi_{n-\alpha}^*$
‡ See also in Chapter 5 the derivation of the Ginzburg–Landau equations in the dirty case.

Reversible Properties

spond to the same energy ξ_n. The perturbation expansion (6.67) can be written as

$$\varepsilon_n = \xi_n + \frac{|\langle n | \Delta(\mathbf{r}) | n \rangle|^2}{2\xi_n} + \sum_{\substack{m \\ m \neq n \\ m \neq Kn}} \frac{|\langle n | \Delta(\mathbf{r}) K | m \rangle|^2}{\xi_n + \xi_m} \quad (6.68)$$

In general the mean value $\langle n | \Delta(\mathbf{r}) | n \rangle$ is not zero so that the expansion diverges for $\xi_n \to 0$. In particular in the B.C.S. case where $\Delta(\mathbf{r}) = \Delta$, the last summation is equal to zero and

$$\varepsilon_n = \xi_n + \frac{\Delta^2}{2\xi_n} \quad (6.69)$$

which is the form to be expected from (6.58b) when $\Delta \ll \xi_n$. Thus when K commutes with T, i.e. when there are no magnetic effects, the system will generally present a gap equal to the average of $\Delta(\mathbf{r})$ as seen by an electron at the Fermi level.

(b) K does not commute with the hamiltonian T. In this case one has to examine the matrix element (6.66) more carefully. Following de Gennes and Tinkham (1964), it is convenient to introduce the power spectrum $I(\omega)$ defined by

$$I(\omega) = \overline{\sum_m |\langle n | \Delta(\mathbf{r}) K | m \rangle|^2 \delta(\xi_m - \xi_n - \omega)}, \quad (6.70)$$

where the average is taken over the states $| n \rangle$ of fixed energy ξ_n. The Fourien transform of $I(\omega)$ has a simple form, namely,

$$I(t) = \int_{-\infty}^{+\infty} I(\omega) e^{i\omega t} d\omega = \overline{\Delta^*[\mathbf{r}(0)] K(0) \Delta[\mathbf{r}(t)] K(t)}, \quad (6.71)$$

where $K(t)$ and $\mathbf{r}(t)$ are the Heisenberg operators which describe the motion of K and $\mathbf{r}$ in the normal state.

At large distances the motion of $\mathbf{r}$ is ruled by a classical transport equation. The motion of $K(t)$ can be obtained from its equation of motion, namely,

$$\frac{dK}{dt} = i[T, K] = ie \left\{ \frac{\mathbf{p} \cdot \mathbf{A} + \mathbf{A} \cdot \mathbf{p}}{mc} \right\} K \quad (6.72)$$

and

$$K(t) = e^{i\phi(t)} K(0), \quad (6.73)$$

where

$$\phi(t) = \frac{2e}{mc} \int \mathbf{A} \cdot \mathbf{p} \, dt = \frac{2e}{\hbar c} \int \mathbf{A} \cdot d\mathbf{l} \quad (6.74)$$

so that

$$I(t) = \langle \Delta^*[\mathbf{r}(0)] e^{i\phi(t)} \Delta[\mathbf{r}(t)] \rangle, \quad (6.75)$$

the average being taken over all classical one electron trajectories.

By studying the behaviour of $I(t)$ for large values of t, it is possible to decide whether there will be a gap in the excitation spectrum or not. Two cases may be distinguished (de Gennes, 1964):

(α) $\lim\limits_{t \to \infty} I(t) = \eta \neq 0$ non ergodic situation,   (6.76a)

(β) $\lim\limits_{t \to \infty} I(t) = 0$ ergodic situation.†   (6.76b)

In case (α), $I(\omega)$ is singular for $\omega = 0$. The perturbation expansion (6.67) diverges for small values of ξ_n. There is a gap in the excitation spectrum. It is clear that the B.C.S. case which was discussed above belongs to case (α).

In case (β), $I(\omega)$ is not singular for $\omega = 0$. The two energies ξ_n and ξ_m appearing in (6.67) can never be equal and the perturbation series converges. The spectrum is not qualitatively different from the normal state spectrum. The superconductor is gapless.‡ We shall now review, without entering into the details, several examples.

(i) Surface superconductivity in parallel fields. In a semi-infinite medium $\Delta(\mathbf{r})$ varies from its maximum value close to the surface to zero at a distance $d \gtrsim \xi(T)$ inside the superconductor.

For $|t| \to \infty$, $\langle \Delta^*[\mathbf{r}(0)] \Delta [\mathbf{r}(t)] \rangle$ tends toward zero since the electron has a high probability of being in the core of the sample where $\Delta = 0$. It is therefore expected that $I(t)$ will tend toward zero when $t \to \infty$. The perturbation expansion (6.67) converges when $H \simeq H_{c3}$ and there is no gap in the energy spectrum.

The above remark is valid in the case of pure and dirty superconductors, but only for a semi-infinite medium. In the case of a slab $[d \ll \xi(T)]$ the situation is somewhat different; $|\Delta|$ is constant throughout the specimen and the problem is to investigate the behaviour of the phase $e^{i\phi(t)}$. This was done by de Gennes and Tinkham (1964) who have drawn the following conclusion:

for a pure superconductor with diffuse surface scattering it is found that

$$\lim_{t \to \infty} \overline{e^{i\phi(t)}} \neq 0 \text{ so that}$$

$$\lim_{t \to \infty} I(t) = \eta |\Delta|^2, \tag{6.77}$$

where

$$\eta = \begin{cases} 1 - \dfrac{\pi}{3} \dfrac{Hd^2}{\phi_0} & H \ll \dfrac{\phi_0}{d^2} \\[2ex] \dfrac{2\phi_0}{\pi H d^2} & H \gg \dfrac{\phi_0}{d^2} \end{cases} \begin{matrix} (6.78\text{a}) \\[2ex] (6.78\text{b}) \end{matrix}$$

H is the applied field and ϕ_0 the flux quantum.

† For a discussion of the stochastic processes in physics, see Chandrasekhar, *Selected papers on noise and stochastic processes*. Wax, Ed., Dover, 1964.

‡ It is, however, necessary to determine the radius of convergence for the perturbation series (6.67).

Reversible Properties

This result arises from geometrical cancellations of the successive contribution to the phase. In the case of a dirty superconductor, however, the ergodic behaviour is restored, and in a certain range of fields, the dirty films are gapless superconductors (Maki, 1964a). In particular for $d < l < \sqrt{(\xi_0 d)}$, where d is the sample thickness, the film should be gapless for

$$0.95\, H_{//} < H < H_{//}$$

(de Gennes and Tinkham, 1964)†.

(ii) Superimposed films of normal and superconducting metals in zero field. This situation has been studied by de Gennes and Mauro (1966). At infinite times if one assumes diffuse reflection and transmission at the interfaces

$$\lim_{t\to\infty} I(t) = |\bar{\Delta}|^2, \tag{6.79}$$

where $|\bar{\Delta}|$ is the average of $\Delta(\boldsymbol{r})$ as seen by an electron at the Fermi level. In general a gap will exist in the excitation spectrum.

(iii) Dirty type II superconductors in high fields (close to H_{c2} or H_{c3}). This is an ergodic case and the calculation can be carried out very simply. If $(H_{\max} - H)/H_{\max} \ll 1$ ($H_{\max} = H_{c2}$ or H_{c3}), $I(t)$ take the following form:

$$I(t) = \langle |\Delta(\boldsymbol{r})|\rangle^2 e^{-|t|/\tau(T)}, \tag{6.80}$$

where $\tau(T)$ is a time which depends only on the temperature. This special behaviour is known as a markoffian ergodic situation.

This remarkable form arises from the fact that, in dirty superconductors, the behaviour of the electron is governed by a diffusion process (cf. Chapter 5). $\tau(T)$ can be obtained from (5.98) for $\Delta(\boldsymbol{r})$ and it is found that

$$\frac{1}{\tau(T)} = \frac{2\pi D H_{c2}(T)}{\phi_0}, \tag{6.81}$$

where $H_{c2}(T)$ is given by (5.117).‡ Thus the measurement of $\tau(T)$ yields a determination of $H_{c2}(T)$, i.e. of the Maki coefficient $\kappa_1(T)$. It is seen that $1/\tau(T)$ tends linearly toward zero when $T \to T_0$ and towards a constant $\Delta_{\text{B.C.S.}}/\hbar$ when $T \to 0$. The eigenvalue (6.67) takes the form

$$\varepsilon_n = \xi_n + \frac{2\langle \Delta^2\rangle \xi_n}{2\xi_n^2 + [\hbar/\tau(T)]^2}. \tag{6.82}$$

† A general study of the gapless situation in films has been made by Guyon, Meunier and Thompson (1966).

‡ In a semi-infinite medium, and close to H_{c3}, $H_{c2}(T)$ must be replaced by 1·69 $H_{c2}(T)$.

This formula is valid as long as $\Delta\tau(T)/\hbar \ll 1$, i.e. $(H_{c2} - H)/H_{c2} \ll 1$. Using the above expression for the excitation spectrum, de Gennes (1964) has obtained the density of states at a point r in the material. It is found that

$$\frac{N(r,\varepsilon)}{N_n(0)} = 1 + \frac{2|\Delta(r)|^2[\tau(T)]^2}{\hbar^2} \frac{[2\varepsilon\tau(T)/\hbar]^2 - 1}{\{[2\varepsilon\tau(T)/\hbar]^2 + 1\}^2}, \quad (6.83)$$

where $N_n(0)$ is the density of states in the normal material at the Fermi

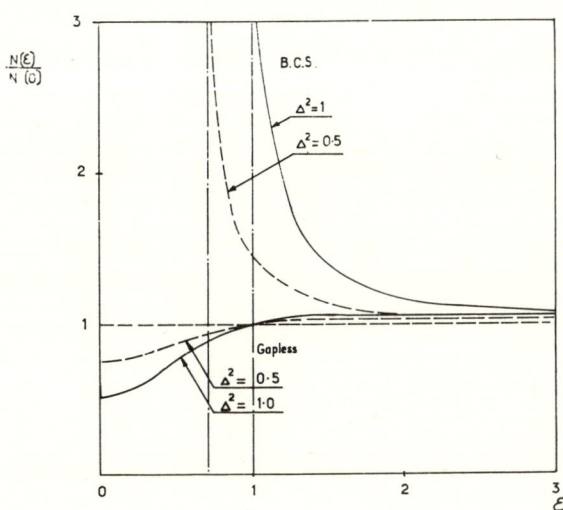

Fig. 6.13. The density of states at point r for a markoffian ergodic gapless superconductor for two different values of the pair potential $|\Delta(r)|$ and for $\hbar/2\tau(T) = 1$. The corresponding density of states for the B.C.S. case with $\Delta = |\Delta(r)|$ is also shown on this diagram

level. The variation of $N(r,\varepsilon)/N_n(0)$ (for a given r) is shown in Fig. 6.13. It is seen that there is no gap, but that, the number of available states in a band $\pm \hbar/2\tau(T)$ close to the Fermi level is strongly reduced. At a finite temperature the effect of the discontinuity of the energy gap in the B.C.S. case is reduced by the existence of thermal excitations. This means that the tunnelling characteristic observed in the B.C.S. case and in the gapless superconductivity mentioned above are not, qualitatively, very different.

Equation (6.83) relates the value of the density of states at point r to the order parameter at the same point. Actually, there are contributions arising from a domain of radius $\xi(T)$ around the point r, but these additions can be expressed in terms of $\Delta(r)$. It is to be remarked that tunnelling experiments in high fields will yield the value of $\Delta(r)$ at the interface of the tunnelling junction. For energies $\varepsilon \gg \hbar/\tau(T)$, $N(\varepsilon)/N_n(0)$ tends towards $(1 + \Delta^2/2\varepsilon^2)$ which is the B.C.S. asymptotic form. The tunnelling characteristic gives an average of $N(\varepsilon)/N(0)$ in a width $k_B T$. Thus if $k_B T \gg \hbar/\tau(T)$ the experiment cannot choose between the B.C.S and the markoffian ergodic gapless situation.

In fields smaller than H_{max} there is still a domain in which a gapless situation exists. However, the markoffian ergodic gapless superconductivity is limited to the vicinity of H_{max}. The boundary between gapless supercon-

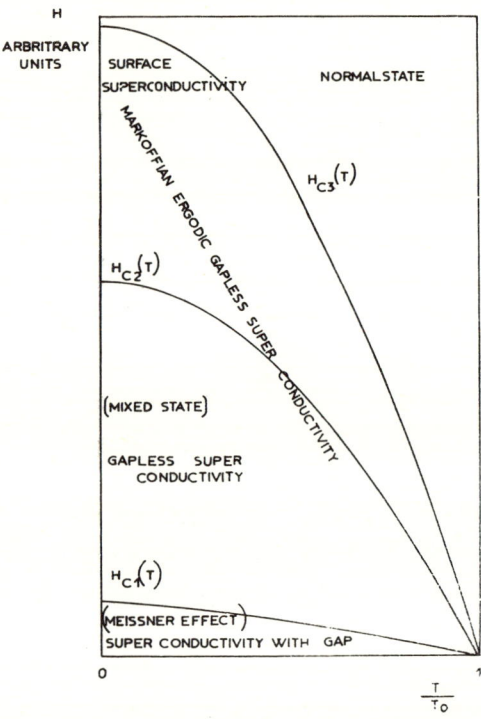

Fig. 6.14. Schematic representation of the domain of existence of the various types of superconductivity in type II superconductors (after Guyon, 1966)

ductivity and superconductivity with a gap is not known. It is probable that this boundary corresponds to the first entrance of vortex lines, i.e. to $H = H_{c1}$. A schematic representation of the various domains is given in Fig. 6.14.

Guyon et al. (1965) and Guyon (1966) have measured the tunnelling characteristics in the markoffian gapless region and the results are shown in Fig. 6.15. It is seen that the agreement between theory and experiment is very good. In particular the experimental points cannot be fitted to a B.C.S. type curve whatever the value of the gap chosen (see also Fig. 6.16).

(iv) Pure type II superconductors in a high magnetic field. This case is difficult to study by tunnelling experiments since it is difficult to prepare tunnel junctions in which the superconductor remains "clean". However it is interesting from a theoretical point of view, because, unlike the dirty case, it leads to gapless superconductivity close to H_{c2} which is not of the markoffian type. De Gennes and Mauro (1965) have shown that, in this case,

$$\lim_{t \to \infty} I(t) = \text{constant} \left[\frac{\xi(T)}{\hbar v_F t} \right]^2 \tag{6.84}$$

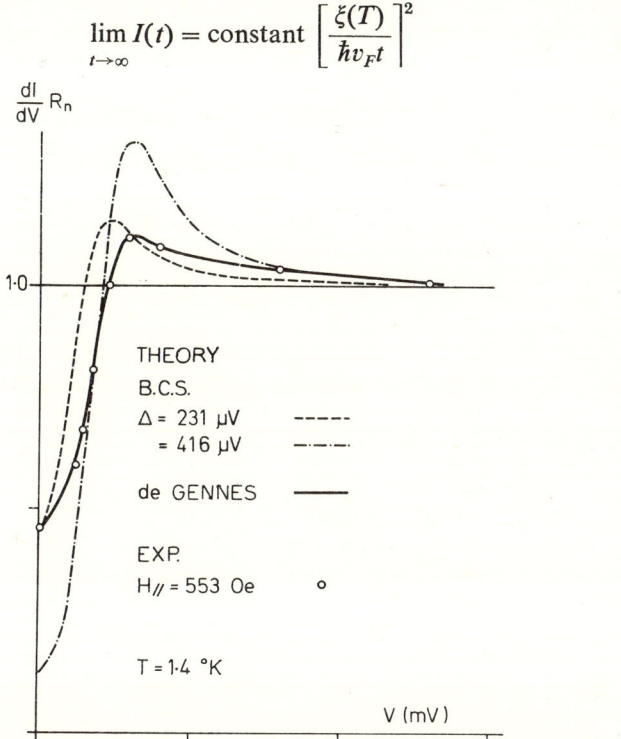

Fig. 6.15. Tunnelling characteristics in the markoffian ergodic gapless region (after Guyon et al., 1965). The points represent the measured ratio of the differential conductivity in the superconductor (Sn – 5% In) to the differential conductivity in the normal state, for an applied field of 553 Oe parallel to the sample at $T = 1\cdot 4°$ K, as a function of the applied voltage. The heavy line represents the theory deduced from (6.83). The light lines represent the B.C.S. tunnelling curves having the same initial slope (a), or the same position for the maximum of dI/dV (b). Note that curve (a) is not in agreement with the voltage at which the maximum occurs. In fact, no value of the effective gap Δ exists for which the B.C.S. characteristic is in reasonable agreement with experiment

where v_F is the Fermi velocity and $\xi(T)$ is the temperature dependent coherence length. This form arises from the fact that the electrons travel in straight lines across the vortex lattice so that the phase of $\Delta(\mathbf{r})$ oscillates in a more or less random fashion and $I(t)$ must therefore tend towards zero. $\xi(T)$ appears

Reversible Properties

in (6.84) because close to H_{c2} the distance between two vortices is of order $\xi(T)$ (cf. Chapter 3). The corresponding density of states has been computed by de Gennes and Mauro (1965) and the reader is referred to their work for further details.

Close to H_{c1} and for $\kappa \gg 1$ the low energy excitations can be calculated exactly (Caroli et al., 1964; Caroli and Matricon 1965). In that case it is also found that the system is gapless.

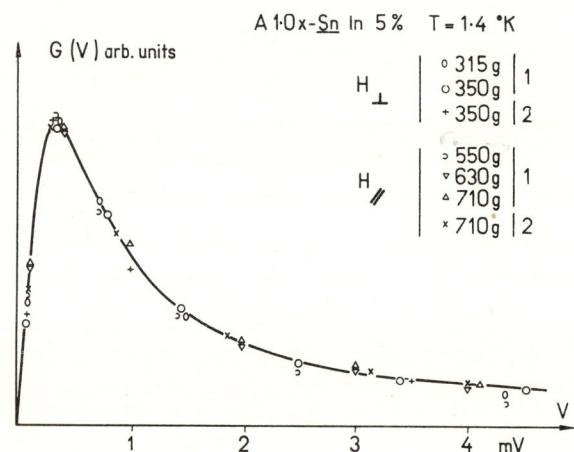

FIG. 6.16. Variation of the function $G(V) = [I_n(V) - I_s(V)] R_n$ where $I_n(V)$ is the tunnelling conductance in the normal state, $I_s(V)$ is this conductance in the superconducting state, and R_n the resistance in the normal state. According to the theory deduced from (6.83) the normalized $G(V)/G(V_{max})$ is independent of both the magnitude and the orientation of the field. The heavy line represents the theoretical variation. Note that the shape of $G(V)$ obtained for various samples of Sn – 5% In, various fields and orientations, is independent of these parameters as predicted (after Guyon et al., 1965)

6.II.3. Conclusions

It has been shown in the preceding section that the existence of a gap in the energy spectrum is not an essential feature of superconductivity and that the main property is the existence of an order parameter, the so-called pair potential $\Delta(r)$. Several types of gapless superconductivity have been studied here in which the gapless situation arises from the effect of the magnetic field but we have not considered the effect of magnetic impurities. In this case the time reversal operator K does not commute with the hamiltonian, since it is necessary to introduce spin dependent contributions into T. The usual situation is one in which $|\Delta|$ is constant and $I(t) = |\Delta|^2 \langle K(0) K(t) \rangle$. The exchange coupling between conduction electrons and localized impurities leads to

$$\langle K(0) K(t) \rangle \neq 1.$$

Type II Superconductors in High Fields

This situation has been investigated by de Gennes and Sarma (1963) and by Abrikosov and Gorkov (1960) and they find that

$$I(t) = |\Delta|^2 e^{-2|t|/\tau_s}. \qquad (6.85)$$

τ_S is the collision time between electrons and impurities due to exchange coupling. Equation (6.85) predicts markoffian gapless superconductivity in which the gapless situation exists in a limited domain of impurity concentration. If C_r is the concentration at which the transition temperature in zero field drops to zero, this domain is

$$0\cdot 95\, C_r < C < C_r.$$

This situation is difficult to analyse because of the impurity–impurity interactions and of the existence of the Kondo effect. Some experiments have been performed on such alloys by Reif and Woolf (1962).

The existence of the gapless region has an influence on several thermodynamic properties of superconductors. Its effect on the thermal conductivity of type II superconductors in high fields has been studied by Maki (1964b), Ambegaokar and Griffin (1965) and by Caroli and Cyrot (1966).

Bibliography for Chapter 6

Part I

ABRIKOSOV, A. A. and GORKOV, L. P. (1962), *Zh. Eksperim. I Teor. Fiz.* **42**, 1088 English translation: *Soviet Phys. JETP*, **15**, 752 (1962)].
ANDERSON, P. W. (1959), *Phys. Rev. Letters*, **3**, 325.
BALTENSPERGER, W. (1958), *Suppl. Physica*, **24**, S, 153.
BERLINCOURT, T. G. and HAKE, R. R. (1963), *Phys. Rev.* **131**, 140.
CAPE, J. A. (1966), *Phys. Rev.* **148**, 257.
CAROLI, C., CYROT, M. and DE GENNES, P. G. (1966), *Solid State Com.* **4**, 17.
CHANDRASEKHAR, B. S. (1962), *Appl. Phys. Letters*, **1**, 7.
CLOGSTON, A. M. (1962), *Phys. Rev. Letters*, **9**, 266.
FERRELL, R. A. (1959), *Phys. Rev. Letters*, **3**, 262.
FULDE, P. and FERRELL, R. A. (1964), *Phys. Rev.* **135** A, 550.
GORKOV. L. P. (1965) *Zh. Eksperim i Teor. Fiz.* **48**. 1772. English translation: *Soviet Phys.* JETP, **21**, 1186.
GRUENBERG, L. W. and GUNTHER, L. (1966), *Phys. Rev. Letters*, **16**, 996.
HAKE, R. R. (1967) *Phis. Rev.* To be published.
KIM, Y. B., HEMPSTEAD, C. F. and STRNAD, A. R. (1965), *Phys. Rev.* **139**, A, 1163.
MAKI, K. (1964), *Physics*, **1**, 127; (1966) *Phys. Rev.* **148**, 362.
SAINT-JAMES, D. (1966), *Phys. Letters*, **23**, 177
SARMA, G. (1963), *J. Phys. Chem. Solids*, **24**, 1029.
SARMA, G. and SAINT-JAMES, D. (1964), *Communication to the Conf. on the Phys. of Type II Superconductivity*, Western Reserve University, Cleveland, (Ohio)
SHAPIRA, Y. and NEURINGER, L. J. (1965), *Phys. Rev.* **140A**, 1638.
STRNAD, A. R. and KIM, Y. B. (1965), *Proceedings of the Symposium on Quantum Fluids*, University of Sussex.
WERTHAMER, N.R., HELFAND, E. and HOHENBERG, P. C. (1966), *Phys. Rev.* **147**. 1, 295.

Part II

ABRIKOSOV, A. A. and GORKOV, L. P. (1960), *Zh. Eksperim. I Teor. Fiz.* **39**, 1781 [English translation: *Soviet Phys. JETP*, **12**, 1243 (1961)].
AMBEGAOKAR, V. and GRIFFIN, A. (1965), *Phys. Rev.* **137** A, 1151.
ANDERSON, P. W. (1959), *J. Phys. Chem. Solids*, **11**, 26.
BARDEEN, J., COOPER, L. N. and SCHRIEFFER, J. R., (1957), *Phys. Rev.* **108**, 1175.
CAROLI, C. (1966), Thesis. Orsay University.
CAROLI, C. and CYROT, M. (1965), *Phys. Cond. Matt.* **4**, 285.
CAROLI, C. and MATRICON, J. (1965), *Phys. Cond. Matt.* **3**, 380.
CAROLI, C., DE GENNES, P. G. and MATRICON, J. (1964), *Phys. Letters*, **9**, 307.
DE GENNES, P. G. (1964), *Phys. Cond. Matt.* **3**, 79; (1966a) *Superconductivity of Metals and Alloys*, Benjamin, New York; (1966b) *Many Body Theory*, Kubo Ed. Benjamin, New York.
DE GENNES, P. G. and MAURO, S. (1965), *Solid State Com.* **3**, 381.
DE GENNES, P. G. and SARMA, G. (1963), *J. Appl. Phys.* **34**, 1380.
DE GENNES, P. G. and TINKHAM, M. (1964), *Physics*, **1**, 107.
GIAEVER, I. (1960), *Phys. Rev. Letters*, **5**, 147 and 464 also: (1961), *Phys. Rev.* **122**, 1101.
GUYON, E. (1966), Thesis. Orsay University.
GUYON, E., MARTINET, A., MATRICON, J. and PINCUS, P. (1965), *Phys. Rev.* **138**, A, 746.
GUYON, E., MEUNIER, F. and THOMPSON, R. S. (1966), To be published in *Phys. Rev.*
LYNTON, E. A. (1964), *Superconductivity*, Methuen and Co., London.
MAKI, K. (1964a), *Prog. Theor. Phys.* **31**, 731; (1964b) *Prog. Theor. Phys.* **31**, 378.
REIF, F. and WOOLF, M. (1962), *Phys. Rev. Letters*, **9**, 315.

CHAPTER 7

Thermal Properties

THE mechanisms of specific heat and thermal conductivity in type II superconductors have received relatively little attention in recent years and are still incompletely understood from a quantitative theoretical view point. The experimental features of the variation of the thermal properties with temperature and magnetic field are well established, however, and this chapter will be confined to a qualitative description of these effects.

7.1. Specific Heat

Early experiments made by Mendelssohn and Moore (1935) established that the variation of the specific heat c_s with temperature of a type II superconductor is similar to the variation seen with type I which is reasonably described, at least at temperatures away from the absolute zero, by the prediction of the Bardeen, Cooper, Schrieffer theory (1957) (B.C.S.) that

$$c_s/\gamma T_0 = a \exp\{-b/t\},\qquad(7.1)$$

where γ is the Sommerfeld constant, T_0 the critical temperature, t the reduced temperature T/T_0, and a and b are constants. Strictly this is an equation for the electronic specific heat but at low temperatures the contribution of the lattice is very small and is usually negligible.

The results of a more recent measurement of the variation of specific heat with temperature of V_3Ga in a number of different externally applied magnetic fields is given in Fig. 7.1. The form of the variation shown is that usually found with type II superconductors, e.g. Nb (Brown *et al.*, 1953), Va (Corak *et al.*, 1956) and V_3Ga (Hake, 1964). The results of Fig. 7.1 are of particular interest since they indicate that a magnetic field as high as 70 kOe has very little effect in reducing the superconductivity of V_3Ga. In this case it was found that the specific heat is independent of the previous magnetic history of the sample and not affected by trapped flux even though the magnetisation of the sample is very irreversible (see Chapter 8). Goodman (1962) has pointed out that since the specific heat is reversible it can be used to estimate the reversible magnetization characteristic of an irreversible sample by applying the equations of reversible thermodynamics.

Reversible Properties

For low κ materials the variation of the specific heat with temperature in a constant magnetic field can be rather different from that illustrated in Fig. 7.1. The variation of specific heat with temperature for a niobium specimen ($\kappa \sim 1$) in a field of 1030 Oe is shown in Fig. 7.2. It is seen here

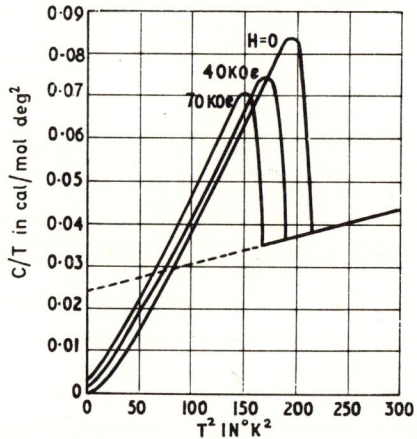

FIG. 7.1. The variation of the specific heat of V_3Ga with temperature in magnetic fields of 0, 40 and 70 kOe (from Morin *et al.*, 1962)

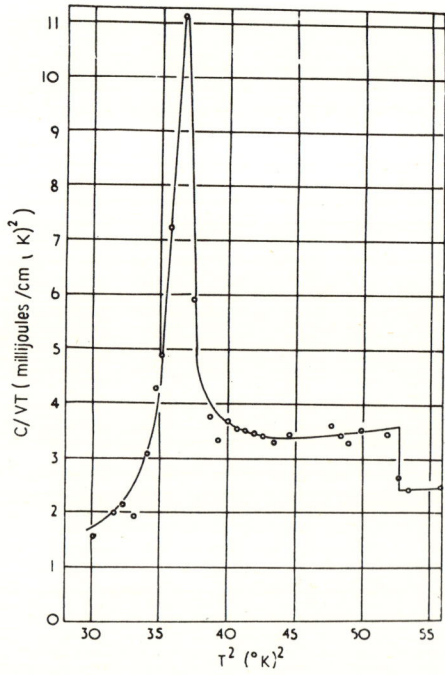

FIG. 7.2. The variation of the specific heat of Nb with temperature in a magnetic field of 1030 Oe (from McConville and Serin, 1965)

Thermal Properties

that as well as the sharp increase in specific heat at the normal-superconducting transition there is, at a lower temperature, a very sharp peak. This specific heat peak is associated with the change in free energy when magnetic flux enters the sample in increasing field and occurs at about the temperature at which the applied field coincides with the lower critical field of the sample. It should, in principle, be possible to decide from the variation of the specific heat close to H_{c1} whether the transition is of first or second order. However the choice between these two possibilities is very difficult and McConville and Serin were not able to decide this question by measurements on the specimens available to them. As quoted in Chapter 1, Serin (1965) has made careful measurements of the magnetization in niobium close to H_{c1} and has proved that the transition at H_{c1} is of second order. McConville and Serin have also reported measurements on Nb–Ta and In–Bi alloys. There is clear evidence for a peak in the specific heat at $H = H_{c1}$ in the latter alloy and they conclude that it is reasonable to infer that all type II superconductors should, ideally, exhibit a λ-point at the lower critical field.

The specific heat data may be used to determine the parameters $\kappa_1(T)$ and $\kappa_2(T)$ in a temperature range where the magnetization measurements are inaccurate. At a given temperature T_H the specific heat discontinuity at the normal-superconducting transition occurs at an applied field $H = H_{c2}(T_H)$. If the thermodynamic field $H_c(T_H)$ is known, the parameter $\kappa_1(T)$ can be obtained from

$$H_{c2}(T) = \kappa_1(T)\sqrt{2}\, H_c(T). \tag{7.2}$$

In order to determine $H_c(T)$, McConville and Serin have used the thermodynamic relation between the specific heat discontinuity, the field and the induction at the transition. According to (1.44), the discontinuity of the specific heat per unit volume at a second order transition is

$$\Delta C = \frac{T}{4\pi}\left(\frac{dH}{dT}\right)^2 \left[\left(\frac{\partial B_i}{\partial H}\right)_T - \left(\frac{\partial B_j}{\partial H}\right)_T\right], \tag{7.3}$$

where the i and j are related to the two different phases. At $T = T_0$, $B \equiv 0$ in the superconducting state and $B = H$ in the normal state so that

$$\left(\frac{\Delta C}{T}\right)_{T_0} = \frac{1}{4\pi}\left(\frac{dH_c}{dT}\right)^2_{T=T_0}, \tag{7.4}$$

where H_c is the thermodynamic critical field. Assuming that the variation of $H_c(T)$ with temperature is†,

$$H_c(T) = H_c(0)\left[1 - \frac{T}{T_0}\right]^2, \tag{7.5}$$

† This assumption leads to an error of a few per cent.

Reversible Properties

one obtains the value of $H_c(0)$, and hence of $\kappa_1(T)$ from the measurement of the jump in the specific heat at $T = T_0$.

The parameter $\kappa_2(T)$ is defined by

$$\left(\frac{dM}{dH}\right)_{H_{c2}} = \frac{1}{4\pi\beta_A[2\kappa_2^2(T) - 1]}, \tag{7.6}$$

where $(dM/dH)_{H_{c2}}$ is the magnetization slope at $H = H_{c2}$, and β_A the geometrical factor related to the symmetry of the vortex lattice introduced in Chapter 3.† The magnetization slope is readily obtained from the specific heat discontinuity at $H = H_{c2}$ since according to (7.3):

$$\left(\frac{\Delta C}{T}\right)_{T_H} = \left(\frac{dH_{c2}}{dT}\right)_{T_H}^2 \left(\frac{dM}{dH}\right)_{H_{c2}} \tag{7.7}$$

and the values of $\kappa_2(T)$ deduced from (7.6). McConville and Serin obtained the values of $\kappa_1(T)$ and $\kappa_2(T)$ in niobium and their results have already been given in Fig. 5.8. The specific heat measurements have yielded the values close to T_0 and the conventional magnetization measurements the values in the low temperature range.

The reader is referred to Chapter 5 for further details on the comparison of the variation of $\kappa_1(T)$ and $\kappa_2(T)$ versus temperature with the existing theories.

7.2. Thermal conductivity

Heat is conducted in metals and alloys by electrons and phonons. In order to understand the behaviour of heat conduction it is necessary to study the way in which these two carriers may be scattered in their motion through the sample. [For more detail, see Rosenberg (1963).] The electrons are scattered both by phonons and by static imperfections (impurities, stacking faults, grain boundaries, etc.). According to Mathiessen's rule, it can be assumed that the contribution of these two mechanisms to the electronic thermal resistivity W_e are additive so that:

$$W_e = W_i + W_0, \tag{7.8}$$

where $W_i = \alpha T^2$ is the phonon scattering contribution and $W_0 = \beta/T$ is the static imperfection contribution.‡ These forms for W_i and W_0 are valid only at sufficiently low temperatures, i.e. T/θ_D small, where θ_D is the Debye

---

† $\beta_A = 1.1596$ for a triangular lattice. McConville and Serin have used the value 1.18 which is correct for a square lattice. This yields a slight error in the κ_2 values.

‡ Note that the variation of W_0 with temperature is the same as that of residual resistivity ρ_0 in agreement with the Wiedeman–Franz law.

Thermal Properties

temperature. It is clear that at low temperatures the dominant term in W_e will be W_0.

The phonons in the sample are scattered by different mechanisms, which in order of increasing effectiveness as the temperature is increased are:

(1) the scattering at the specimen boundaries which gives a contribution of the form BT^{-3};
(2) the scattering by electrons which gives a contribution of the form ET^{-2};
(3) the scattering by dislocations which is of the form DT^{-2};
(4) the scattering by point defects which is of the form PT;
(5) the scattering by the other phonons which is of the form $gT^n e^{-\theta/mT}$

[for more detail see Klemens (1958)].

The total phonon resistivity is then

$$W_g = BT^{-3} + ET^{-2} + DT^{-2} + PT + gT^n e^{-\theta/mT}. \qquad (7.9)$$

In well-annealed metals and alloys at low temperature the main contribution comes from the electron–phonon scattering and the variation of W_g with temperature will be of the form γT^{-2}. The total thermal conductivity of a metal or an alloy is then written as

$$K = K_e + K_g, \qquad (7.10)$$

where

$$K_e = \frac{1}{W_e} \quad \text{and} \quad K_g = \frac{1}{W_g}.$$

It is clear that, at low temperatures, in normal metals and alloys the main contribution arises from the electrons so that $K \simeq T/\beta$.

7.2.1. *Variation of the thermal conductivity with temperature*

In superconductors, the electrons involved in the condensation process are no longer able to carry heat and, unlike the electrical resistance, the thermal resistance of the electrons should tend towards infinity, i.e. $K_e \to 0$. Thus in the superconducting state only the phonon contribution to the thermal conductivity should be observed.

At first sight it would appear that the thermal resistance of a superconductor should be far higher in the superconducting than in the normal state and this has been observed in several pure metals. Indeed, superconducting wires are used as thermal switches which are "open", i.e. non-conducting in the superconducting phase, and "closed" when superconductivity is destroyed by means of a magnetic field (Reese and Steyert, 1962).

Reversible Properties

However, in the condensation process the scattering between electrons and phonons is also strongly reduced so that the phonon conductivity is modified and, in principle, increased. It is not at present possible to predict the extent to which each resistivity mechanism will affect the thermal conductivity in any particular sample so that no quantitative theoretical description of the net thermal conductivity exists as yet.

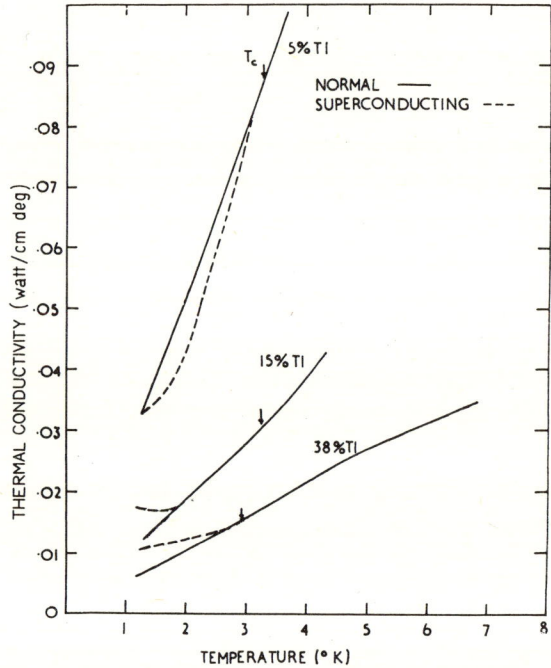

FIG. 7.3. The variation of thermal conductivity with temperature for In – 5% Tl, –15% Tl and –38% Tl in the normal and superconducting states (from Sladek, 1955)

Different behaviours can be observed which are illustrated in Fig.7.3, where the variation of thermal conductivity of In–Tl alloys with temperature and concentration is shown. Increasing alloy concentration decreases the thermal conductivity and at the same moment there is a gradual change of the relative position of the normal and superconducting curves. In high concentration alloys the thermal conductivity in the superconducting state may even exceed that of the normal state. This was observed by de Haas and Bremmer in Pb–Tl (1932), by Mendelssohn and Olsen (1950) in Pb–Bi, and by Lindenfeld (1961) in Pb–In, and seems to be a general feature of alloys. This is roughly explained by the fact that, in high concentration alloys, the phonon contribution to the thermal resistance in the normal state is no longer negligible and that in the superconducting state the phonon conductivity is increased.

Thermal Properties

The electronic contribution to the thermal conductivity in the superconducting phase, though possibly small, is not negligible. It can be calculated in the frame of the B.C.S. theory (1957). This was done by Bardeen, Rickaysen and Tewordt (1959) (B.R.T.) who have used a simple Boltzmann equa-

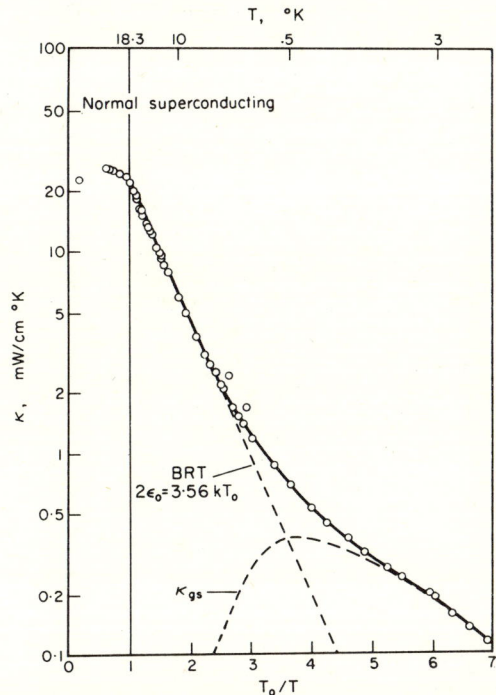

FIG. 7.4. The thermal conductivity of Nb_3Sn as a function of the inverse reduced temperature (from Cody and Cohen, 1964)

tion to describe the transport properties of the superconductor. The ratio of the electronic thermal conductivity in the superconducting phase to that in the normal state is given by

$$\frac{K_{es}}{K_{en}} = \frac{\int_{\varepsilon_0}^{\infty} \varepsilon^2 (\partial f/\partial \varepsilon)\, d\varepsilon}{\int_{0}^{\infty} \varepsilon^2 (\partial f/\partial \varepsilon)\, d\varepsilon}, \qquad (7.11)$$

where f is the Fermi function and ε_0 is the gap in the quasi particle excitation spectrum. The comparison of the B.R.T. formula with experiment is not always easy because it is necessary to remove the phonon contribution from the experimental results. However, Cody and Cohen (1964) have observed that close to T_0 in Nb_3Sn, the thermal conductivity follows the B.R.T. prediction fairly well. At low temperatures, it is found that the phonon part

Reversible Properties

is predominant and that the variation of K with temperature is closely described by a T^3 law indicating that the phonon conductivity is limited by crystal boundaries (Fig. 7.4).

7.2.2. *Variation of the thermal conductivity with magnetic field*

The situation is less simple when a magnetic field is applied and the observations as well as the theories become much more involved. We shall

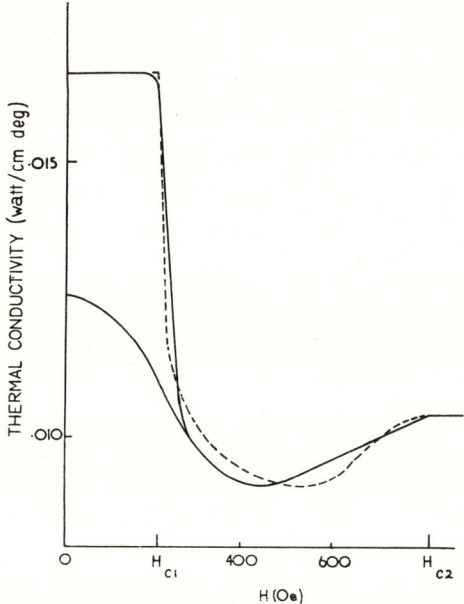

FIG. 7.5. The variation of the thermal conductivity of In -3% Bi with magnetic field at $2 \cdot 16°$K. The solid lines show the experimental results in increasing and decreasing field. The dashed line is that calculated from the magnetization (from Dubeck *et al.*, 1963)

concentrate here on the behaviour of type II superconductors. The usual form of the variation of the thermal conductivity is shown in Fig. 7.5. It is seen that in an increasing field there is a very sharp drop at H_{c1} followed by a gradual rise to the normal state value which is reached at H_{c2}. In a decreasing field the same curve is followed until close to H_{c1} where it breaks away, rising to a lower thermal conductivity value than the initial one showing the presence of trapped flux.

In order to interpret this variation Dubeck *et al.*, (1963 and 1964) have used a tempting and simple procedure which consists of retaining the B.R.T. equation (7.11) and inserting a field dependent gap $\varepsilon_0(H)$. This is the "effective gap approximation". Actually, Dubeck *et al.*, have assumed that $\varepsilon_0(H)$ is proportional to the square root of the magnetization, since the mean square order parameter in the Ginzburg, Landau, Abrikosov theory is proportional to the magnetization in the vicinity of H_{c2} (see sections 3.3 and 3.4). The resulting variation of the electronic thermal conductivity with field as predicted by this calculation is shown by the dotted line in Fig. 7.5 and is seen to be in quite good agreement with the experimental results.

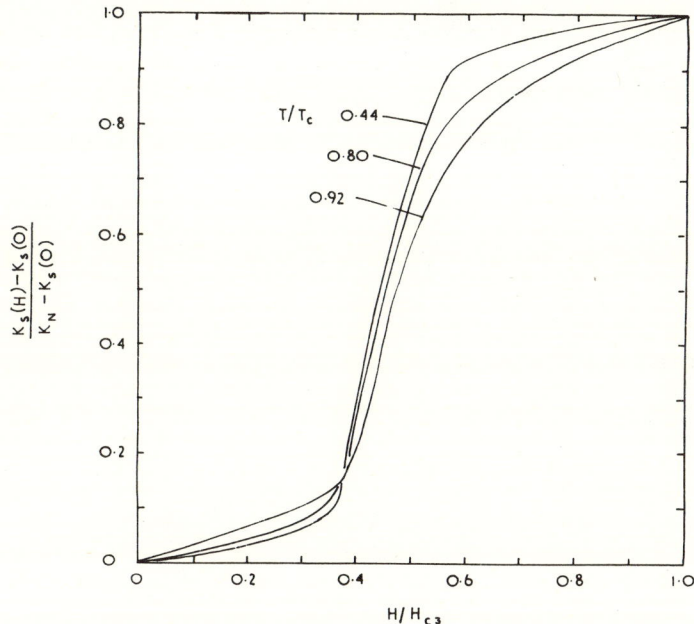

FIG. 7.6. The reduced thermal conductivity versus parallel magnetic field for a 15,200 Å film of In −5% Pb film for various temperatures ($T_0 = 3.75°$K). Note that the slope close to H_{c3} is finite, (from Mochel and Parks, 1966)

This agreement, however, is somewhat deceptive since the effective gap approximation is open to serious objections. As was shown in the second part of Chapter 6, type II superconductors in high fields are actually "gapless" superconductors and, for example, the density of states cannot be calculated in an effective gap model. It must therefore be suspected that the agreement between the theory of Dubeck et al. and experiment is fortuitous.

The behaviour of K can be understood qualitatively by considering the phonon and electron contributions. At $H = H_{c1}$ vortex lines appear in the sample. As was shown in Chapter 3 (sections 3.3 and 3.4) their density per unit surface area increases very quickly in the vicinity of H_{c1}. This results in a scattering of phonons by vortices and in a rapid decrease of the thermal conductivity associated with phonons. This is the so-called "phonon dip". For higher values of the field the phonon thermal resistivity increases steadily with the density of lines and reaches the normal state value at $H = H_{c2}$.

Reversible Properties

At the same moment the electronic resistance should decrease with the entry of flux into the sample. The combination of the two effects yields qualitatively the observed behaviour. It may be remarked that close to the critical field the phonon contribution to the conductivity is expected to be negligible in comparison to the electronic one.

The validity of this explanation is indirectly demonstrated by Mochel and Parks (1966) who did not observe the phonon dip in the vicinity of H_{c1} when studying films of In–7% Pb and Sn–6% In alloys (Fig. 7.6). Due to the very fine structure of the films studied, the phonon mean free path is reduced to a point where the electron–phonon scattering is unimportant so that no phonon dip effect is seen. This means that, in the superconducting state, the variation of the thermal conductivity with magnetic field can be attributed only to excitations from the superconducting ground state.

The experiments were performed by Mochel and Parks to check the theoretical predictions of Caroli and Cyrot (1965). These two authors have computed the electronic thermal conductivity in the gapless region. Their calculation is restricted to the vicinity of the critical field (H_{c2} or H_{c3}) where the order parameter $\psi(r)$ is small, and K_{es} is calculated to order $\langle |\Delta|^2 \rangle$ for *dirty* superconductors. As shown in section 5.4e, the pair potential $\Delta(r)$ still obeys a linearized equation of the form

$$\frac{1}{\tau K} \Delta = D \left(\nabla - \frac{2ieA}{\hbar c} \right)^2 \Delta, \qquad (7.12)$$

where D is the diffusion coefficient. τ_K may be considered as a lifetime of electron pairs and is given by

$$\log \frac{T}{T_0} = \psi\left(\frac{1}{2}\right) - \psi\left[\frac{1}{2} + \frac{\hbar}{4\pi k_B T \tau_K}\right], \qquad (7.13)$$

where $\psi(Z)$ is the digamma function. The ratio of the electronic thermal conductivities in the superconducting and the normal state is found to be

$$\frac{K_{es}}{K_{en}} = 1 - \frac{6\hbar^3}{\pi^2} \beta^3 \int_0^\infty d\omega \, \omega^3 \frac{1}{\cosh^2(\hbar\beta\omega/2)} \frac{\langle |\Delta|^2 \rangle \tau_K^2}{(4\omega^2 \tau_K^2 + 1)^2}, \qquad (7.14)$$

where $\beta = 1/k_B T$ and $\langle |\Delta|^2 \rangle$ is the space average of the squared pair potential.

The most interesting prediction of this calculation is that, close to H_{c2} (or H_{c3}), the slope of the thermal conductivity dK/dH is finite, unlike the effective gap approximation which predicts a vanishing slope. The former was actually observed close to H_{c2} by Dubeck *et al.*, and can be seen in Fig. 7.5.† Mochel and Parks have made a more systematic study of this property

† It should be noted that close to the critical field the phonon contribution is negligible.

close to H_{c3}. Their results as well as the theoretical curve calculated from Caroli and Cyrot theory are shown in Fig. 7.7. It is seen that the agreement is fairly good. Incidentally, these results give new evidence of the gapless nature of type II superconductors. Lindenfeld *et al.* (1966) have also measured the thermal conductivity of indium and lead–bismuth alloys. In In–2% Bi the agreement with theory is very good while for other concentrations, the variation of the thermal conductivity slope at $H = H_{c2}$, if it follows the gener-

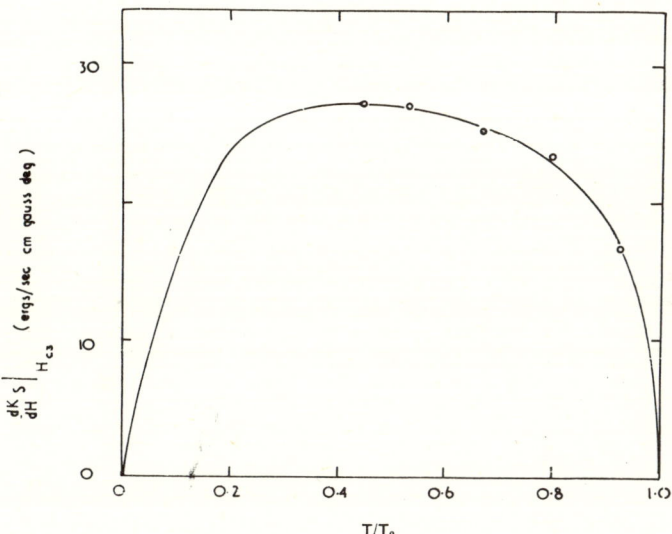

FIG. 7.7. Comparison of the experimental results for the temperature dependence of $(dK_s/dH)_{H_{c3}}$ for a 15,200 Å In –5% Pb film [circles] compared with the theory of Caroli and Cyrot [full line] (from Mochel and Parks, 1966)

al behaviour predicted by the theory, differs appreciably from the absolute value computed by Caroli and Cyrot (Fig. 7.8). No explanation of this discrepancy exists at the moment.

At lower fields the situation is very complicated. It is expected that the vortex cores are better thermal conductors than the outer regions. This implies an anisotropic effect depending on the respective orientation of the field and of the thermal gradient.† However, the observation of this anisotropy should be made in the absence of phonon contribution which can mask the phenomenon of interest. This was probably the case in the experiments by Dubeck *et al.*, where an effect of 3% or less was observed when the heat flow was tilted from parallel to perpendicular to the vortex lines direction.

For fields lower than H_{c1}, the type II superconductor is in the Meissner state and the B.R.T. relation should be applicable.

† Note that a small anisotropic effect may also exist for the phonons.

Reversible Properties

Some experiments have been carried out in thin films, mainly by Morris and Tinkham (1964). They have found an increase of the conductivity with field, the variation being proportional to H^2. In that case the effective gap approximation is not in good agreement with the observed data.

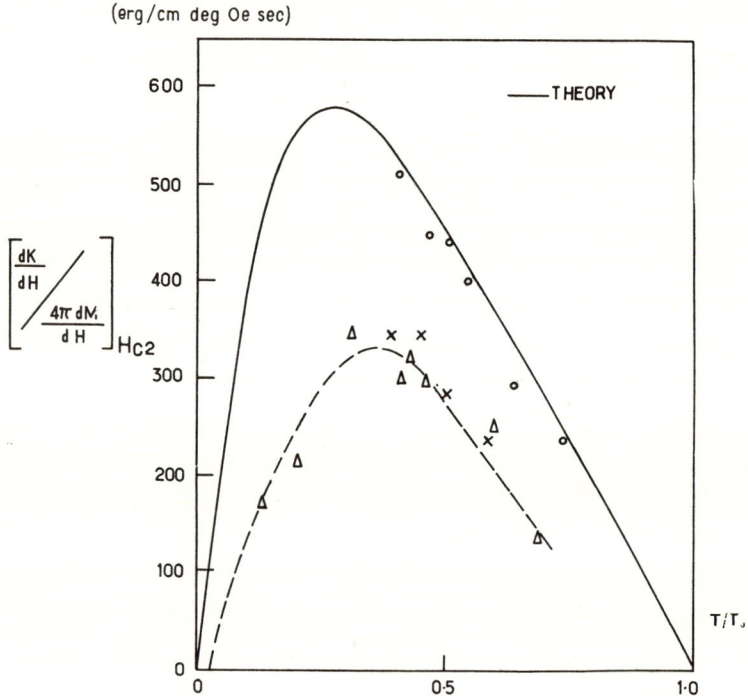

FIG. 7.8. Variation of the slope of the thermal conductivity in In–Bi alloys (after Lindenfeld et al.). The points for the 2% Bi alloy (o) are almost exactly on the theoretical line, while for more impure specimens ($\triangle$, $\times$), if they show the trend of the theoretical curve, they are lower in magnitude by about a third

Maki (1964) has calculated in detail the thermal conductivity for a "dirty" thin film. The results are different from the effective gap approximation. Unless the pair potential can be considered as constant in the specimen, there is no gap in the excitation spectrum for certain field values. However, as the order parameter is different from zero the thermal conductivity differs from the normal value. No comparison of Maki's calculation with experiment exists at the moment. A different, but related, situation has been considered by Ambegaokar and Griffin (1965). In this case, instead of a magnetic field, the perturbing agent is a set of magnetic impurities. As pointed out in Chapter 6 this leads to a gapless situation even if the pair potential Δ is constant throughout the specimen. As in the Caroli and Cyrot treatment one can introduce an electron pair lifetime τ_K and the thermal conductivity is a universal function of Δ and τ_K.

Bibliography for Chapter 7

ABRIKOSOV, A. A. (1957), *Sov. Phys. JETP* **5**, 1174.
AMBEGAOKAR, V. and GRIFFIN, A. (1965), *Phys. Rev.* **137A**, 1151.
BARDEEN, J., COOPER, L. N. and SCHRIEFFER, J. R. (1957), *Phys. Rev.* **108**, 1175.
BARDEEN, J., RICKAYSEN, G. and TEWORDT, T. L. (1959), *Phys. Rev.* **113**, 982.
BROWN, A., ZEMANSKY, M. W. and BOORSE, H. A. (1953), *Phys. Rev.* **92**, 52.
CAROLI, C. and CYROT, M. (1965), *Phys. Cond. Matt.* **4**, 285.
CODY, G. D. and COHEN, R. W. (1964), *Rev. Mod. Phys.* **36**, 121.
CORAK, W. S., GOODMAN, B. B., SATTERTHWAITE, C. B. and WEXLER, A. (1956), *Phys. Rev.* **102**, 656.
DE HAAS, W. J. and BREMMER, H. (1932), Leiden Comm. 220c (1932).
DUBECK, L., LINDENFELD, P., LYNTON, E. A. and ROHRER, H. (1963) *Phys. Rev. Letters*, **10**, 98; (1964) *Rev. Mod. Phys.* **36**, 110.
GOODMAN, B. B. (1962), *Phys. Letters*, **1**, 215.
GORKOV, L. P. (1960), *Sov. Phys. JETP*, **10**, 998.
HAKE, R. R. (1964), *Rev. Mod. Phys.* **36**, 124.
KLEMENS, P. G. (1958), *Solid State Physics*, vol. 7, p. 1. Seitz and Turnbull, Eds., Academic Press, New York.
LINDENFELD, P. (1961), *Phys. Rev. Letters*, **6**, 613.
LINDENFELD, P., LYNTON, E. A. and SOULEN, R. (1966), L. T. X Conference
LYNTON, E. A. (1964), *Superconductivity*, Chapter 9, Methuen.
MAKI, K. (1964), *Prog. Theor. Phys.* **31**, 378.
MCCONVILLE, T. and SERIN, B. (1965), *Phys. Rev.* **140A**, 1169.
MENDELSSOHN, K. and MOORE, J. R. (1935), *Proc. Roy. Soc.* **A152**, 39.
MENDELSSOHN, K. and OLSEN, J. L. (1950), *Proc. Phys. Soc. (Lond.)*, **A63**, 2.
MOCHEL, J. M. and PARKS, R. D. (1966), *Phys. Rev. Letters*, **16**, 1156.
MORIN, F. J., MAITA, J. P., WILLIAMS, H. J., SHERWOOD, R. C., WERNICK, J. H. and KUNZLER, J. E. (1962), *Phys. Rev. Letters*, **8**, 275.
MORRIS, G. and TINKHAM, M. (1964), *Phys. Rev.* **134A**, 1154.
REESE, W. and STEYERT, W. A. (1962), *Rev. Sci. Inst.* **33**, 43.
ROSENBERG, H. M. (1963), *Low Temperature Solid State Physics*, Chapter 5, Oxford University Press.
SERIN, B. (1965), *Phys. Letters*, **16**, 112.
SLADEK, R. J. (1955), *Phys. Rev.* **97**, 902.

PART II

IRREVERSIBLE PROPERTIES

FOR
ERICA
AND
KATHERINE GRACE

E. J. THOMAS

CHAPTER 8

Flux Trapping and the Irreversible Magnetization Characteristic

It has already been mentioned (Chapter 3) that when the external magnetic field is raised above the lower critical value, H_{c1}, of a type II superconductor the full Meissner effect no longer occurs and lines of magnetic flux penetrate into the bulk of the material. In general, when the magnetic field is removed again some of the flux remains trapped within the bulk. The lowest field at which flux trapping occurs may be identified with the field at which flux first penetrates [H_{c1}, if the surface of the superconductor is not smooth (see section 8.3)]. Finding the minimum field at which flux is first trapped may be used as a convenient way of observing H_{c1} experimentally (Hecht, 1964).

The trapping of flux in rings of both type I and type II superconductors is an expected and familiar phenomenon but it must be emphasized that, in type II superconductors, flux trapping occurs also in singly connected specimens. The possibility that these may contain a multiconnected mesh which causes the trapping is considered in section 8.2.

As a consequence of the ability to trap flux, the magnetic behaviour of type II superconductors is generally irreversible. This is shown particularly in the variation of the magnetization with field and will be considered in detail later in the chapter.

8.1. Observation of trapped flux

An idea of the flux distribution within a sample may be gained by examining the flux which emerges at the surface. This can be done in a number of ways (see Alers, 1957) but undoubtedly the most satisfactory method is that proposed by Alers using glass containing paramagnetic cerous nitrate. This glass causes a particularly marked rotation of the plane of plane polarized light in the presence of a magnetic field (Faraday effect). A sheet of the glass is placed on the smooth surface of a superconductor and illuminated with plane polarized light. This surface is viewed through a Nicol set to extinction with respect to the incident light so that the surface is dark.

Irreversible Properties

The arrangement is shown in Fig. 8.1a. If a magnetic field is now applied, flux penetrates into the superconductor, and at the points on the surface where the flux emerges, the plane of polarization of the light is rotated and a bright region appears as shown in Fig. 8.1b. Because of local fields in the glass the resolution using this method is limited to about 0·2 mm and

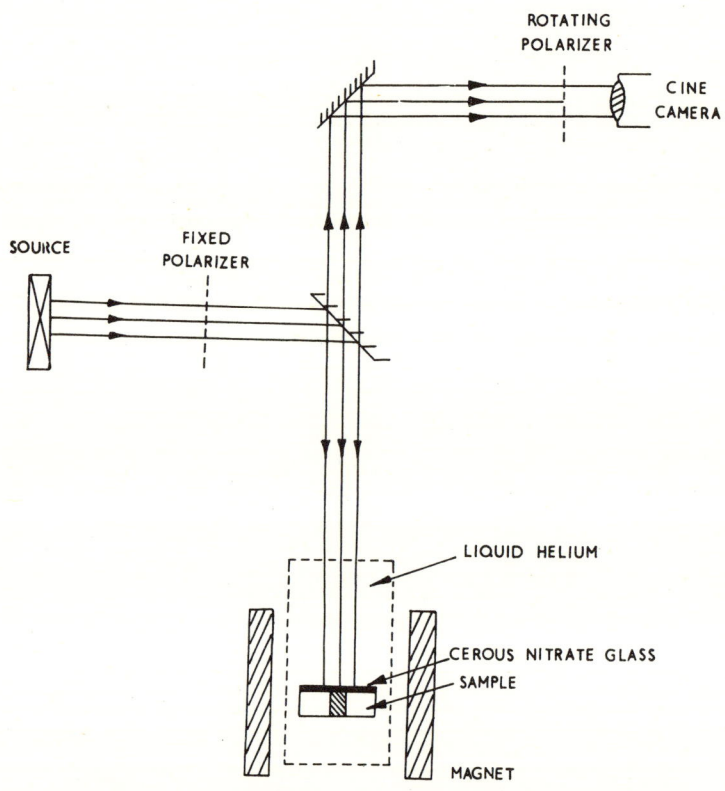

FIG. 8.1a. Arrangement of the apparatus used to examine the penetration of flux into a superconductor using the Faraday effect. The rotating polarizer is set to extinction when the sample is superconducting. Bundles of flux penetrating into the sample are seen as white spots on the sample surface

it is not possible to observe single flux quanta or the structure of the mixed state using this technique. However, these experiments do show that at least some of the flux trapped inside the material is present in bundles.

The presence of single flux quanta (vortex lines) trapped in the bulk may be detected using an interferometric technique which depends upon the macroscopic quantum interference effect which occurs in a Josephson junction of the type shown in Fig. 8.2a. The supercurrent flowing through an arrangement of this type depends upon the magnetic flux within the area S

Flux Trapping

and varies with a period ($hc/2e$), the quantum of flux. By monitoring this current as a type II superconducting wire containing trapped flux is moved past, a trace showing the pattern of the flux in the wire as a function of position is obtained. An actual trace is shown in Fig. 8.2b (Zimmerman and

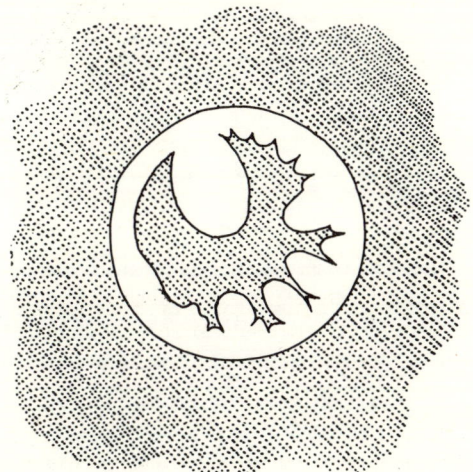

FIG. 8.1b. The appearance of a well-annealed Nb –10% Ta disc in a field of 500 Oe (drawn from De Sorbo and Healey, 1964). See also frontispiece.

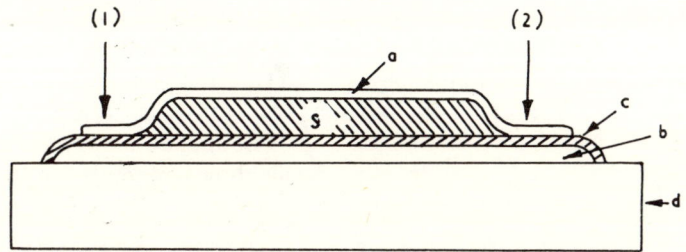

FIG. 8.2a. Cross-section of a Josephson junction. Junction area S is enclosed between thin ($\sim$ 1000 Å) tin films, a and b, separated by a thin oxide layer, c. Current flow is measured between films a and b (from Jaklevic et al., 1964)

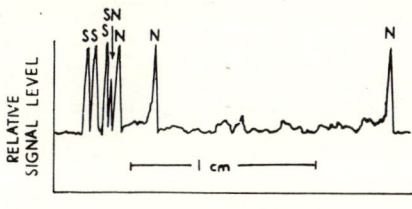

FIG. 8.2b. Junction current as a function of wire position for a niobium wire. S and N indicate the direction of the emerging field (from Zimmerman and Mercereau, 1964)

Mercereau, 1964). The maxima in this curve correspond to the size expected when a single vortex line threads the junction.

It is, therefore, seen that, in irreversible type II superconductors, the flux is trapped both as isolated vortex lines† and in flux bundles.

8.2. Trapping in a filamentary system

The property of a superconducting ring that flux is trapped within it in a changing external field suggests that there may be a multiconnected system of thin filaments with critical field greater than that of the surrounding material inside the body of the superconductor (Mendelssohn, 1935). Many people have identified this *sponge structure* with the dislocation network which exists in the sample.

A prediction of the variation of magnetization with field for a material of this type may be made assuming a filamentary structure capable of sustaining lossless currents up to the critical current density, J_c (Bean, 1962a). Within the mesh is a type I superconductor with critical field H_c. For simplicity it is assumed that the critical current is independent of the field (which is equivalent to assuming that the applied fields are much less than the critical field of the filaments). The magnetization M is defined by the equation

$$4\pi M = \frac{\int (h-H) d\tau}{\int d\tau}, \tag{8.1}$$

where both integrals are over the volume of the specimen, h is the internal field, H the external applied field and τ is the volume of the specimen. To perform the integration h must be stated in terms of H and r for the different field regions of interest. For mathematical convenience the specimen is taken to be a long rod of radius a_c with field applied along its axis. Then, if the applied field is less than H_c, the shielding is complete if a_c is much greater than the penetration depth and

$$h = 0; \quad 0 \leq r \leq a_c, \quad 0 \leq H \leq H_c. \tag{8.2}$$

As the field is raised above H_c the soft superconductor on the outside becomes normal and a current is induced in the outer filaments. By Ampère's law, this will flow to a depth δ from the surface where

$$\delta = 10(H - H_c)/4\pi J_c,$$

i.e. a macroscopic, field dependent, penetration depth exists. With increasing field δ becomes larger and reaches a maximum when $\delta = a_c$, the radius of the cylinder. The external field is then $(H^* + H_c)$ where $H^* = 4\pi J_c a_c/10$

---

† Individual flux lines have recently observed using an electron microscope technique [U. Essman and H. Traüble, *Phys. Letters* 24A, 526 (1967). These results show that the flux line lattice contains both "elastic" and "plastic" defects. Among the latter type seen are the flux line equivalents of dislocations, stacking faults and point defects.

Flux Trapping

and the internal field in the next stage of the magnetization is

$$h = 0; \quad 0 \leq r \leq a_c[1 - (H - H_c)/H^*]$$
$$h = H - H^*(1 - r/a_c); \; a_c[1 - (H - H_c)/H^*] \leq r \leq a_c \quad \Bigg\} \; H_c \leq H \leq H^* + H_c$$
(8.3)

Here $(H - H_c)/H^* = \delta/a_c$ so that the first condition is over the range $r = 0$ to $r = (a_c - \delta)$ and the second is over the range $r = (a_c - \delta)$ to $r = a_c$; these are written in terms of field so that they may be used in the integration. At this stage all the bulk properties have been destroyed so that only filamentary currents flow in the superconductor and the condition is

$$h = H - H^*(1 - r/a_c); \quad 0 \leq r \leq a_c, \; H^* + H_c \leq H. \quad (8.4)$$

Putting (8.2), (8.3) and (8.4) into (8.1) and integrating gives

$$4\pi M = -H \quad\quad\quad\quad\quad\quad\quad\quad\quad\quad\quad\quad\quad 0 \leq H \leq H_c$$

$$4\pi M = -H + \left(\frac{H^2 - H_c^2}{H^*}\right) + \frac{[H_c^2(3H - 2H_c) - H^3]}{3H^{*2}} \quad H_c \leq H \leq H^* + H_c$$

$$4\pi M = -H^*/3 \quad\quad\quad\quad\quad\quad\quad\quad\quad\quad\quad\quad H^* + H_c \leq H$$

Figure 8.3a shows the magnetization curves predicted by this theory for different values of the filamentary critical current, i.e. different values of H^*. The case $H^* = 0$ corresponds to a type I superconductor.

The theory shows that in the low field region ($0 \leq H \leq 2 H_c$) the shapes of the magnetization curves are similar to those found by experiment. No agreement is expected, or found, at higher values of the applied field where the assumption of a field independent critical current no longer holds.

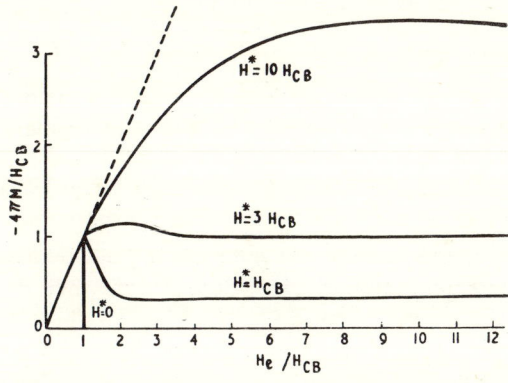

FIG. 8.3a. Magnetization curves for a low cylindrical filamentary superconductor predicted by Bean's theory. H_{cB} is the critical field of the material in the interstices and H^* is a measure of the critical current of the filaments (from Bean, 1962a)

Irreversible Properties

When Vycor glass is leached at an intermediate stage in its manufacture a porous body of silica is left with interconnected pores having a mean radius of about 30 Å. If these pores are evacuated and then mercury forced into them, samples are produced which have a form similar to that assumed in the previous calculation. Measurements of $M(H)$ on such samples (Fig. 8.3b) show the characteristic variation of an irreversible type II superconductor (Bean, 1962b).

While the filamentary model is almost certainly valid for this special case, it fails to account for the initial field penetration at fields less than H_c which occurs in type II superconductors and also for the flux trapping caused by point defects when no multiconnected structure exists.

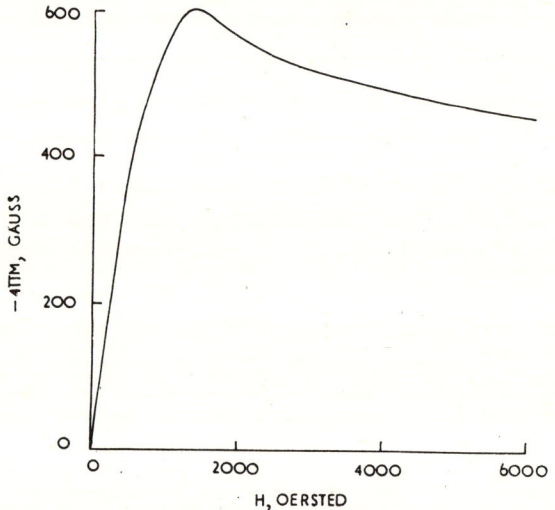

Fig. 8.3b. Experimental magnetization curve of a long cylindrical filamentary superconductor made by pressing mercury into Vycor glass (from Bean, 1962b)

8.3. Flux trapping in a singly connected system

Experimental magnetization characteristics differ at least slightly from that predicted by Abrikosov (see Section 3.3) for some hysteresis occurs in the magnetisation of all type II superconductors in the region H_{c1} to H_{c2}. For low κ materials hysteresis may also be detected in some samples in the field region between H_{c2} and H_{c3} (Sandiford and Schweitzer, 1964). This hysteresis is caused by the persistent current induced in the superconducting sheath by the changing external magnetic field (Jones and Park, 1966). The effect can be reduced significantly if the surface is scratched to impede the current flow.

One cause of hysteresis in large samples is the size effect: the area under the magnetization curve, $\int M dH$, increases with increasing cross-section area of the sample.

Flux Trapping

The primary cause of hysteresis, however, is the presence of lattice defects (dislocations, point defects and particles of a second phase) within the body of the superconductor. This is clearly illustrated in Fig. 8.4 for a Pb–8.23 wt% In specimen (Livingston, 1963). In the annealed state the sample shows little hysteresis and practically no trapped flux. As a result of swaging, however, in which a high dislocation density and a high density of different point defects are produced, there is considerable hysteresis and trapped flux

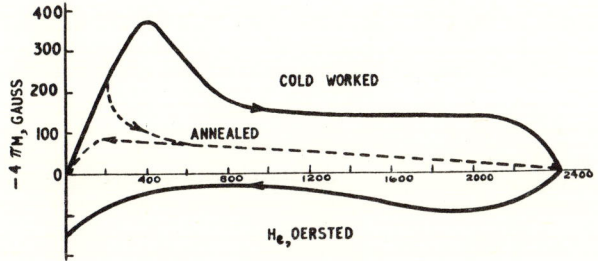

Fig. 8.4. The result of extreme cold work on the magnetization of Pb −8·23 wt.% In alloy (from Livingston, 1963)

when the field is reduced to zero. Subsequent annealing to remove the defects removes also the hysteresis and the initial characteristic can be regained.

By making measurements on metallurgical systems for which details of particular defect structures are known, it is possible to examine the effects of some specific lattice defects on the magnetisation characteristic. The

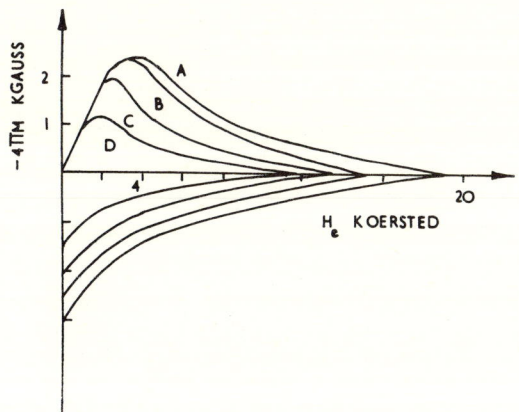

Fig. 8.5. The variation of the magnetization of the Pb − Bi eutectic at 4·2°K showing the effect of ageing at room temperature for various times. (a) as quenched (b) 1 day, (c) 5 days, (d) 19 days (from Evetts *et al.*, 1964)

Irreversible Properties

effect of changing concentration and size of a normal precipitate has been followed using chill-cast Pb–Bi eutectic samples (Evetts *et al.*, 1964). In the as-cast form there is a very fine random distribution of normal bismuth particles in a type II matrix for which $\kappa \sim 12$. Ageing the specimens causes spheroidation of the bismuth with the number of particles decreasing and an increase in the average size. The variation of magnetization with ageing is shown in Fig. 8.5. It is seen that considerable flux trapping occurs even though no multiconnected system exists in the bulk.

Fast neutron irradiation, which produces a number of different types of point defects but which has little effect on the dislocation density also causes an increase in the hysteresis which is approximately proportional to the neutron dose (Swartz *et al.*, 1964).

Even with the most highly annealed samples, however, it is not possible to remove the hysteresis near H_{c1} completely and it is probable that the lack of reversibility in this region has another cause. Close to H_{c1} there is a net force on the vortex line which opposes the penetration of the flux and the flux will not enter the surface until an extra field to overcome this barrier is applied. A simple calculation of the order of magnitude of this barrier field for high κ materials has been performed by Bean and Livingston (1964) and by de Gennes (1965). Considering a single flux line parallel to the surface at a distance x_L outside it, the London equation can be applied everywhere, except within the vortex core of radius $\xi(T)$. The line is subject to two forces due to:

(a) the external field H which has penetrated a distance λ within the superconductor. This repels the flux line with a force of magnitude

$$F_1 = \frac{1}{c} \phi_0 J(x_L),$$

where J is the current density induced by the external field and ϕ_0 the flux quantum. In the London approximation this current density is

$$J(x) = \frac{cH}{4\pi\lambda(T)} e^{-x/\lambda(T)};$$

(b) the image of the line (located at $x = -x_L$) tending to attract the line towards the surface:

$$F_2 = -\frac{1}{c} \phi_0 J_L(2x_L).$$

$J_L(x)$ is the current density due to a line at a distance x from the core. For small x the superelectron velocity is

$$v = \frac{\hbar}{2mx}.$$

where m is the mass of an electron and the current density is

$$J_L(x) = n_s ev = \frac{e\hbar n_s}{2mx}.$$

The smallest distance, x_L, at which the equations may be applied is $\xi(T)$. The line will enter if $F_1 \geq |F_2|$ and the barrier field H_s is given when $F_1 = |F_2|$, i.e. $(\phi_0/c) J[\xi(T)] = (\phi_0/c) J_L[2\,\xi(T)]$ at $H = H_s$.
If $\kappa \gg 1/\sqrt{2}$, $\lambda(T) \gg \xi(T)$ and one obtains

$$\frac{cH_s}{4\pi\lambda(T)} = \frac{e\hbar n_s}{4m\xi(T)},$$

i.e.

$$H_s = \frac{\phi_0}{4\pi\lambda(T)\xi(T)} \simeq H_c.$$

Equations (1.15) for $\lambda(T)$ and (3.13) for ϕ_0 have been used in deriving the expression for H_s. It is clear that the coefficient of proportionality between H_s and H_c is uncertain since the choice of $x_L = \xi(T)$ is rather arbitrary. Matricon (1966) has considered the more realistic case of a "half loop" instead of an infinite vortex line. His results are similar to the above conclusions. Experimentally, the barrier field H_s seems to be very close to H_c in a highly polished Nb–Ta alloy investigated by de Blois and de Sorbo (1964) ($H_c = 310$ Oe, $H_s = 320$ Oe at $4\cdot2°$K).

It may be remarked that when H is of the order of H_c the value of the order parameter is strongly reduced near the surface, even before the vortex has appeared. This effect has been discussed by de Gennes (1965). It is relatively easy to study in the Ginzburg–Landau scheme with $\kappa \gg 1$ since the problem is one dimensional. For a bulk superconductor it is found that $H_s = H_c$ exactly.† The predicted behaviour of a thin film, however, is quite different.

Surface irregularities and end effects cause local field concentrations along the surface and at these "weak" points flux may penetrate so that, in an increasing field (i.e. flux moving into the specimen) the effect of the barrier is reduced. When the external field is lowered, however, the flux inside the specimen is in the same direction as in the applied field and these same irregularities will oppose flux escape and the proportion of trapped flux will be increased.

These predictions have been confirmed using smooth and scored samples of Pb–Tl alloys (Joseph and Tomasch, 1964) and Nb and a Nb–Ta alloy (de Blois and de Sorbo, 1964).

† *Note added in proof.* Matricon and Saint-James have recently calculated the barrier field as a function of κ [Phys. Letters 24A, 241 (1967)] The barrier field is found to decrease from a value of the order of H_c for $\kappa = \dfrac{1}{\sqrt{2}}$ toward H_{c1} when κ increases.

Irreversible Properties

8.4. Pinning points

The fact that flux is trapped within a type II superconductor containing point defects and no multiconnected structure has led to the idea of flux *pinning points* throughout the body of the superconductor. The physical reason for the pinning is not known for certain. One possible cause of the pinning by inclusions in the lattice is that an attractive image force exists at the superconductor-inclusion boundary. Order of magnitude calculations also suggest that the stress field associated with lattice defects is sufficient to explain the observed pinning.

The relative sizes of the flux line and the pinning points will be important in the determination of the actual pinning force since the effect of the core extends over a region of the order of the coherence length ξ within the superconductor. It is probable, therefore, that pinning points with a dia-

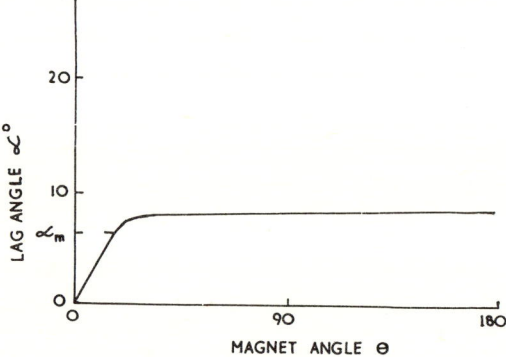

Fig. 8.6a. The angle between m and H measured as a function of the magnet angle for a field of 5 kOe (from Heise, 1964)

meter less than ξ will have little pinning effect. ξ varies from over 1000 Å in well-annealed niobium (Cribier *et al.*, 1964) to ~ 50 Å in the compound V_3Ga (Goodman, 1962), decreasing with increasing κ. The localized defects can vary from a few angstrom units diameter for a precipitate to within an order of the specimen size for a duplex structure. Because of the variation of the size of ξ the effectiveness of different defects to trap flux will vary in different materials.

An experimental estimate of the mean energy associated with pinning points can be made using the method suggested by Heise (1964). If a type II superconducting sample is suspended on a torsion wire and a magnetic field applied and then removed again flux will be trapped in the sample. A small field is reapplied and the magnet rotated through an angle θ. If the sample is now released, it too will rotate and come to rest at an angle α to its original position. The variation of α as a function of θ for a field of 5 kOe is shown in Fig. 8.6a. The value α_m, at which α first begins to deviate from θ, may be interpreted as the angle at which sufficient torque is exerted on the flux lines to overcome the pinning forces.

Flux Trapping

If the magnetic moment of the specimen is m then the torque in field H is

$$\tau = mH \sin \alpha$$

where α is the angle between the sample and the field. In terms of magnetic induction for a sample of volume V this is

$$\tau = BVH \sin \alpha. \tag{8.4}$$

The total number of vortex lines in the specimen of cross-sectional area S is

$$n = BS/\phi_0 \tag{8.5}$$

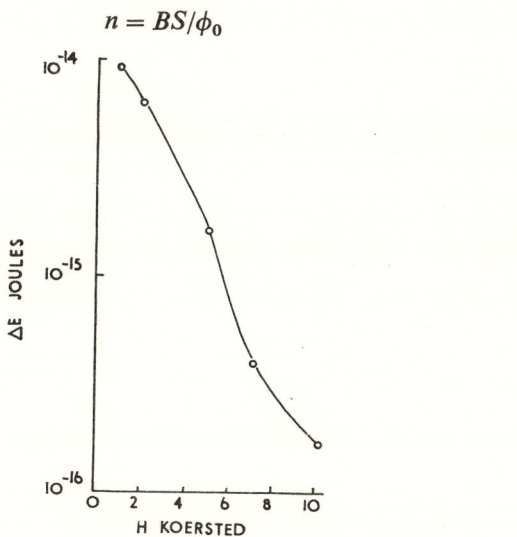

FIG. 8.6b. The variation of the energy available to pin one vortex line in a Nb–25%Zr wire with applied magnetic field (from Heise, 1964)

and the energy of all the pinning points in the specimen

$$E = \int_0^{\alpha_m} \tau d\tau .$$

So, using (8.4) and (8.5), the pinning energy per fluxoid

$$E_f = \phi_0 H l [1 - \cos \alpha_m] \tag{8.6}$$

for a sample of length l. The variation of E_f with field in Nb–25% Zr wire calculated using (8.6) is shown in Fig. 8.6b.

The force exerted on the vortices by the stress field round a spherical inclusion in the lattice is also a possible mechanism causing pinning (Toth and Pratt, 1964). If the critical temperature in this region is less than that of the rest of the lattice (or if the region is normal), the free energy in the region of the inclusion will be higher than in the rest of the lattice and it will have a repulsive effect on the superconducting electrons. Hence this position will be one of stable equilibrium for a vortex (which has a normal core). The elastic energy density u in the matrix surrounding a spherical inclusion of radius a in a lattice with shear modulus μ is

$$u = 6\mu a^4(a - r_s)^2/r^6,$$

where r_s is the radius the sphere would have if it contained the same number of stress free atoms with the same composition as the matrix. The interaction energy in the region arises from the change in energy due to the presence of the vortex line. Integrating over the volume of a vortex, the change in energy in the matrix in this volume is

$$U_m = 9\pi \Delta\, S_{44} \mu^2 a(a - r_s)^2,$$

where ΔS_{44} is the change in the elastic compliance between the normal and the superconducting region. There is also an energy contribution due to the strain in the sphere (given by Friedel, 1954),

$$U_s = \{6\pi a(a - r_s')^2/\chi\} \cdot (\Delta\chi/\chi)$$

where r_s' is the radius the sphere would have if it were in a stress free state, χ is the compressibility within the sphere and $\Delta\chi = \chi_n - \chi_s$.

Then the total energy U is

$$U = U_m + U_s.$$

The force F on a fluxoid is given by

$$F = dU/dx$$

and an approximate value of the mean force $\bar{F}$ can be found in this case by defining

$$\bar{F} \doteq \{U(b = 2a) - U(b = 0)\}/2a.$$

Substituting reasonable values for niobium [$\chi = 2 \times 10^{-12}$ cm^2/dyne; $\Delta\chi/\chi = -1 \times 10^{-6}$ (Alers and Waldorf, 1961), $|(a - r_s')/a| = \frac{1}{2}|(a - r_s)/a|$ and $a = 40$ Å] leads to $F \sim 10^{-2}$ dyn/cm and a strain energy of about 10^{-20} joules/pinning point.

A similar approach has been taken by Webb (1963) to estimate the interaction energy and force on a fluxoid due to dislocations. Using the general formulae given by Seraphim and Marcus (1962) for the change in critical field with stress, the changes with the tensor stress components of the free energy difference per unit volume between the normal and the superconducting state can be calculated. Treating this in the same way as the elastic energy density in the case of the point defects and integrating over the volume of a vortex line of radius R at distance w away from the dislocation gives the total interaction energy. In the case where the flux is perpendicular to the dislocation and in the limit $w \gg R$

$$U = - \Delta S_{44}(\mu b/2\pi)^2 \pi^2 R^2/2w \quad (w \gg R).$$

Here μ is the shear stress and b the Burgers vector. Using as an approximation for the differential (dU/dw) the average slope in U in the region $w = 0$ to $w = R$ (in order to avoid the nonphysical cut-off at the core of the dislocation)

$$F \sim - \Delta S_{44}(\mu b/2\pi)^2 \pi^2/d,$$

where d is the spacing of the pinning dislocations. The orders of magnitude of the pinning force and interaction energy are about the same as those predicted for pinning by inclusions in the lattice.

It will be seen in Chapter 9 that, when a current flows in the mixed state of a type II superconductor with pinning points, a Lorentz-type force is exerted on the vortex lines and the critical current corresponds to the force which is sufficient to unpin the vortices. Hence an order of magnitude estimate of the validity of F can be made by equating it to the Lorentz force and putting in a known value of the critical current density. In this case

$$J_c = 10F/\phi_0$$

and taking $J_c = 10^5$ A/cm^2 for Nb–25% Zr gives $d \sim 10^{-5}$ cm and, therefore, a dislocation density $\sim 10^{10}$ cm^{-2}. The dislocation density measured in these materials is $\sim 10^{11}$–10^{12} cm^{-2}. It is possible, however, that only a small proportion of the dislocations are in a suitable configuration to cause flux trapping.

Comparing these results with the experimental values found by Heise, assuming that these are the interactions which cause the pinning in his experiment, since the pinning energy of a single filament varies between 10^{-14} and 10^{-16} joules and the energy of a single pinning point is $\sim 10^{-20}$ joules, the number of pinning points on each flux line varies from 10^4 to 10^6. The effective length of Heise's sample was about 30 cm, so the spacing of these pinning points varied from 3×10^3 to 3×10^5 Å. Assuming the points themselves to be an order of magnitude smaller than this then it is seen that they vary in size from the order of ξ upwards as expected.

Since the interaction energy between a vortex line and a dislocation is negative we must expect that an elongated mesh of dislocations transverse to the fluxoid is necessary to provide a suitable barrier. This is certainly true in niobium where tangles of dislocations are much more effective in pinning flux (Narlikar and Dew-Hughes, 1964).

Irreversible Properties

The free energy of one vortex line in a field H is $-\phi_0 H/4\pi$ so the x component of the force per unit length

$$F_x = -\phi_0(\partial H/\partial x)/4\pi = -\phi_0(\partial H/\partial B)(\partial B/\partial x)/4\pi, \quad (8.7)$$

where $(\partial B/\partial x)$ is the gradient of the flux in the specimen and to a first approximation $(\partial H/\partial B)$ is the reciprocal of the gradient of the reversible $B(H)$

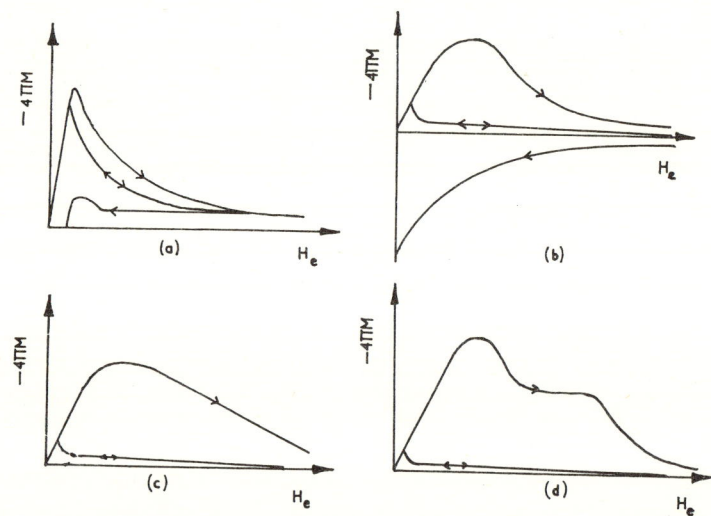

FIG. 8.7. Magnetization curves for four types of pinning centres calculated by Campbell *et al.* (1964): (a) for weak pinning; (b) for strong pinning; (c) closely spaced pinning centres; (d) a mixture of strong widely spaced centres and weak closely spaced centres

curve. Knowing F_x, the flux in the specimen can be calculated as a function of position for different values of external field and then integrated over the volume of the specimen to find the contribution $M(H)$ due to the presence of pinning points. A reasonable estimate of the field dependent hysteresis may be made by assuming that the pinning force is inversely proportional to the magnetic induction (Campbell *et al.*, 1964)

$$F = F_0 \phi_0 / Bl^2,$$

where F_0 is the pinning force on an isolated vortex line and l is the spacing between pinning points. Putting this into (8.7) and integrating leads to

$$B^2 - B_0^2 = (8\pi m F_0/l^2)x = \beta x, \quad (8.8)$$

where $m = (\partial H/\partial B)$, β is a constant and the boundary condition for the integral is $B = B_0$, the reversible value of the induction, at the surface of the specimen. Assuming reasonable values for the reversible $M(H)$ curve, the form of the hysteresis may be calculated. The irreversible $M(H)$ curves predicted from this equation for four cases of interest are shown in Fig. 8.7 together with the reversible curve from which the calculations were made.

These predictions have been compared with the experimental results on Pb–Bi already shown in Fig. 8.5. The resulting fit is shown in Fig. 8.8 where the values of the constants are chosen so that the maxima in the observed and calculated curves have the same value of magnetization. The fit at low field is seen to be quite good but presumably the assumption of an

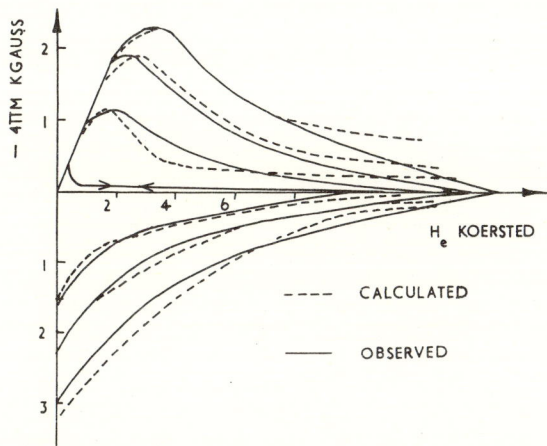

Fig. 8.8. The variation of magnetization with field for Pb–Bi alloys after annealing for various times compared with the theory due to Campbell *et al.* (1964). The reversible curve assumed as the basis of the calculation is that shown with the twin arrowheads (from Campbell *et al.*, 1964)

inverse proportionality between the pinning force and the induction is not valid when the vortex line density is high.

A similar result to (8.8) for the induction due to the hysteresis can be obtained by considering the interaction energy between the vortex lines in a triangular lattice when they are acted upon by pores in the lattice (Silcox and Rollins, 1963). Using this approach it is not necessary to assume the $F \propto 1/B$ relationship as a starting point in the calculation but a number of other simplifying assumptions (nearest neighbour interactions only, the decrease in the free energy of the lattice due only to the presence of the pores and that all the pores contribute to the pinning for all values of field between H_{c1} and H_{c2}) need to be made.

8.5. Variation of trapped flux with temperature

A systematic study of the variation of trapped flux with temperature has been made by Pippard (1955) using long rod samples of tin containing up to 3 wt% In. Characteristic results showing the effect of crystallographic

Irreversible Properties

perfection and of annealing are given in Fig. 8.9. Similar experiments have also been performed by Budnick *et al.* (1956) with tin (-Sb, -Bi, -In) samples. None of these latter results showed the drop in trapped flux to zero at $t = 1$. The samples used were well annealed, had bright surfaces and consisted of a few crystallites similar to Pippard's and it is not known why they failed to show the behaviour. A similar variation of trapped flux with temperature (with large values only at $t = 1$) is also found in rhodium (Doulat *et al.*, 1959). In this sample the proportion of trapped flux increased in a regular way with the neutron irradiation dose.

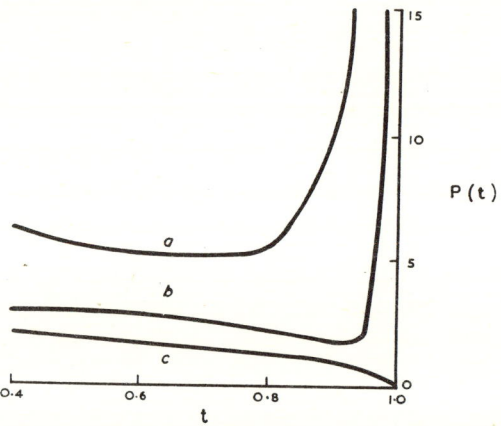

Fig. 8.9. The variation of the percentage of trapped flux with temperature in a Sn–1% In sample showing (1) the change in the characteristics between a polycrystalline specimen (a) and a monocrystalline specimen (b) and (2) the effect of annealing: (a) unannealed; (b) annealed for 44 hours; (c) annealed for 44 days (from Pippard, 1955). This material is type I but the changes which occur are similar to those in a type II material

No completely satisfactory explanation of the variation of the trapped flux with temperature exists but a simple prediction may be made assuming that the flux is present in a series of normal channels in the bulk of the material (Jurisson and Oakes, 1962). The effective trapping area of the channel will be increased by the penetration of the trapped field into the surrounding region and decreased by the coherence effect between electrons across the channel interface. If C is the circumference of the channel, then the change in its effective cross-sectional area due to the combined effects is of order $C(\lambda - \xi)$ and the effective trapping area of the channel is

$$A_e = A_c + C(\lambda - \xi),$$

where A_c is the actual channel cross-sectional area. This implies that the percentage flux trapped in a given specimen will be proportional to $(\gamma + \lambda - \xi)$ where γ is a characteristic dimension of the channel. There is no reason for γ to be a function of temperature but λ and ξ are known to vary so the percentage flux trapping as a function of temperature $P(t)$ may be written as

$$P(t) = P(0)\left\{\frac{\gamma + \lambda(t) - \xi(t)}{\gamma + \lambda(0) - \xi(0)}\right\}.$$

The temperature dependence of the penetration depth in tin is reported by Schawlow and Devlin (1958) to be approximately $\lambda(t) = \lambda(0)(1 - t^4)^{-1/2}$. No direct measurements of the coherence length are available but Faber

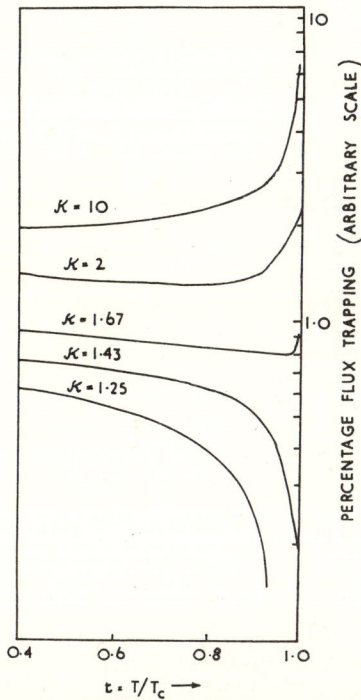

FIG. 8.10. Variation of the percentage flux trapped as a function of temperature for $\kappa = 1\cdot25$ to $\kappa = 10$ predicted by Jurisson and Oakes (1962)

(1958) gives results for the variation of surface energy with temperature for pure tin and, because of the high purity, these will be dominated by the coherence length. The temperature variation of the coherence length is, therefore, assumed to be the same as that of the surface energy in this case and may be written

$$\xi(t) = \xi(0)(1 - t^{3/2})^{-1/2}.$$

The variation of percentage flux trapping with temperature predicted by this model for different values of κ is shown in Fig. 8.10. The main features of the flux trapping characteristic and its change with κ are reproduced although, as might be expected from so simple a model, quantitative agreement with experiment is rather poor.

Bibliography for Chapter 8

ALERS, G. A. and WALDORF, D. L. (1961), *Phys. Rev. Letters*, **6**, 677.
ALERS, P. (1957), *Phys. Rev.* **105**, 104.
BEAN, C. P. (1962a), *Phys. Rev. Letters*, **8**, 250; (1962b), *Phys. Rev. Letters*, **9**, 93.
BEAN, C. P. and LIVINGSTON, J. D. (1964), *Phys. Rev. Letters*, **12**, 14.
BUDNICK, J. I., LYNTON, E. A. and SERIN, B. (1956), *Phys. Rev.* **103**, 286.
CAMPBELL, A. M., EVETTS, J. E. and DEW-HUGHES, D. (1964), *Phil. Mag.* **10**, 333.
CRIBIER, D., JACROT, B., MADHAV RAO, L. and FARNOUX, B. (1964), *Phys. Letters*, **9**, 106.
DE BLOIS, R. W. and DE SORBO, W. (1964), *Phys. Rev. Letters*, **12**, 499.
DE GENNES, P. G. (1965), *Solid State Comm.* **3**, 127.
DE SORBO, W. and HEALEY, W. A. (1964), *Cryogenics*, **4**, 257.
DOULAT, J., GOODMAN, B. B., RENARD, M. and WEIL, L. (1959), *Comptes Rendus*, **249**, 2017.
ESSMAN, U. and TRAUBLE, H. (1967) *Phys. Letters*, **24A**, 526.
EVETTS, J. E., CAMPBELL, A. M. and DEW-HUGHES, D. (1964), *Phil. Mag.* **10**, 339.
FABER, T. E. (1958), *Proc. Roy. Soc.* **A248**, 460.
FRIEDEL, J. (1954), *Advan. Phys.* **3**, 446.
GOODMAN, B. B. (1962), *IBM J. Res. Develop.* **6**, 63.
HECHT, R. (1964), Cleveland Conference IV, 98.
HEISE, B. H. (1964), *Rev. Mod. Phys.* **36**, 64.
JAKLEVIC, R. C., LAMB, J., SILVER, A. H. and MERCEREAU, J. E., (1964), *Phys. Rev. Letters.* **12**, 159.
JONES, D. P. and PARK, J. G. (1966), *Phys. Letters*, **20**, 111.
JOSEPH, A. S. and TOMASCH, W. J. (1964), *Phys. Rev. Letters*, **12**, 219.
JURRISON, J. and OAKES, J. A. (1962), *Phys. Letters*, **2**, 187.
LIVINGSTON, J. D. (1963), *Phys. Rev.* **129**, 1943.
MATRICON, J. (1966), Thesis (unpublished: Private communication).
MENDELSSOHN, K. (1935), *Proc. Roy. Soc. (Lond.)*, **A152**, 34.
NARLIKAR, A. V. and DEW-HUGHES, D. (1964), *Physica Status Solidi*, **6**, 383.
PIPPARD, A. B. (1955), *Phil. Trans. Roy. Soc. (Lond.)*, **A248**, 97.
SANDIFORD, D. J. and SCHWEITZER, D. G. (1964), *Phys. Letters*, **13**, 98.
SCHAWLOW, A. L. and DEVLIN, G. E. (1959), *Phys. Rev.* **113**, 120.
SERAPHIM, D. P. and MARCUS, P. M. (1962), *IBM J. Res. Develop.* **6**, 94.
SILCOX, J. and ROLLINS, R. W. (1963), *Appl. Phys. Letters*, **2**, 231.
SWARTZ, P. S., HART, H. R. and FLEISCHER, R. L. (1964), *Appl. Phys. Letters*, **4**, 71.
TOTH, L. E., and PRATT, I. P. (1964), *Appl. Phys. Letters*, **4**, 75.
WEBB, W. W. (1963), *Phys. Rev. Letters*, **11**, 191.
ZIMMERMAN, J. E. and MERCEREAU, J. E. (1964), *Phys. Rev. Letters*, **13**, 125.

CHAPTER 9

Flux Movement within Type II Superconductors

IN GENERAL the flux inside a type II superconductor in the mixed state is in a metastable equilibrium. Any change in the external conditions of the superconductor may cause the distribution of this flux to change. The ways in which these changes take place are described in this chapter.

A qualitative study of flux movement within the superconductor may be made using the Faraday effect technique described in section 8.1 (de Sorbo, 1957; de Sorbo and Healey, 1964). These results show that flux may move smoothly into a type II superconductor (section 9.1) or as a series of sudden movements, called *flux jumps* (section 9.2) depending on the density and also the arrangement of the imperfections in the sample. The latter behaviour is more common in samples with a high defect density.

The motion of flux lines causes energy to be dissipated in the sample and, therefore, heating (section 9.3). Because of the energy dissipation the movement of flux in type II superconductors is irreversible even in a sample free from pinning points.

9.1. Steady movement of flux lines

The flux lines in the body of a superconductor in the mixed state may be subject to a number of forces: when there is more than one vortex line in the material the long range Coulomb interaction will produce a force between two vortex lines which is positive or negative depending on whether they are in opposition or in the same sense; in the presence of a transport current density J in the sample there is a Lorentz-type force $F_L = J \wedge \phi_0/c$ on each vortex line perpendicular to the direction of the current; observations of the potential difference between two points along a sample in the mixed state show that there is a steady voltage between these points when the field and current are kept constant which may be interpreted as a steady movement of flux across the sample under the action of the Lorentz force together with a resistive "viscous" force; if there are suitable pinning points in the material these will impede the movement of flux lines and may

Irreversible Properties

stop the flow altogether; finally, localized regions of higher temperature in the superconductor may give the flux lines sufficient energy to overcome the pinning points.

The size of the Coulomb forces between the flux lines depends upon the distribution of the lines and, in general, this is irregular throughout the sample. For this reason an exact calculation of the magnetic forces on the fluxoids can rarely be made. Approximations to the flux distribution can be

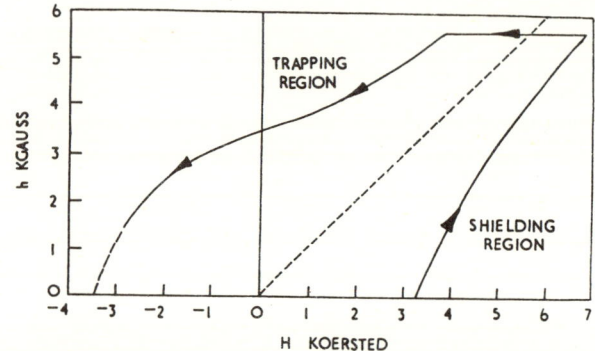

FIG. 9.1. The variation of the magnetic field inside a Nb–25% Zr tube with changing external field (from Kim et al., 1963)

taken for particular cases, however. In this case if the local flux density is $B(x)$, the energy per unit length of the line is $[\phi_0 B(x)/4\pi]$. Hence the force per unit length on the flux line in the x direction is

$$F_x = \frac{\phi_0}{4\pi} \frac{\partial B(x)}{\partial x}. \tag{9.1}$$

The suggestion that there is a Lorentz force on the fluxoids when a transport current is flowing was first made by Gorter (1962) and has been used by Kim et al. (1962) to account for their experiments on the magnetisation of tubes of type II material. The variation of the internal field, h, with external field, H_e for a Nb–25% Zr hollow tube is shown in Fig. 9.1. In the field region, h positive, H_e positive (and also in the region, h negative, H_e negative, not shown) the curves are rectangular hyperbolae while in the other two quadrants the curves are circles.

For a tube a magnetization, M, can be defined by:

$$M = h - H_e.$$

Also the average field in the tube wall is

$$H^* = \frac{1}{2}(h + H_e).$$

Flux Movement Within Type II Superconductors

The systematic law which these results obey can be seen by plotting $1/M$ against H^*. A series of these results at different temperatures for sintered niobium tubes is shown in Fig. 9.2. For field regions not too close to H_{c1} and H_{c2} a family of straight lines is produced. Representing these straight lines by

$$W\alpha/M = H^* + B_0, \qquad (9.2)$$

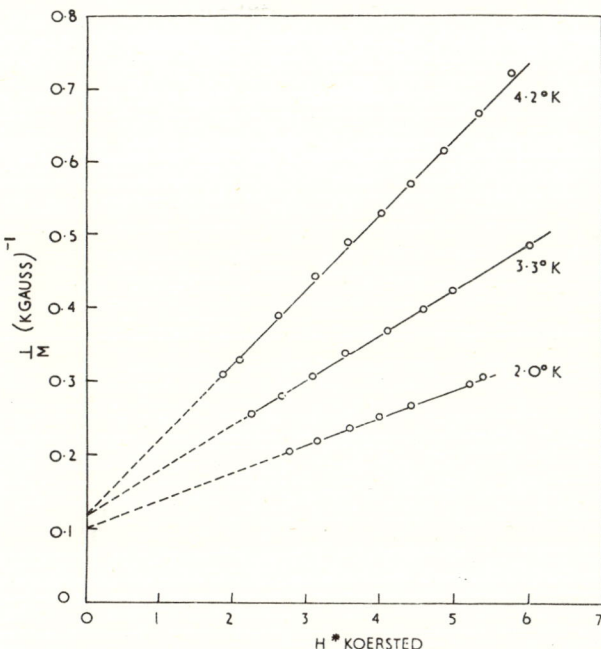

Fig. 9.2. $1/M$ versus H^* plots for a niobium powder sintered tube as a function of temperature (from Kim *et al.*, 1963)

where W is the wall thickness, two constants α and B_0, the slope and intercept of the line, which characterise any particular material appear. The hyperbolae and circles of the critical state curve, $h(H_e)$, are obtained by substituting the expressions for M and H^* into (9.2.).

The field difference $(h - H_e)$ is due to the current induced in the cylinder walls. The internal and external field may, therefore, be related by the expression

$$h = H_e + k \int_0^W JB(r)dr,$$

where r is the radial variable measured inwards from the tube surface, J is the current flowing in the cylinder walls and $B(r)$ is the magnetic induction at r and k is a constant. Eliminating r gives the integral relation

$$kW = \int_{H^*-M/2}^{H^*+M/2} \frac{dB}{J(B)}.$$

Irreversible Properties

This leads to (9.2) if the current-induction relation is of the form

$$\alpha_c = J(B + B_0).$$

Away from H_{c1}, $B \sim H$ (see Fig. 9.1) so that α_c is proportional to ($JH +$ + constant) showing that the relation between the field and the current is determined by the Lorentz force. The constant B_0 varies for different materials and may even be negative and its significance is not fully understood.

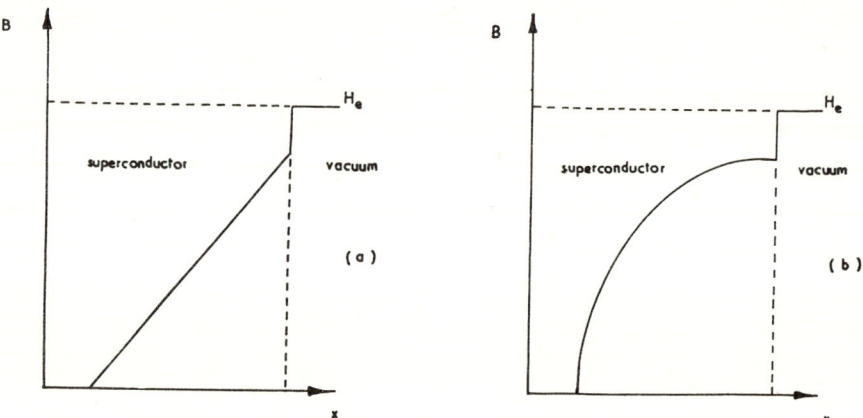

FIG. 9.3. The variation of the magnetic induction within the superconductor according to (a) Bean's model and (b) Kim and Anderson's model

The most likely possibility is that it is a composite term representing all the other forces acting on the fluxoids under the conditions of the experiment:

$$\sum F = JB_0.$$

It may be remarked here that the Bean model discussed in section 8.2 assumes that $J = \alpha$, i.e. J is independent of B and yields the linear variation of B shown in Fig. 9.3a. The Kim–Anderson model, Fig. 9.3b, implies a parabolic variation of the field within the superconductor.

Measurements of the resistive transition show that a voltage and, therefore, an apparent resistance appears along the sample while it is still in the mixed state (Druyvesteyn and Volger, 1964; Strnad et al., 1964). The normalised mixed state resistivity as a function of field at a number of different temperatures and as a function of temperature at different field values for a Nb–90% Ta strip is shown in Fig. 9.4. As the reduced temperature t approaches zero the results approximate to

$$\rho_m/\rho_n = H_e/H_{c2},$$

where ρ_m and ρ_n are the resistivities in the mixed and the normal state respectively.

It has been suggested (Anderson, 1962) that the potential difference is caused by the movement of flux across the sample under the action of the Lorentz force and this is shown to be true by the thin film experiments of

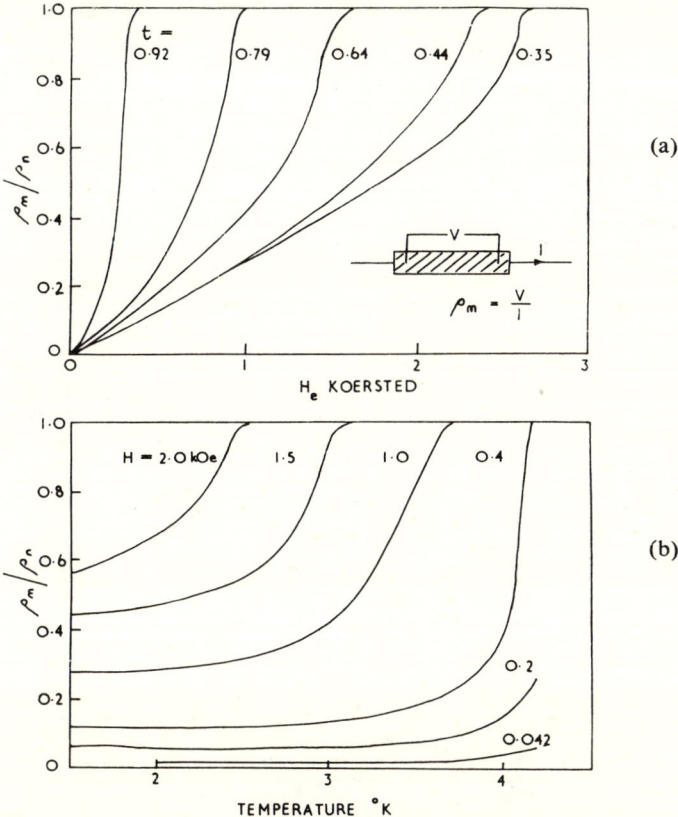

FIG. 9.4. Flow resistivity in a Nb–90% Ta strip in the mixed state (a) as a function of field at constant temperature values and (b) as a function of temperature at constant field values (from Strnad et al., 1964)

Giaever (1965). The cause is not completely straightforward however, since a steady movement of flux through the sample would cause flux to build up in the measuring circuit while the alternative possibility of flux creation at one surface of the sample and annihilation at the other would not, by itself, produce the steady voltage (Jones et al., 1965). The moving flux lines, however, have the effect of changing the chemical potential along the superconductor and this chemical potential difference is seen as a voltage difference between any two points in the mixed state (Josephson, 1965). The result of this is exactly the same as if the voltage were produced directly by the Lorentz force

Irreversible Properties

and we may write that the potential difference E across two points in the superconductor in the mixed state as

$$E = -(v/c) \wedge B$$

where v is the velocity of the flux lines and B is the local magnetic induction

The resistive force on the flux lines which prevents them reaching an infinite velocity may be written as a viscous term in the form

$$F_v = -\eta v_e,$$

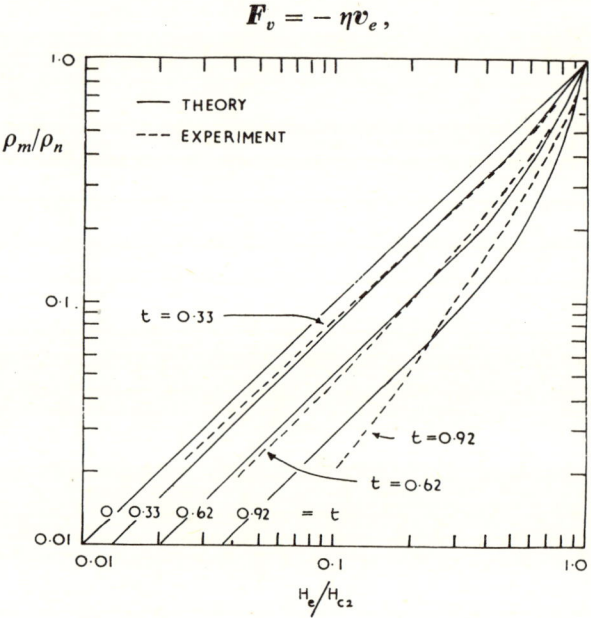

FIG. 9.5. Comparison of Tinkham's theory with experimental values for the flow resistivity as a function of field at a number of different temperatures (from Tinkham, 1964)

where η is the viscosity coefficient and v_e is the velocity of the flux lines relative to their surroundings. The exact nature of the viscosity coefficient is not completely clear. One possibility is that the dissipation occurs due to eddy currents induced in the core of the flux line which is almost normal (Volger *et al.*, 1964). However, the experiments of Strnad *et al.*, (1964) show that the dissipation found experimentally is much larger than that predicted by this theory. Tinkham (1964) has calculated the power dissipation in the superconductor phenomenologically by taking into account an assumed relaxation time for the superconducting wave function at a point as it returns to its equilibrium form when a flux line has passed. Adding the eddy current term and with a number of approximations this leads to an expression for ρ_m/ρ_n of the form

$$\rho_m/\rho_n = b/(1+t)^2(1-b^2) + b^2 + (b-b^2)t,$$

where $b = B/H_{c2}$ and t is the reduced temperature. The variation of (ρ_m/ρ_n) predicted by this expression is shown as a function of (H_e/H_{c2}) in Fig. 9.5. More recent work (Tinkham, 1966) has indicated that the relaxation time assumed here is probably too long in the case of dirty materials. However, it still provides a reasonably good empirical description of the experimental results. In particular the equation agrees reasonably agrees reasonably well with the experimental results already shown (with a linear scale) in Fig. 9.4. The expression also agrees very well with results on Pb–In wire (Druyvesteyn, 1965).

Kim et al., have shown by measurements of the longitudinal voltage in the mixed state that, over a wide range of temperature and composition, the viscosity coefficient is given by the empirical relation

$$\eta = \pi \hbar H_{c2} \sigma / ec,$$

where σ is the normal state conductivity. When a flux line is moving, the superconducting electrons near the edge of the core, at any point in the lattice, are accelerated as the flux line moves past. Hence an electric field is generated in the core. If the power loss due to this field is equated to (ηv_l), the dissipation in the core per unit length (Stephen and Bardeen, 1965), then the expression for η is

$$\eta = \frac{1}{2} \pi \hbar H_{c2} \sigma / ec.$$

This represents the viscosity coefficient due to dissipation inside the core. Dissipation outside the core may be expected to be of an equivalent amount so doubling this value. According to Stephen and Bardeen this mechanism is responsible for the major part of the energy loss in an extreme type II superconductor ($\xi \ll \lambda$).

The suggestion that the motion of the flux lines can be thermally activated was put forward by Anderson (1962) to explain the results of Kim et al. (1962) on tube magnetization, already mentioned. The effect must be analysed in terms of the movement of bundles of flux rather than of single fluxoids because of the observation that the voltage induced in a coil placed inside or outside a superconducting tube when flux moves in or out corresponds to a movement of 20–50 fluxoids, n say, rather than just one. The Lorentz force on a bundle of flux lines of effective diameter d is $\int \mathbf{J} \wedge \mathbf{B} \, d\tau$ so the free energy is

$$\mathcal{F}_l = -Jn\phi_0 \, dx$$

The size of the effective flux pinning points is $\geq \xi$ (see section 8.4) and if the pinning points are considered to be normal the free energy of the barrier is $\sim (H_c^2/8\pi)\xi^3$. Assuming that only a fraction p of this is effective, the total energy of the barrier to the motion of a flux line in the presence of a transport current is

$$\mathcal{F}_b = (pH_c^2/8\pi)\xi^3 - Jn\phi_0 d^2.$$

Irreversible Properties

Due to thermal activation, a particular barrier will allow flux lines through at a rate

$$R = R_0 \exp\{-\mathscr{F}_b/kT\}. \tag{9.3}$$

Here R_0 is a characteristic vibration frequency of a bundle, $\sim 10^5$–10^{10} per second.

This reasoning will fail in the two cases when the field is near to H_{c1} and H_{c2}. Near H_{c1} there is a large difference between B and H (see Fig. 9.1, where $B \sim h$ and $H \sim H_e$). The larger this difference, the larger is the current induced in the tube wall and this will complicate the results. Also in this limit the number of vortex lines becomes small so the force on the vortex will tend to $\mathbf{J} \wedge \boldsymbol{\phi}_0/c$, independent of the external field. As H_e approaches H_{c2}, the vortices are forced to overlap so that the interactions between them are greatly increased. In this case the magnetic pressure becomes comparable with the Lorentz force causing the results to depart from the simple Lorentz-type behaviour.

One consequence of forces on the vortex lines is that, unlike a type I superconductor, the flux trapped in a type II superconducting ring does not remain constant but decays exponentially. This follows from (9.3) when it is written for a ring shaped specimen. If $h > H_e$, the flux creep causes flux bundles to move out of the ring at a rate RN_w/N_p where N_w is the number of bundles in the wall and N_p is the number of effective pinning points that the bundle sees in traversing the wall. Now

$$N_w \sim 2\pi a\, WH^*/\phi_0$$

and $N_p \sim pW/d$, where a is the radius of the tube so, when H_e is zero the creep rate is given by

$$dh/dt = -(2dR_0 H^*/pa)\exp(-\mathscr{F}_b/kT). \tag{9.4}$$

Putting

$$K = (2dR_0/pa)\exp(-pH_c^2 d^3/8\pi kT)$$

$$\alpha(t) = J(t)[B(t) + B_0],$$

then (9.4) becomes

$$d\alpha/dt = -H^*|d\alpha/dh|\,K\exp\{\alpha d^4/kT\}.$$

Assuming that the total variation in α is small and allowing the initial transient to die out this leads to

$$\delta\alpha \sim \text{constant} - (kT/d^4)\ln t. \tag{9.5}$$

The value of α found from experiment is the quasi constant value α_c, which is attained once the initial rapid decay of the exponential has taken place. From (9.5) it follows that

$$h = (dh/d\alpha)(kT/d^4)\ln t = -[k'W/(h+B_0)](kT/d^4)\ln t, \text{ where } k' \text{ is a constant.}$$

Field decay, exponential with time, has been detected in type II superconducting tubes by Kim *et al.* (1962) with decay rates of up to 10 Oe/decade. If the current in the sample walls continued to decay at this rate, then it would be reduced effectively to zero in about 3×10^{92} years. Hence, in any practical sense, the current in the walls is persistent and the flux trapped remains constant once the initial rapid decay period has passed. Because of this decay, the curve shown in Fig. 9.1 represents only a quasi-equilibrium state in which the change in h at any particular value of H_e can be followed experimentally.

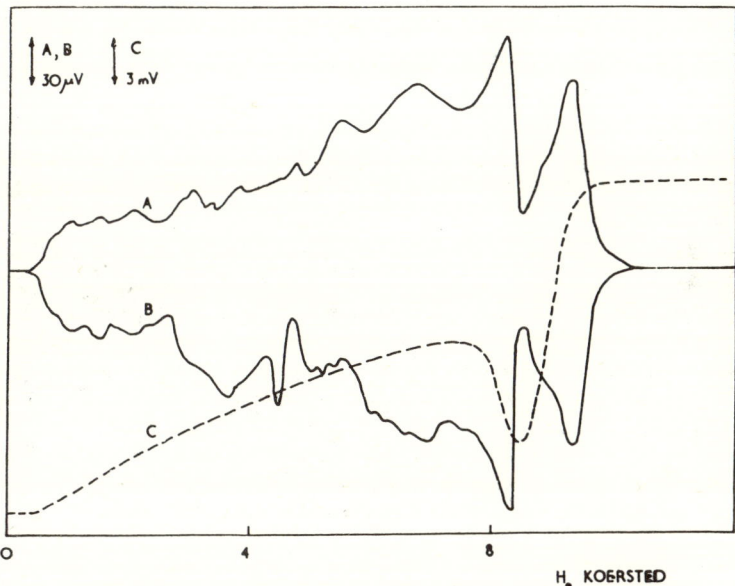

FIG. 9.6a. Transverse and longitudinal voltages as a function of magnetic field in a Nb–50% Ta sheet at a temperature of 1·3°K and with a current density of 3×10^3 A/cm² in the sample (from Niessen and Staas, 1965). A, field parallel to current; B, field antiparallel to current; C, field perpendicular to current. The fine structure in A and B and the dip in C disappear on annealing

Flux lines may have a longitudinal component to their velocity as they move through the sample. In this case a transverse voltage will be induced, reversible with magnetic field and transport current, i.e. a Hall voltage. The Hall effect has been measured in the mixed state for the type II superconductors Nb (Reed *et al.*, 1965) and Nb–50% Ta (Niessen and Staas, 1965) (and also in the intermediate state of the type I superconductor, In). Unless the surfaces of the sample are very regular, large non-reversible transverse voltages are also produced across the sample and these may mask the Hall effect. It is probably for this reason that other workers have failed to detect the Hall voltage. The irregular transverse voltages are found to correlate with slight corrugations in the surface of the sheet samples which

Irreversible Properties

form when they are rolled and it is apparent that these corrugations cause *guided* motion of the vortex lines; the flux can move along the valleys in the corrugations but is not able to move across them. This behaviour can be simulated on a macroscopic scale by producing the variation in the potential energy of flux lines in the sample by modulating the external magnetic field using corrugated pole pieces in the magnet providing the external field. These experiments on guided flux motion were originally analysed in terms

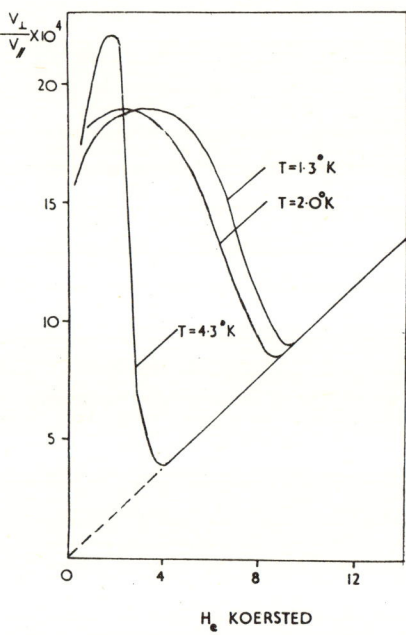

FIG. 9.6b. Hall angle as a function of magnetic field for Nb–50% Ta at various temperatures (from Niessen and Staas, 1965)

of the Magnus force on the vortices by analogy with the classical theory of vortices in hydrodynamics. [Bardeen (1964) has suggested that, for an infinite coupling between the electrons and the lattice, i.e. in the limit of short collision time, the Magnus force on the vortex of superconducting electrons is exactly cancelled by the presence of the positive ions of the lattice so that the net Magnus force is zero. In this limit the Hall effect also vanishes.] The effect of the Hall voltage and the irreversible guided-motion voltage together are shown for a Nb–50% Ta sample at 1·3°K in Fig. 9.6a where the current is reversed with respect to A in curve B. The difference between these curves is twice the Hall voltage. The fine structure in the curves is determined by the sequence of pinning points encountered by the flux in the material at different field values. This is shown by the facts that the fine structure is repeated if both the current and the field around the sample are reversed and also that if the samples are well annealed the structure disappears (Staas *et al.*, 1965).

Curve C of Fig. 9.6 shows the longitudinal voltage detected along the sample. The variation of the Hall angle $\tan^{-1}(V_\perp/V_{//})$ as a function of field for three different temperatures is shown in Fig. 9.6b. The Hall angle is very small indicating that the fluxoid motion is nearly perpendicular to the transport current.

In a very pure superconductor in which $\omega_c\tau \gg 1$ and $\xi_0 \gg l$ (i.e. in which the lattice defect density has little effect) it is possible that the array of vortex lines in the mixed state will undergo collective oscillations (de Gennes and Matricon, 1964). Considering an array of flux lines in a field $H(//Oz)$ each displaced by the variable amount $s_x(r)$, $s_y(r)$, then the potential may be expanded in an isotropic approximation:

$$G = G_0 + \frac{1}{2}K_1[(\partial s_x/\partial x)^2 + (\partial s_y/\partial y)^2] + K_2[(\partial s_x/\partial x)^2 + (\partial s_y/\partial y)^2$$

$$+ \frac{1}{2}(\partial s_x/\partial y + \partial s_y/\partial x)^2] + \frac{1}{2}K_3[(\partial s_x/\partial z)^2 + (\partial s_y/\partial z)^2].$$

Here K_1 and K_2 are the Lamé coefficients of the two-dimensional line lattice. The force on the flux line calculated from this equation may be equated to the usual electrodynamic force on a moving flux line:

$$F = \left(\frac{n_s e v}{c}\right)\phi_0.$$

Values of the Lamé coefficients can be derived from the general formula

$$G = G_0 + \frac{1}{8\pi}\left\{\frac{[B_z - B(H)]^2}{\mu_z} + \frac{B_x^2 + B_y^2}{\mu}\right\},$$

where $\mu_z = dB(H)/dH$ and $\mu = B(H)/H$. When $\lambda \gg \xi$, $\mu_z \sim \mu \sim 1$ so

$$\frac{1}{2}K_1 + K_2 = B^2(H)/4\pi\mu_z \sim B^2/4\pi,$$

$$K_3 = HB(H)/4\pi$$

and the shear modulus K_2 is small so $K_2 \sim K_1(\xi/\lambda)^2$. Putting these values of K into the force equation and looking for eigenmodes $s = s_0 \exp i(kr - \omega t)$ leads to the dispersion relations

$$\omega = \frac{eH}{m^*c}k^2\lambda^2 \quad (k//Oz) \tag{9.6}$$

and

$$\omega \sim \frac{eB}{m^*c}k^2\lambda d \quad (k \perp Oz), \tag{9.7}$$

Irreversible Properties

where d is the average distance between the flux lines and m^* is the effective mass of the electrons. The mode given by (9.6) is circularly polarized while that of (9.7) is elliptically polarized and when k is along Ox, $s_x/s_y \sim \xi/\lambda \ll 1$.

This behaviour has been calculated without taking into account the complicating presence of pinning points in the lattice. The effect of these will be to add terms s_x^2 and s_y^2 to the potential expression. Well defined long wavelength plane wave modes will still exist in this case but there will be a gap in the frequency spectrum.

From the time dependent Ginzburg–Landau theory it follows that each flux line has an inertial mass per unit length (Suhl, 1965). An idea of the relaxation time of a resonating vortex line may be obtained by equating the acceleration term resulting from the inertial mass with the damping term found experimentally by Strnad et al. (1964). This indicates that with the Nb–50% Ta sample used in that case the relaxation time will be less than 10^{-12} sec so that the collective modes of oscillation will be very difficult to observe.

9.2. Flux jumping

The rapid movement of flux within the bulk of a superconductor is a widespread property of superconductors and occurs in rings of type I materials in the intermediate state (Corsan and Thomas, 1962) as well as in type II. It has been detected in artificially made, filamentary superconductors formed by pressing mercury or lead into the pores of leached Vycor glass (Bean and Doyle, 1962) and in solid and hollow cylinders and wire wound solenoids of type II materials.

Flux jumping may be understood in terms of the *critical state* model of the flux in a type II superconductor proposed by Kim *et al.* (1963). A tubular specimen with a random distribution of pinning points in an external field between H_{c1} and H_{c2} in the critical state is shown schematically in Fig. 9.7. The flux lines penetrate in the region $r_1 < r < r_2$ and here the pinning points are each holding their maximum number of flux lines. The sample is in this condition when it produces the curve, Fig. 9.1, and for this reason this curve is often called the critical state curve.

In this condition any increase in the external field will cause r_1 to decrease and may initiate a flux jump. The flux jump is detected as a discontinuous change in the magnetisation of the sample or by a transient voltage induced in the turns of a search coil wound round the sample. If the flux jump is particularly severe some flux will appear suddenly at the centre of the tube and so may be detected with a search coil or with a field measuring device using the magneto-resistance principle or the Hall effect. When the sample is in the form of a wire wound solenoid the winding itself becomes its own search coil and the occurrence of flux jumps can be detected by monitoring the voltage developed across the coil.

The initial localized flux movement can be caused by the external forces already mentioned in section 9.1. Once the sample is in the critical state, the flux jump may be initiated by a change in the magnetic pressure, by a change in the Lorentz force (by varying the external field or the transport current) or by an increase in the temperature. The viscous forces and the forces due to pinning points are not controlled by external conditions and so do not play any part in setting off the flux jump. The effect of an initial flux jump may also be produced by a mechanical shock in the sample. The flux line has an inertia (Suhl, 1964) so the shock will have the effect of moving the lattice relative to the fluxoids.

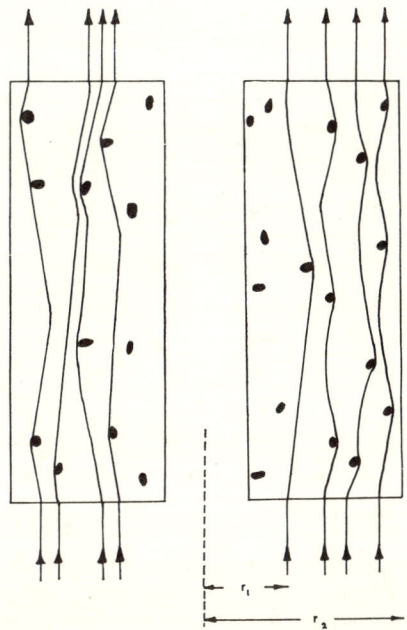

FIG. 9.7. A type II superconductor in the critical state. At an intermediate stage in the penetration of field into the superconductor, flux is being forced in by the external field and is held back by the pinning points. In the figure the pinning points in the region $r_1 \leq r \leq r_2$ are holding the maximum number of flux lines that they can while no flux has penetrated into the region $r < r_1$

The initial flux movement causes only a small flux jump. This may build up, producing an avalanche of flux, or it may die away again. The two possible behaviours are shown in Fig. 9.8. For the non-catastrophic flux jump the voltages induced across the end of a superconducting coil and across a coil wound outside it are relatively small and the voltage appearing in the solenoid is due to the movement of flux (i.e. it has the same form as the voltage in the search coil). For a catastrophic jump, however, the search coil voltage indicates that the flux leaves the sample very quickly while a voltage is detected across the superconducting coil after the flux has left indicating that a normal region is produced. If a suitably large transport current is flowing in the coil, the joule heat produced will be sufficient to maintain the coil above the critical temperature of the wire even though it is in the liquid helium. If the transport current is below this value, however, the coil will cool again and the voltage will eventually return to zero.

The appearance of normal resistance in the coil is an indication that large scale flux movement causes heating. This has been shown directly (Goede-

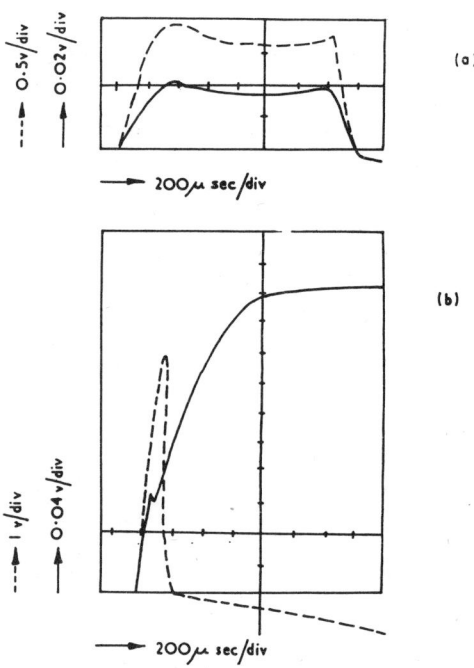

FIG. 9.8. Oscilloscope traces taken when a flux jump occurs in Nb–25% Zr wire carrying a current for (a) a non-catastrophic flux jump and (b) a catastrophic flux jump: the dotted trace is the voltage across a pick up coil surrounding the sample and shows the movement of flux; the full line trace is the uncompensated voltage which appears between the two ends of the superconducting coil and represents the sum of the voltages due to the normal resistance and to flux movement (drawn from Le Blanc and Vernon, 1964). Note that the scale of (a) is twice that of (b). The transport current is slightly different in the two cases [(a) 30 amps, (b) 25 amps]

moed *et al.*, 1964; Zebouni *et al.*, 1964) by measuring the temperature of the sample; discontinuous increases in the sample temperature are observed when a flux jump takes place. The thermal nature of flux jumping is also supported by the observation that porous tubes of sintered Nb_3Sn are much less likely to show flux jumping than samples of the same material without pores: the interior of the porous material is much better cooled owing to the penetration of liquid helium into the pores, preventing the formation of localized hot-spots. (Goldsmid and Corsan, 1964). If these pores are filled even with a good thermal conductor such as silver or copper flux jumping becomes more likely (Smith *et al.*, 1965) while cooling the helium to below the λ-point, where the thermal conductivity of the liquid helium is much larger, prevents flux jumping in Nb_3Sn samples in which the porosity is not sufficient to prevent the flux jumping completely at higher temperatures (Lange, 1965).

The requirement for the build up of the initial flux movement into a large scale flux jump is that a thermal wave large enough to unpin the flux should

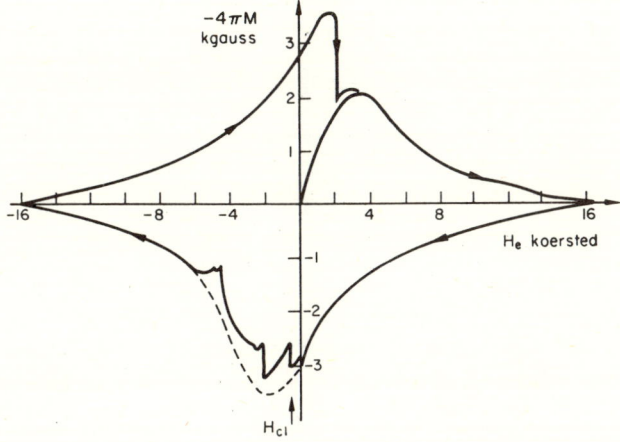

FIG. 9.9a. Magnetization curve of Pb–Bi eutectic at 4·2°K showing flux jumps at $-H_{c1}$. The dashed curve shows the magnetization the sample would have if it remained in the critical state (from Evetts *et al.*, 1964)

spread through the sample. Whether this will happen or not is determined by the specific heat and thermal conductivity of the superconductor and also by the rate at which heat is produced in the bulk. The main source of heat in most samples is from the energy dissipation caused by the viscous motion of the vortex lines.

The possibility of the liberation of further heat by the annihilation of vortex lines of opposite sign is suggested by the experimental observation shown in Fig. 9.9a (Evetts *et al.*, 1964): there is no major flux jumping here until the point on the magnetisation curve is reached at which flux of opposite direction begins to penetrate, $-H_{c1}$. The energy released when two flux lines coalesce is $2(\phi_0 H_{c1}/4\pi)$ equivalent to a local temperature rise at 4°K of about 1°K. This will give some of the neighbouring vortices enough energy to release them from the pinning points and the Coulomb interaction will then cause lines of opposite sign to move together and annihilate liber-

Irreversible Properties

ating more heat while the remaining lines in the region move apart dissipating energy due to the viscous motion as they move. If there are enough vortex lines with opposite signs present, the heat produced will be sufficient to unpin all the vortices. This type of instability has been confirmed by calculating the field dependence of the instability region when a transport current flows through the sample (Beasley et al., 1965). The transport current has the effect of reducing the magnetization. Very good agreement is found between the predicted field values at which flux jumping is expected to begin for particular transport currents flowing through the wire and

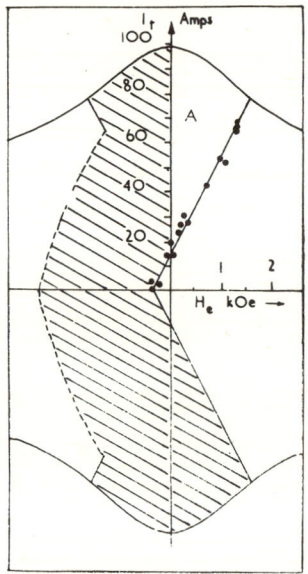

Fig. 9.9b. The instability region (shaded part and region A) calculated by Beasley et al. for a flat plate, shown on the magnetisation curve of Nb–25% Zr. The points are experimental results of the field at which the instability first occurs with different transport currents in the sample (from Beasley et al., 1965)

the values found experimentally. This is illustrated in Fig. 9.9b where the calculated region of instability is that shaded plus the region labelled **A**.

Large flux jumps due to the Lorentz force on the fluxoids have been observed in experiments on Nb–25% Zr by Lubell et al. (1963, 1964). Voltage spikes are observed at regular intervals across the ends of a coil when either the transport current in the coil is kept constant and the external field is increased or when the external field is kept constant and the transport current is increased.

It is expected that flux jumps will occur more frequently in specimens with a low specific heat, low density, low thermal conductivity and in which no special precautions are taken to cool the inside of the material. Flux jumps are also more likely in the sample as the rate of heating is increased. This is determined partly by the rate of change of field (Corsan, 1964a) and also by the density of pinning points. This latter has been confirmed by measurements on neutron irradiated samples of Nb_3Sn (McEvoy et al., 1964). Before irradiation and with the external field changing steadily at a rate

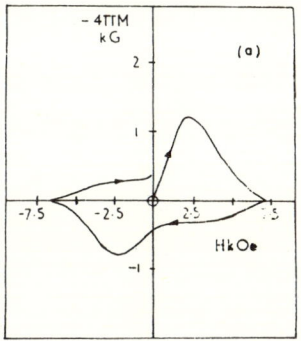

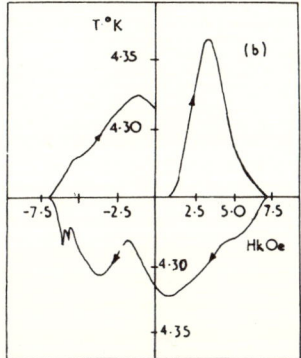

Fig. 9.10. The variation of (a) the magnetisation and (b) temperature with field for a niobium sample near 4·2°K in a varying field of about 1000 Oe/sec (from Goedemoed, 1966 The quadrants in the temperature curve (b) correspond to the quadrants of the magnetisation curve (a). The temperature of the sample never falls below 4·25°K.

of 10 Oe/sec the specimen remains in the critical state throughout the field sweep with no major flux jumps taking place. After irradiation with a density of $\sim 10^{18}$ neutrons/cm^2, however, although the critical state curve extends to a slightly higher field it is only occasionally reached by a sample; instead large scale flux jumping occurs.

It is difficult to make a quantitative prediction of the amount of flux which will move in any flux jump. It has been found experimentally, how-

Irreversible Properties

ever, (Corsan, 1964b) for a sintered Nb_3Sn cylinder that the amount of flux which remains in the cylinder, H_f, after a flux jump has occurred and once the temperature of the sample has returned to that of the helium bath is related to the initial rate of change of external field by the exponential law:

$$H_f \propto \exp\left\{\frac{\partial H_e}{\partial t}\right\}$$

when the field is swept at rates varying from 5 to 20 Oe/sec.

9.3. Magneto-thermal effects

Since a moving flux line dissipates energy, heat is liberated in the body of a type II superconductor when flux lines move through and a magneto-thermal effect is expected. This has been observed by Goedemoed et al. (1964) in an impure niobium specimen. The variation of the temperature of the sample in a swept field of ~ 1000 Oe/sec at 4·2°K measured using a carbon resistance thermometer is shown in Fig. 9.10 together with the magnetization characteristic of the sample. When the superconductor was heated by passing a current pulse through it, a very steep rise in temperature occurred, followed by an exponential decay with an equilibrium time of about 0·02 sec showing that the heat was carried away very effectively by the helium exchange gas which surrounded the sample and, therefore, that the temperature curve recorded is an equilibrium curve. It is seen that flux does not enter the sample until H_{c1} and that, after this, the temperature continues to change to near H_{c2}. When the field is swept in the opposite direction, however, the unusual structure shown in the third quadrant appears, with very sharp temperature fluctuations near the end of the sweep which correlated with flux jumps observed with a search coil. At 4·2°K the total amount

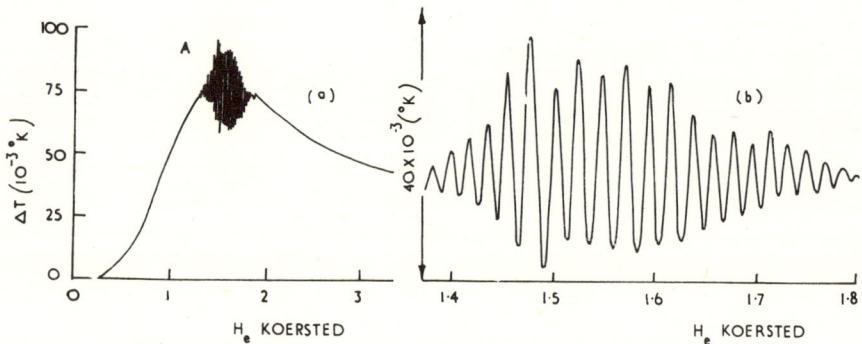

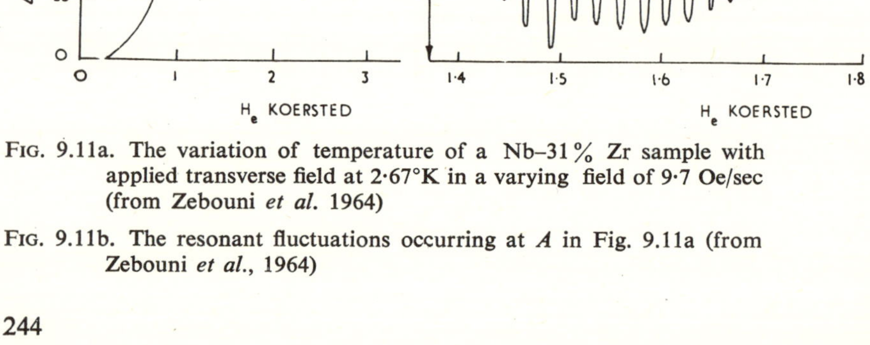

Fig. 9.11a. The variation of temperature of a Nb–31% Zr sample with applied transverse field at 2·67°K in a varying field of 9·7 Oe/sec (from Zebouni et al. 1964)

Fig. 9.11b. The resonant fluctuations occurring at A in Fig. 9.11a (from Zebouni et al., 1964)

of heat developed in the second and third quadrants is equal to $-\int MdH$ in these quadrants within the limits of the experimental precision.

Similar experiments on Nb and Nb–31% Zr have been made by Zebouni et al. (1964) with much lower sweep rates (0·8–13·5 Oe/sec). These measurements show the temperature rising at H_{c1} falling again towards H_{c2} and also discontinuous changes in temperature which correlate with flux jumps. In these experiments a further feature is an oscillatory fluctuation of the temperature with a period of 1–10 sec occurring at the maximum of the heating curve as shown in Fig. 9.11a. The expanded portion of the curve, Fig. 9.11b, shows the resonant character of the fluctuations. The period of the fluctuations increases rapidly with increasing temperature.

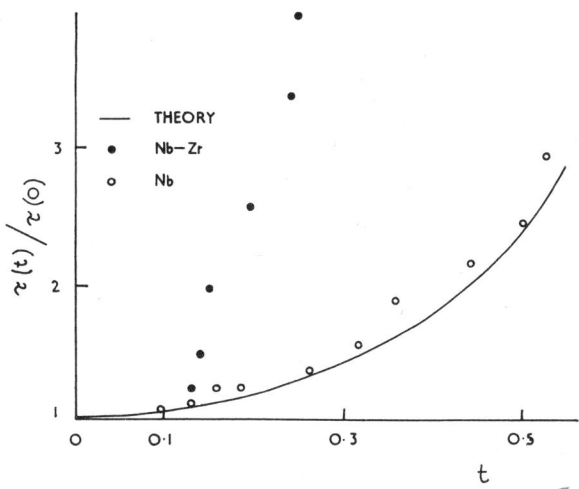

FIG. 9.12. Normalized period of the magneto-thermal oscillations $\tau(t)/\tau(0)$ as a function of temperature for Nb and Nb –31% Zr samples (from Zebouni et al., 1964)

The temperature dependence of the period of oscillation of these fluctuations may be compared with that of the elliptical mode of the collective oscillation predicted by de Gennes and Matricon (1964) which has the dispersion relation already given in (9.7).

$$\omega \sim \frac{eB}{m^*c} k^2 \lambda d \quad (k \perp H).$$

Taking $k = \pi/L$, where L is some unknown characteristic distance in the oscillating system, then, for niobium, $m^* \sim 50$ electron masses, $B \sim 8700$ gauss, $\lambda \sim 10^{-5}$ cm, $\omega\tau(0) \sim 2\cdot 1$ sec^{-1} leading to $L \sim 0\cdot25$ cm which is of the order of the radius of the niobium rod. The temperature dependence of $\tau \;(= 2\pi/\omega)$ follows from the facts that $d \propto (1-t)^{-\frac{1}{2}}$, $\lambda \propto (1-t^4)^{-\frac{1}{2}}$ and

Irreversible Properties

$L \propto (1-t^2)^{-1}$, the temperature dependence expected for the macroscopic, Bean, penetration depth (see section 8.2) as

$$\tau(t) = \tau(0)\left[(1-t^4)/(1-t)(1-t^2)^4\right]^{1/2}.$$

This prediction is compared with the experimental results for Nb and Nb–Zr in Fig. 9.12: the agreement is quite good with the results from the niobium sample but not with those of Nb–Zr. This latter disagreement is expected since the theory will certainly not apply to an alloy in which $l \ll \xi_0$ and $\omega_c \tau \ll 1$.

Bibliography for Chapter 9

ANDERSON, P. W. (1962), *Phys. Rev. Letters*, **9**, 309.
BARDEEN, J. (1964), *Phys. Rev. Letters*, **13**, 747.
BEAN, C. P. and DOYLE, M. V. (1962), *J. Appl. Phys.* **33**, 3334.
BEASLEY, M. R., FIETZ, W. A., ROLLINS, R. W., SILCOX, J. and WEBB, W. W. (1965), *Phys. Rev.* **137** A, 1205.
CORSAN, J. M. (1964a), unpublished; (1964b), *Phys. Letters*, **12**, 85.
CORSAN, J. M. and THOMAS, E. J. (1962), unpublished.
DE GENNES, P. G. and MATRICON, J. (1964), *Rev. Mod. Phys.* **36**, 45.
DE SORBO, W. (1957), *7th International Conference on Low Temperature Physics*, Toronto.
DE SORBO, W. and HEALEY, W. A. (1964), *Cryogenics*, **4**, 257.
DRUYVESTEYN, W. F. (1965), Thesis, p. 76.
DRUYVESTEYN, W. F. and VOLGER, J. (1964), *Philips Research Reports*, **19**, 359.
EVETTS, J. E., CAMPBELL, A. M. and DEW-HUGHES, D. (1964), *Phil. Mag.* **10**, 339.
GIAEVER, I. (1965), *Phys. Rev. Letters*, **15**, 825.
GOEDEMOED, S. H., VAN KOLMESCHATE, C., DE KLERK, D. and GORTER, C. J. (1964), *Physica*, **30**, 1225.
GOEDEMOED, S. H. (1966), Private communication.
GOLDSMID, H. J. and CORSAN, J. M. (1964), *Phys. Letters*, **10**, 39.
GORTER, C. J. (1962), *Phys. Letters*, **1**, 69.
JONES, R. G., RHODERICK, E. H. and ROSE-INNES, A. C. (1965), *Phys. Letters*, **15**, 214.
JOSEPHSON, B. D. (1965), *Phys. Letters*, **16**, 242.
KIM, Y. B., HEMPSTEAD, C. F. and STRNAD, A. R. (1962), *Phys. Rev. Letters*, **9**, 306.
KIM, Y. B., HEMPSTEAD, C. F. and STRNAD, A. R. (1963), *Phys. Rev.* **131**, 2486.
LANGE, F. (1965), *Cryogenics*, **5**, 143.
LE BLANC, M. A. R. and VERNON, F. L. (1964), *Phys. Letters*, **13**, 291.
LUBELL, M. S., CHANDRASEKHAR, B. S. and MALLICK, G. T. (1964a), *Appl. Phys. Letters*, **3**, 79.
LUBELL, M. S., MALLICK, G. T. and CHANDRASEKHAR, B. S. (1964b), *J. Appl. Phys.* **35**, 956.
McEVOY, J. P., DECELL, R. F. and NOVAK, R. L. (1964), *Appl. Phys. Letters*, **4**, 43.
NIESSEN, A. K. and STAAS, F. A. (1965), *Phys. Letters*, **15**, 26.
REED, W. A., FAWCETT, E. and KIM, Y. B. (1965), *Phys. Rev. Letters*, **14**, 790.
SMITH, P. F., SPURWAY, A. H. and LEWIN, J. D. (1965), *Brit. J. Appl. Phys.* **16**, 947.
STAAS, F. A., NIESSEN, A. K. and DRUYVESTEYN, W. F. (1965), *Phys. Letters*, **17**, 231.
STEPHEN, M. J. and BARDEEN, J. (1965), *Phys. Rev. Letters*, **14**, 112.
STRNAD, A. R., HEMPSTEAD, C. F. and KIM, Y. B. (1964), *Phys. Rev. Letters*, **13**, 794.
SUHL, H. (1965), *Phys. Rev. Letters*, **14**, 226.
TINKHAM, M. (1964), *Phys. Rev. Letters*, **13**, 804.
TINKHAM, M. (1966), Private communication.
VOLGER, J., STAAS, F. A. and VAN VIJFEIJKEN, A. G. (1964), *Phys. Letters*, **9**, 303.
ZEBOUNI, N. H., VENKATARAM, A., RAO, G. N., GRENIER, C. G. and REYNOLDS, J. M. (1964), *Phys. Rev. Letters*, **13**, 606.

CHAPTER 10

Critical Current Characteristics

The behaviour of the critical current of type II superconductors is well established although a detailed description of the basic current carrying mechanisms does not yet exist. As with type I superconductors, there is an upper limit to the transport current density which a type II superconductor will carry without losing its property of zero resistance. This critical current I_c is a function of the external field and the temperature. With pure materials free from lattice defects the change from the superconducting to the normal state can be extremely sharp and occurs at a well-defined current value. Type II superconductors, with the exception of niobium and vanadium, are all alloys or compounds and usually contain an appreciable defect density and for these materials the resistance transition is spread over a finite current range.

Good measurements of the critical current are difficult to make on many type II materials since the critical current of a sample of diameter 0·010 in. can be up to 450 amp. It is difficult to pass a current of this size without producing a large amount of heat in the current leads and at the points where they make contact with the sample. If this heat is not removed, it will raise the sample temperature and reduce the critical current value measured. This problem can be overcome by inducing the current in a ring specimen of the material (Kamper, 1963) so that no current leads or contacts are involved but in this case the critical current can only be measured in the presence of a magnetic field and this will also change the critical value. The measured critical current is also lowered if the current is applied too rapidly (Le Blanc, 1962; Lesensky and Neurath, 1963) since this will move flux lines rapidly in the wire and lead to energy dissipation and an increase in the wire temperature.

It has been mentioned in Chapter 9 that, in the mixed state, the transport current produces a Lorentz-type force on the lines of magnetic flux within the bulk of the material. This may be seen particularly in its effect on the magnetization of a sample when a transport current is passed through (Le Blanc, 1963; Yasukochi *et al.*, 1964; Druyvesteyn, 1965a). Flux which was originally pinned is moved out of the sample by the current and the magnetization becomes more reversible. This is shown in Fig. 10.1 where the

Critical Current Characteristics

continuous line is the magnetization curve of an In–8·7 at.% Pb alloy before the current is applied and the vertical lines show the effect of increasing current on the magnetization at different field values (Druyvesteyn, 1965a). The current needed to produce the near reversible characteristic is much smaller than the critical value so the effect on H_{c2} is small. It is too small to be seen in Fig. 10.1.

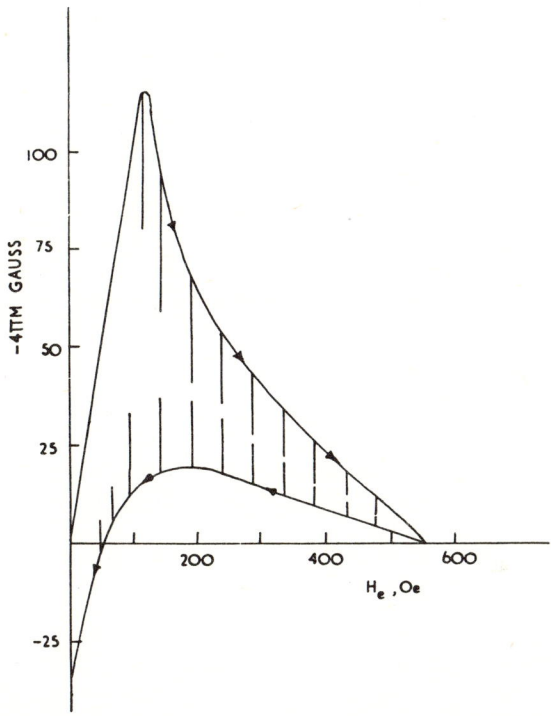

Fig. 10.1. The influence of a transport current on the magnetization curve of a type II superconductor (In–8·7% Pb) at 2·65°K. The continuous curve is the magnetization with zero current. On applying the transport current at a constant field value the magnetization follows the vertical paths (from Druyvesteyn, 1965a)

10.1. Variation of the critical current with field

The critical current of a type II superconductor is usually measured using the well-established four-probe technique which eliminates the effect of contact resistances appearing across the voltage terminals: the arrangement is shown in Fig. 10.2a. Suitable contacts which will carry 100 amps or more can be made to many samples by soldering on stout copper wires using indium, and to most other materials by clamping them firmly between

Irreversible Properties

copper blocks. The magnetic field is applied and the current increased until the first measurable voltage appears across the terminals. This current is usually taken to be the critical value. It depends upon the sensitivity of the measuring apparatus, usually of the order of 10^{-7}–10^{-6} volts and is a measure of the first detectable viscous flux flow (see section 9.1) in the material. Nevertheless, it is a suitable criterion for most critical current measurements.

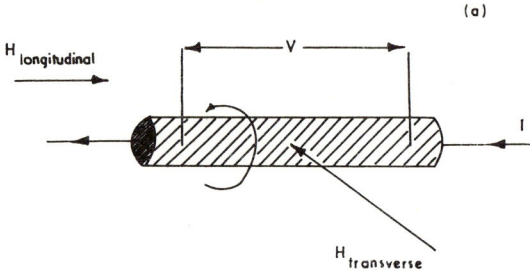

Fig. 10.2a. The usual arrangement of current and potential leads on a sample to measure critical current in a magnetic field

Investigation of the resistance transition when the current is increased shows that its shape depends upon the temperature, upon the current in the sample and on the angle between the current and the field. A shape quite often found for the variation of resistance when the applied magnetic

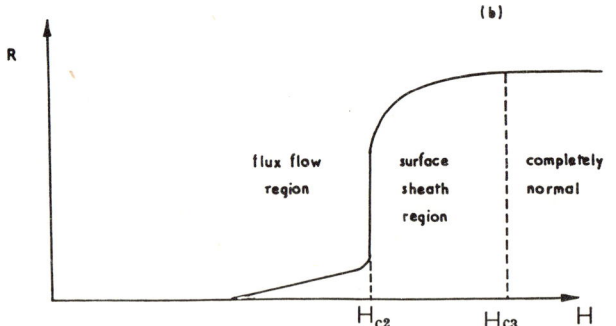

Fig. 10.2b. An idealized picture of the variation of resistance with magnetic field parallel to the measuring current

field is parallel to the current is shown in Fig. 10.2b. By comparing the magnetization characteristic of the sample with the resistance transition, the field value where the curve is vertical (measured using a very small transport current so that the self-field of the wire is negligible) has been identified as H_{c2} (Druyvesteyn, 1965b). The small voltage detected below this value

250

Critical Current Characteristics

when the superconductor is in the mixed state is ascribed to the viscous motion of flux in the mixed state, mentioned above, while the failure to reach the completely normal state at fields higher than H_{c2} is due to the presence of the superconducting sheath at the surface of the sample (see Chapter 4) which exists up to H_{c3}. Hence, ideally, H_{c1}, H_{c2} and H_{c3} should all be obtainable from the resistance transition as the field at which a voltage first appears (H_{c1}), the field of the abrupt rise (H_{c2}), and the field at which the final trace of superconductivity disappears (H_{c3}). In practice, however,

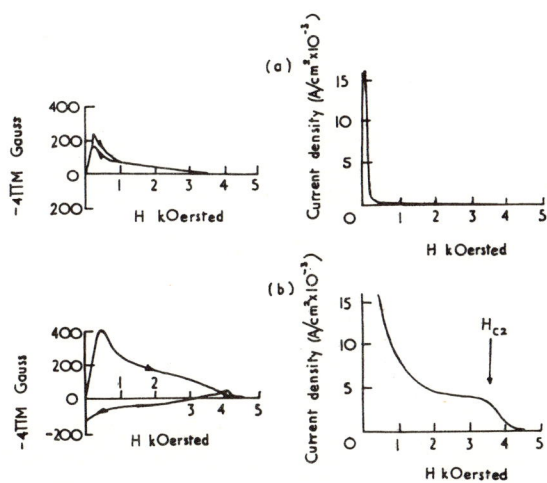

Fig. 10.3. The critical current and magnetisation characteristics of (a) annealed and (b) cold worked Nb–45% Ta wires at 4·2°K (from Heaton and Rose-Innes, 1964)

the entry of flux into the superconductor is delayed by the surface barrier (section 8.3) until the external field reaches a value greater than H_{c1}. Also the low field tail can run into the high field region so that no vertical portion appears. Hence, only H_{c3} may be found with certainty and this only in a longitudinal field.

10.1.1. *Critical current in a transverse field*

The critical current (I_c)–transverse field (H_t) characteristic of a number of materials has been studied in great detail since it is of importance in predicting the commercial usefulness of a hard superconducting material in producing high magnetic fields (see section 11.1). In this case the most important requirement of the wire is that it should carry a large transport current in a high magnetic field and the relative properties of different materials can, therefore, be compared directly from their $I_c(H_t)$ characteristic.

Irreversible Properties

It has already been shown (section 1.8) that the critical current of a type II material with a reversible magnetisation characteristic is very small. This is illustrated in Fig. 10.3a for a well-annealed Nb–45% Ta wire (Heaton and Rose-Innes, 1964) where, except for the very low current tail and taking into account the demagnetization coefficient of the sample, the critical current is found to be limited by H_{c1} (see section 1.8). The high critical

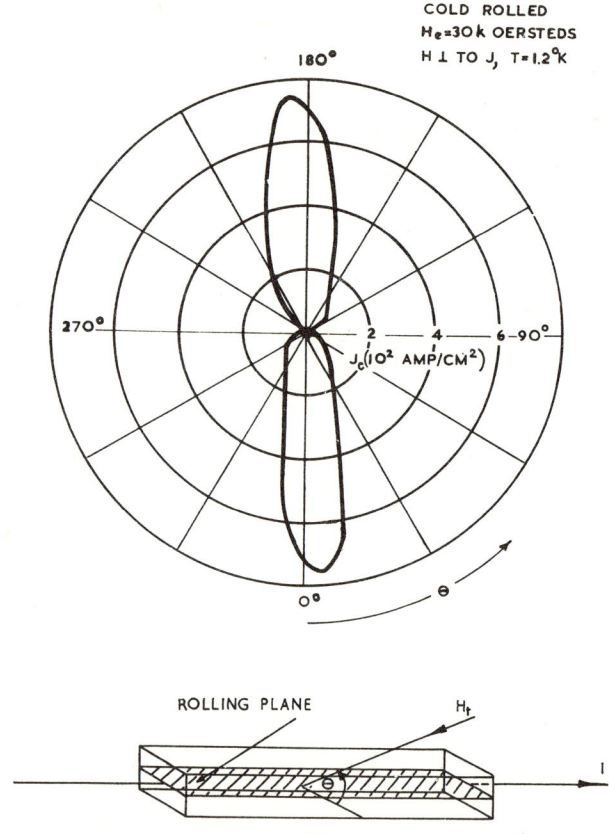

Fig. 10.4. A polar plot of the critical current density against θ, the angle between the transverse field and the rolling plane (defined in the inset) for a Ti–10% Mo specimen at 1·2°K in a field of 30 kOe (from Hake and Leslie, 1963)

currents which are measured in some type II materials are due to the presence of defects in the lattice of the material. This is illustrated in Fig. 10.3b for the same Nb–Ta specimen when it is cold drawn before annealing. In this case the current density remains large up to H_{c2}. The presence of a large defect density is shown by the irreversible magnetization curve.

Most of the work on the effects of lattice defects in high field superconductors (Nb–Zr alloys, Nb_3Sn, V_3Ga and V_3Si compounds) has been empirical, changing the variables in the preparation of the material, e.g. reaction temperature and time, rather than examining the effects of specific imperfections since all these materials have high melting points and are difficult to handle metallurgically.

A systematic study has been made, however, of the effect of lattice imperfections on the critical current – transverse field characteristic of the type II superconductor Nb (Tedmon et al., 1965). The starting material was a niobium single crystal produced using an electron beam, floating zone technique (Calverley et al., 1957) and was annealed for several hours in a vacuum of 5×10^{-6} torr at 2300°C. The effect of a known number of dislocations can be studied by bending the crystal round a known radius of curvature and then bending it straight again. Vacancies are also formed in the bending but these can be removed again by annealing at 160°C leaving only randomly spaced dislocations. Annealing the bent crystal at 900 °C causes dislocations to come together to form walls leaving only a low dislocation density in the rest of the material (polygonization). The effect of interstitial oxygen is studied by adding the oxygen in increments of 5–100 ppm. Finally, anisotropy of the critical current can be studied by rotating the field around the single crystal sample. In all cases the effect of the lattice imperfections is to raise the critical current and, in the case of the oxygen, to raise the upper critical field as well. In well-annealed samples no significant anisotropy is observed even when they are strained up to 50%. For samples containing a small amount of oxygen, however, maxima in the $[\bar{1}11]$ and $[1\bar{1}1]$ directions and minima in the $[1\bar{1}0]$ and $[001]$ directions are observed in strained and unstrained samples. Since niobium is b.c.c. these represent the directions of closest and most open packing respectively. There is no theory at present which relates the critical current to the direction of the applied field relative to the lattice.

Measurements on cold rolled niobium also show a marked anisotropy of the $I_c(H_t)$ characteristic even though the samples, in this case, are not single crystals and lattice orientation can no longer affect the measurements (Le Blanc and Little, 1960). In this case it is found that the critical current is near a maximum when the applied magnetic field is parallel to the rolling plane. The sample arrangement and a polar plot of the critical current as the transverse field direction is varied is shown in Fig. 10.4. Misalignment of the voltage leads may account for the fact that the maxima of the J_c–θ plot do not fall on the 0–180° axis.

This type of anisotropy has been observed in a large number of different type II superconductors (Hake and Leslie, 1963) but a few alloys (Ti–6% Mo, Ti–9% Mo, Ti–20% Nb) do not show it. A possible explanation of this latter behaviour is that a martensitic phase transformation occurs during the cold rolling which is sufficient to suppress the usual cold rolled defect structure.

Irreversible Properties

Measurements in the normal state on materials which show anisotropy indicate that there is an increase in the electrical resistivity, i.e. an increase in the defect density, across the specimen in the field directions in which there are maxima in the critical current. It is therefore probable that cold rolling produces an oriented defect structure in the sample which provides more flux pinning when the flux motion is perpendicular to the plane of the defects. This increases the critical current density when the field is in this direction as will be seen later.

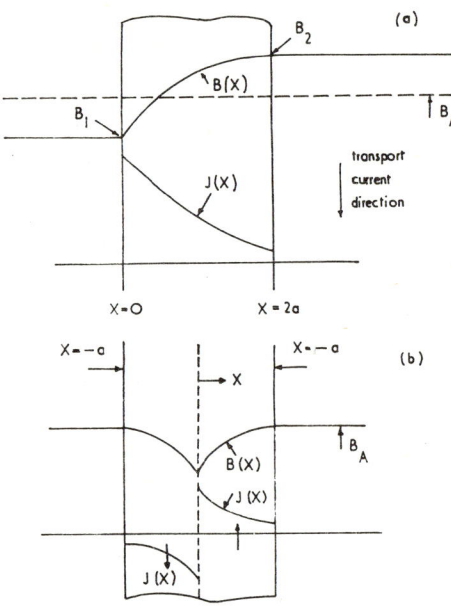

Fig. 10.5. The current and field distribution in a slab (a) with a transport current flowing in the slab and (b) with no current flowing (from Riemersma, 1964). The field is applied out of the paper

An elementary calculation of the critical current in the mixed state as a function of transverse field may be made in the field region where the Lorentz force is the dominant one on the fluxoids in the material. In this case

$$\boldsymbol{J} \wedge \boldsymbol{B} \sim \alpha$$

and, also

$$\nabla \boldsymbol{B} = \mu \boldsymbol{J},$$

where μ is the magnetic permeability of the parts of the superconductor which are carrying the current. Considering the sample to be in the form of a thin strip so that there is no demagnetization effect, then in the one-dimensional geometry shown in Fig. 10.5 (Riemersma, 1964), these equations become

$$J(x) \cdot B(x) \sim \alpha$$

and

$$J(x) = (1/\mu)\, dB(x)/dx\,.$$

Substituting for $J(x)$ and integrating

$$B^2(x) = 2\mu\alpha x + C\,. \tag{10.1}$$

The constant C is determined from the boundary condition of the sample when $x = 0$:

$$B(0) = B_1, \qquad \text{so}\quad B^2(x) = 2\mu\alpha x + B_1^2 \tag{10.2}$$

$$B(2a) = B_2, \qquad \text{so}\quad (B_2^2 - B_1^2) = 4\mu\alpha a\,. \tag{10.3}$$

Thus

$$I_t = \int_0^{2a} J(x)\, dx = \alpha \int_0^{2a} dx/B(x)\,.$$

Evaluating this using (10.2) and (10.3) leads to

$$I_t = \frac{B_2 - B_1}{\mu}\,.$$

By symmetry, $(B_1 + B_2) = 2 B_A$, the average field in the lamina, so from (10.3)

$$I_t = \frac{2a\alpha}{B_A}\,. \tag{10.4}$$

A low field contribution must be added to this effect to take account of the Silsbee-type current flow below H_{c1} and the effect of any surface transport properties must also be added (including that of the Saint-James – de Gennes sheath). This will be mentioned later in the section. The first two effects which contribute to the critical current, together with their sum are shown in Fig. 10.6 and it can be seen that, if the discontinuity at H_{c1} is smoothed out, the general shape which is predicted agrees reasonably well with the characteristics found experimentally up to near H_{c2}.

Two conclusions may be drawn from (10.4): the maximum transport current increases both with α (the maximum force which the pinning points

Irreversible Properties

can exert, which increases with the defect density),† and with the thickness of the strip. The effects are shown in Fig. 10.7 of increasing α on (i) the critical current density – transverse field characteristic, and (ii) the corresponding changes to the magnetization characteristic. Increasing a also has the effect of increasing the amount of hysteresis but in this case the effect is not mirrored in a corresponding increase in the critical current density. The fact that the field

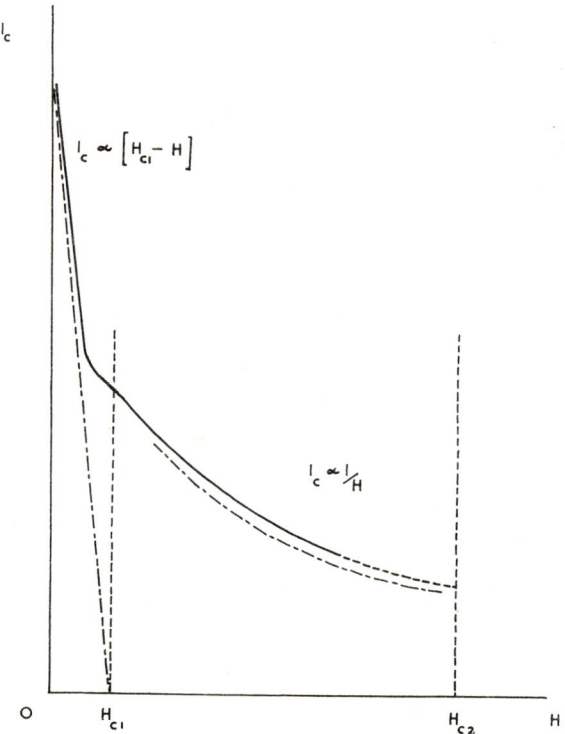

FIG. 10.6. The variation of critical current with field showing the Silsbee behaviour below H_{c1} and the Lorentz force limited behaviour in the mixed state. The continous line, smoothed at the discontinuity at H_{c1}, shows the effect of the two mechanisms together. It is assumed that the sample has a zero demagnetization coefficient.

is distributed throughout the sample cross-section (see Fig. 10.5) implies that the transport current is no longer limited to the surface of the material as in a type I superconductor but also flows within the bulk. This theory predicts the current distribution will be approximately uniform across the slab at high fields, but it does not take into account the effect of the surface sheath

† It has recently been shown [H. R. Hart, Jr. and P. S. Swartz, *Phys. Rev. 156*, 403, (1967)] that this conclusion is also true for *surface sheath* critical transport currents as well as for the critical currents of material in the mixed state.

Critical Current Characteristics

and, as will be seen later, this plays a major role in carrying current in the high field region.

Figure 10.7 also illustrates the close relation between the critical current and the magnetization characteristic. The magnetization may be defined from the field distribution in the sample in the absence of a transport current and, using the notation shown in Fig. 10.5b,

$$-4\pi M = 2B_A a - \int_{-a}^{+a} B(x)\, dx$$

so that

$$-4\pi M = 2B_A a - \alpha \int_{-a}^{+a} dx/J(x),$$

where $J(x)$ in this case is the current density induced in the sample when the external field is applied. Once again this prediction is only valid in the field region where the Lorentz force is dominant.

The critical current – transverse field characteristic has also been calculated assuming that the superconductor is composed of a network of straight line type I superconducting filaments each with diameter much less than the penetration depth (Blaisse and de Jong, 1965). Reasonable agreement is found between this theory and the experimental results of Druyvesteyn *et al.* (1964) on Pb–6·6% In.

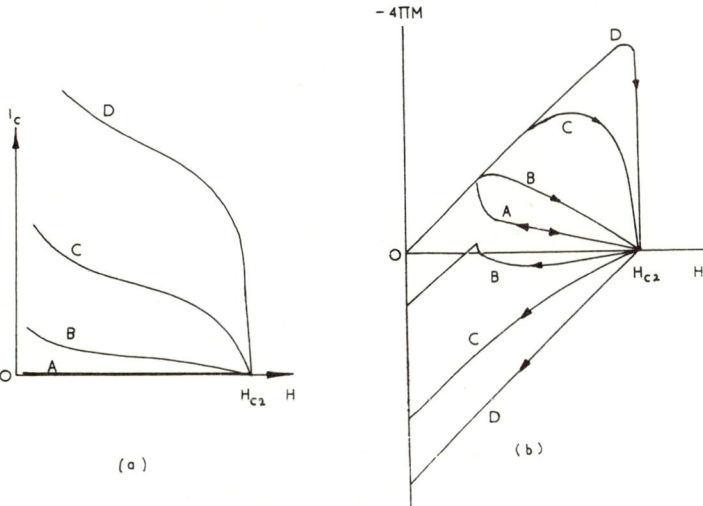

FIG. 10.7. The increase in (a) critical current and (b) the equivalent increase in magnetization which takes place from A to D with increasing defect density in the material. The increase in the hysteresis of the $M(H)$ curve occurs with increasing specimen diameter as well as with increasing defect density

Irreversible Properties

An estimate of the critical current characteristic can also be made by assuming that the current is carried in a layered series of thin laminae (El Bindari and Litvak, 1965). This predicts a current variation of the form:

$$I_c(H)/I_c(0) = 1 - (H/H_{c2})^{1/3},$$

where $I_c(0)$ is the critical current in zero field. This equation agrees well with experiments on a number of high κ, type II materials.

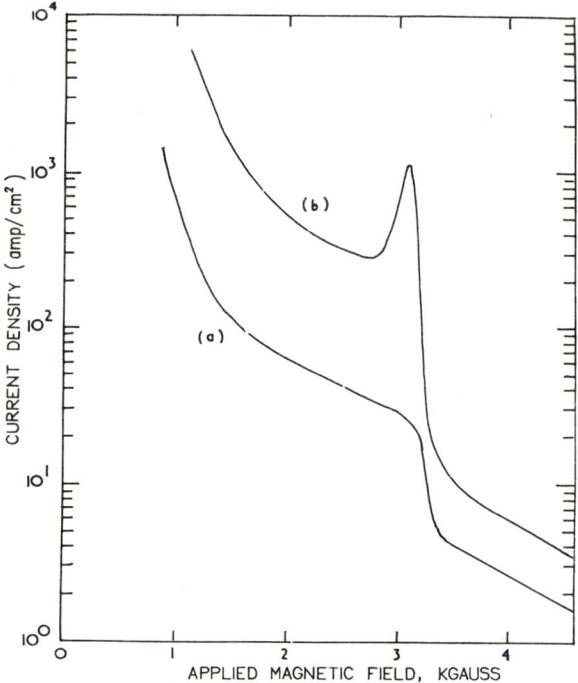

FIG. 10.8. The effect of plastic deformation on the critical current–field characteristic of a niobium single crystal at 4·2°K: (a) unstrained, (b) strained 50% (from Tedmon et al., 1965)

The $I_c(H_t)$ characteristics so far mentioned have all been monotonically decreasing curves in an increasing external field. In contrast to this general behaviour, however, characteristics measured for Nb (Le Blanc and Little, 1960; de Sorbo, 1964; Tedmon et al., 1965), Nb–Zr alloys (Berlincourt et al., 1961), Mo–Re alloys (Hauser and Treuting, 1963) and Pb–Tl alloys (Swartz and Hart, 1965) show that, in the higher field region the critical current rises to form a peak, shown in Fig. 10.8, curve b, which exists in both increasing and decreasing field. The peak is reproducible and is mirrored by a dip in the resistance field transition as shown in Fig. 10.9.

The peak effect is associated with the imperfections in a type II superconductor. This is illustrated in Fig. 10.8 where it is shown that plastic de-

Critical Current Characteristics

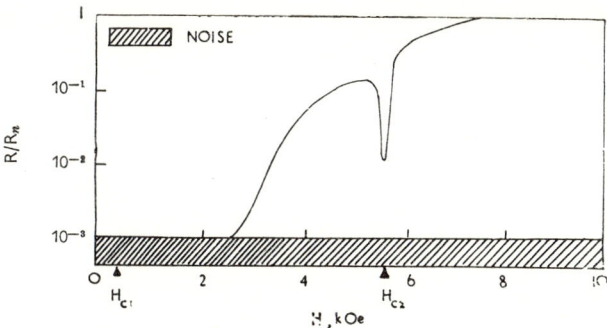

Fig. 10.9. The variation of resistance with magnetic field of a niobium wire at 4·2°K carrying a transport current density of 450 amps/cm² (from Autler *et al.*, 1962)

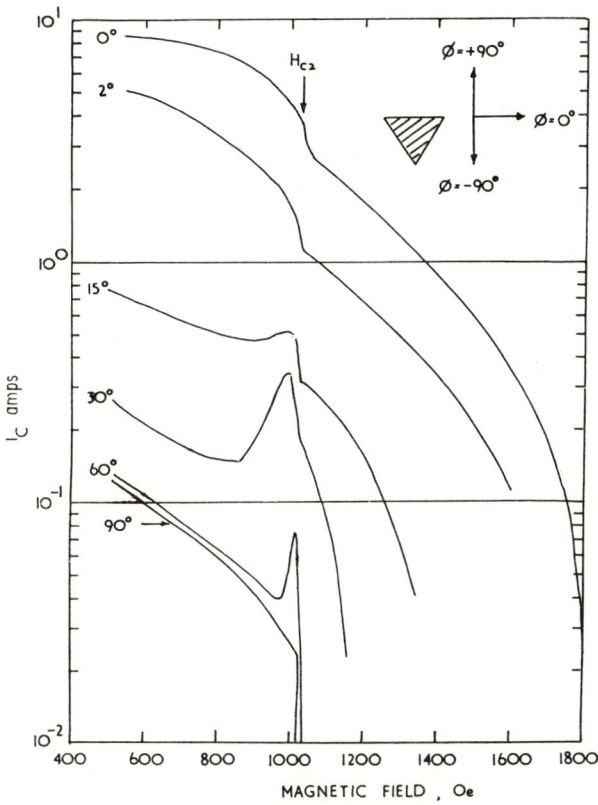

Fig. 10.10a. The critical current–field characteristics of a Pb–5% Tl prism with a triangular cross-section at 4·2°K when the transverse field is set at different angles defined in the insert (from Swartz and Hart, 1965)

Irreversible Properties

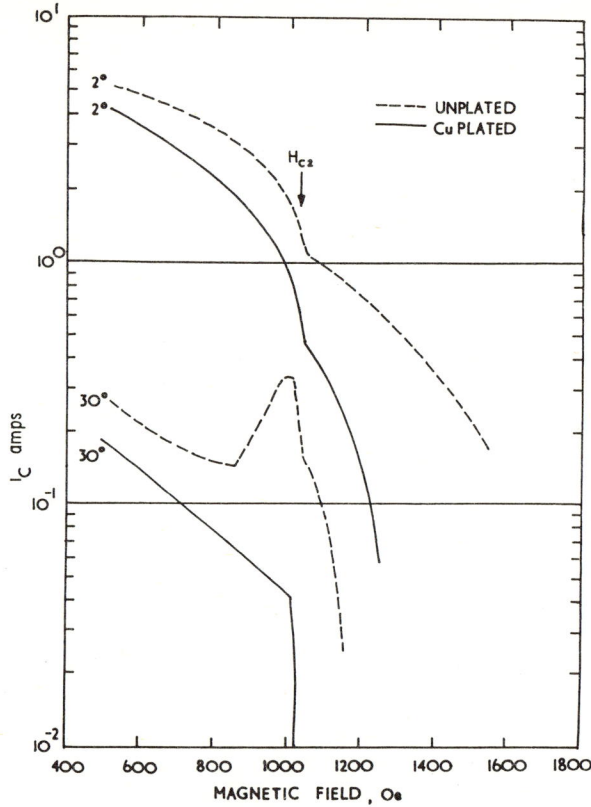

Fig. 10.10b. The critical current–field characteristic of an annealed Pb −5% Tl prism with triangular cross-section before and after electroplating (from Swartz and Hart, 1965)

formation producing dislocations near the surface causes the peak to appear close to H_{c2} as well as increasing the current density at lower field values. The peak effect is also associated with the surface of the sample. This is shown very clearly by the critical current measurements made on a prism of triangular cross-section shown in Fig. 10.10a where the angles refer to different orientations of the field around the sample (Swartz and Hart, 1965). Here a peak in the critical current–field curve appears as the field becomes parallel with one of the sides. That the peak is associated with the Saint-James – de Gennes sheath is illustrated by copper plating the sample. The metal film reduces the effect of the sheath and in this case the peak is also reduced as shown in Fig. 10.10b.

These results indicate that there are two surface current carrying mechanisms for type II materials in the higher field region of the mixed state having the field dependence shown in Fig. 10.11. The current associated with mechanism A decreases approximately exponentially and cuts off sharply

Critical Current Characteristics

at H_{c2} and may be associated tentatively with orientated imperfections parallel with each surface which form during the construction of the prism, an effect already mentioned; mechanism B, which disappears when a 1000 Å layer of copper is deposited on the surface may be linked more surely with the Saint-James – de Gennes surface sheath. When the two families of critical currents shown in Fig. 10.11 are added, a series of curves results which is nearly identical with that of Fig. 10.10a.

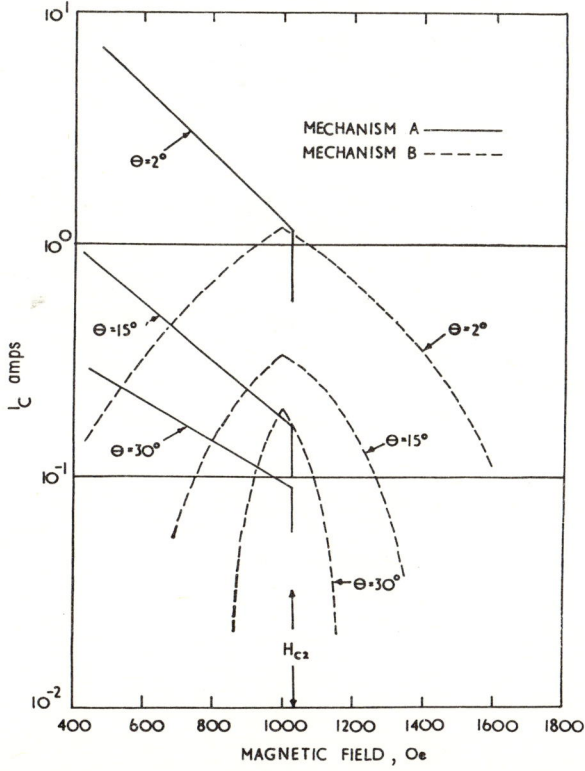

FIG. 10.11. Two families of constructed critical current curves for well annealed Pb–5% Tl. These have the property that when added together they result in a family of curves nearly identical to those of Fig. 10.10a (from Swartz and Hart, 1965)

10.1.2. Critical current in longitudinal field

Most measurements of the critical current characteristic have been made with the magnetic field transverse to the sample. A number of experiments, however, have been made in a longitudinal field with samples of Nb_3Sn (Cody et al., 1964), Nb–Ta alloys (Heaton and Rose-Innes, 1964); Pb–In alloys (Druyvesteyn et al., 1964); Pb–Bi alloys (Grassman and Rinderer, 1959) and Mo–Re and Ta–Ti alloys (Sekula et al., 1963). The critical current behaviour of four cold drawn Nb–Zr alloys is shown in Fig. 10.12a.

Irreversible Properties

As for all the other materials, the Nb–Zr characteristics show a broad maximum and have similar shapes. In this case the height of the peak is a function of the zirconium content. It must be stressed that this peak does not

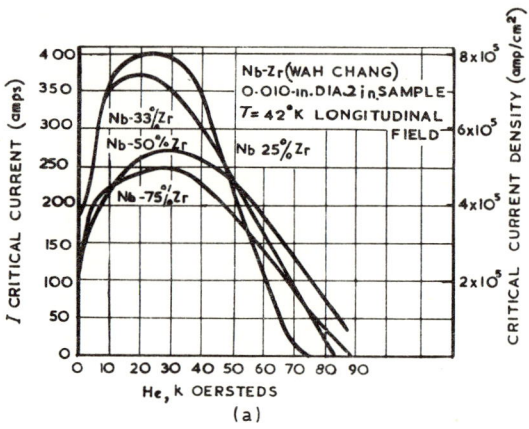

FIG. 10.12a. The critical current–longitudinal field characteristic for four different Nb–Zr alloys at 4·2°K (from Sekula *et al.*, 1963)

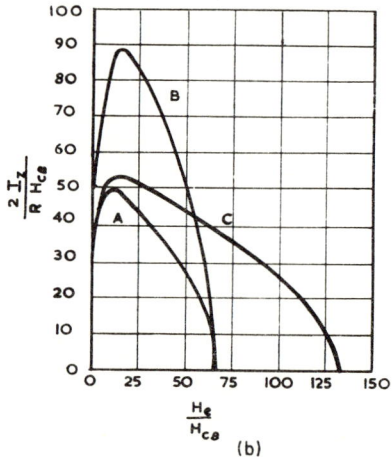

FIG. 10.12b. Normalized critical current – longitudinal field curves resulting from (10.5) (from Bergeron, 1963a):

	$q =$	0·01	$a/\lambda =$	10
A				
B		0·01		20
C		0·005		10

appear for the same reason as the one in the critical current–transverse field characteristic.

A prediction of the critical current–longitudinal field characteristic has been made by Bergeron (1963a, b) assuming that the material consists of a structure of laminar superconducting filaments embedded in a normal matrix. Assuming that the current will take up a force-free configuration (i.e. a helical path) as it flows along the sample, then the longitudinal critical current density J_l is related to the external field H_e by the relation

$$4\pi J_l = c_s H_e,$$

where c_s is a spatial constant. In this case the equilibrium condition is given by the parametric relations:

$$\left. \begin{array}{c} 2I_l/aH_c = (q/A)^{1/2} J_1(c_s a), \\ H_e/H_c = (q/A)^{1/2} J_0(c_s a), \\ c_s \lambda = (A/q - \tanh^2 p)^{1/2}, \end{array} \right\} \qquad (10.5)$$

where I_l is the critical current with the field longitudinal to the wire, a is the wire radius, H_c the thermodynamic critical field, $q = \Delta/\lambda$ (where λ is the London penetration depth and the energy associated with unit area of the filament surface is $\Delta H_c^2/8\pi$), $A = q(H_c/H_e)^2$, $p = d/2\lambda$ where d is the thickness of the superconducting lamina and J_0 and J_1 are zero and first order Bessel functions respectively. Some typical curves generated by this calculation are shown in Fig. 10.12b and it is seen that the general shapes are in good agreement with experiment. An attempted fit has been made between these curves and the experimental curve found for Nb–25% Zr (neglecting the high field tail which is probably due to the surface sheath and which is not considered in this calculation). The λ-value which must be assumed for the fit is too large but no close quantitative agreement with experiment is expected in the calculation since, as shown in section 2.5, the free energy of a laminar structure is greater than that of a tubular filamentary one, so that the laminar free energy expression which is the starting point of this calculation will not be quite right.

The critical current – longitudinal field characteristic has been calculated for an arrangement of anchored superconducting filaments in a normal medium (Blaisse and de Jong, 1965). This agrees with experiment only in the field region near to H_{c2} where the filamentary structure is stabilized by the repulsive Coulomb interaction between the fluxoids.

10.1.3. *Critical currents at intermediate field angles*

The critical current in the mixed state at intermediate angles is predicted by the Lorentz force relation:

$$\mathbf{J} \wedge (\mathbf{H} + \mathbf{B}_0) = \text{constant}$$

in the field region in which this applies and may be written as

$$I_c \propto 1/(H \sin \theta + B_0'),$$

Irreversible Properties

where θ is the angle between the field and the current and $B_0' = B_0 \sin \theta$. The results for strip samples of Nb_3Sn in a constant field at two different temperatures are shown in Fig. 10.13. It is seen that the equation is accurately obeyed except at low θ-values for both temperatures. This is also found to be the case for Pb–Tl alloys (Swartz and Hart, 1965).

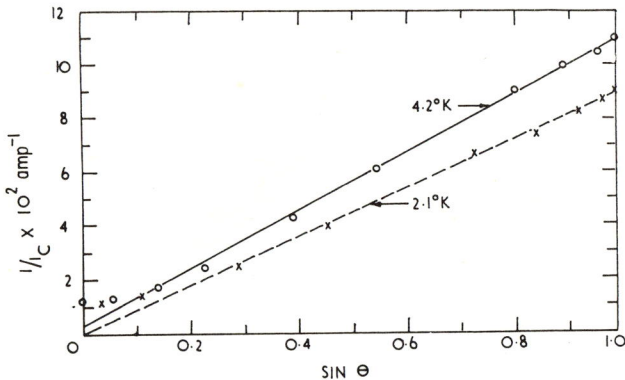

FIG. 10.13. The reciprocal of the critical current of a strip sample of Nb_3Sn as a function of the angle between the strip axis and the field at 4·2°K and 2·1°K in a field of 7·4 kOe (from Cody et al., 1964) Futher results show that the reciprocal critical current varies linearly with current up to about 100 kOe, the highest field measured.

10.2. Training

The previous history of a specimen during an experiment can influence its critical current–transverse field characteristic (Le Blanc, 1962): in particular it is found that if the transport current is increased until the superconductivity is quenched, the current taken to zero and then increased again, the quench will occur at a higher current value the second time. This behaviour is illustrated in Fig. 10.14 for a coil of Mo–49% Re wire at 1·5°K and is called *training*. The solid circles represent transitions which occur when the sample is held in a fixed transverse field and driven normal by increasing the current. The notable feature about each of these series of transitions is that they progress towards higher current densities until a maximum critical current value is obtained. It is also found that an in-increase in the critical current can take place without the wire going normal if the current is increased slowly enough in the range in which the characteristic is changing. The wire can also be trained by keeping the current constant and increasing the field again from zero after each transition to the normal state as shown by the series of open circles.

Training can be induced or annihilated in wire wound solenoids (Lubell and Mallick, 1964). The behaviour of a solenoid depends upon whether or not flux antiparallel to the self-field of the solenoid is trapped in the superconducting wire. Flux of opposite sign in the wire can result either from

Critical Current Characteristics

cycling a field external to the coil or cycling the field of the coil itself in the mode $0 \rightarrow -H \rightarrow 0$. When antiparallel flux is present no flux jumps are detected in the coil until a major flux jump occurs at about half the final critical current value. This causes the superconductivity to be quenched entirely. Subsequent measurements of the critical current value yield a small increase in the previous result until the maximum is reached. Running the solenoid in the field cycle $0 \rightarrow +H \rightarrow 0$ and then increasing the solenoid

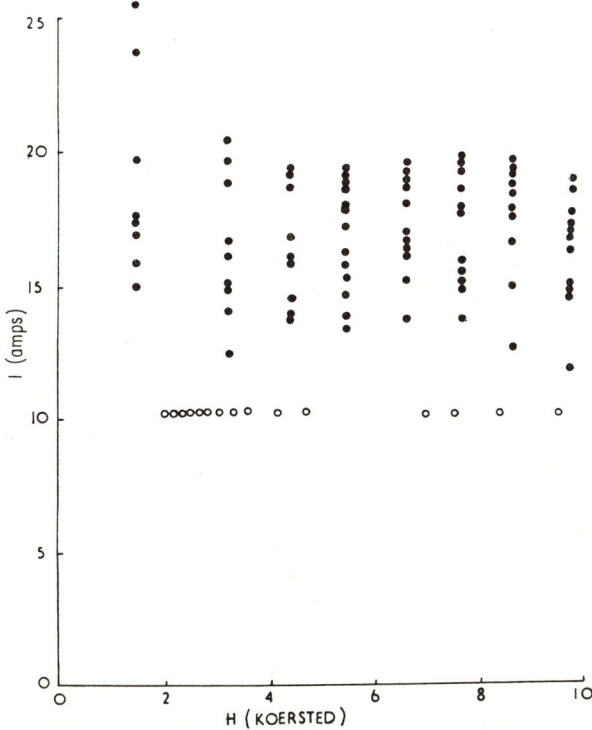

Fig. 10.14. Training in an unannealed Mo–49% Re wire at 1·5°K (from Le Blanc, 1962)

current produces a number of minor flux jumps but no training and the coil does not go normal until the previous maximum value is reached.

These results may be interpreted in terms of the flux moving within the superconducting wire. When flux is moved in the wires of the solenoid by the Lorentz force, minor flux jumps are recorded but, since these are small, only a small amount of heat is liberated in the sample. When flux of opposite sign is present in the wire, however, at the field value at which the flux produced by the current in the wire meets the fluxoids of opposite sign trapped there, they will mutually annihilate one another and a large amount of heat will be liberated locally within the wire. This will cause a catastrophic flux jump and lead to the coil going normal.

Irreversible Properties

A possible explanation of training is that when the sample becomes normal and the magnetization returns to its zero field value, it does so in the presence of a current and, although flux will be trapped opposite to the direction of the coil field, it will be less than the previous amount. On raising the transport current in the coil a higher field value will be reached than that of the previous quench and, because of the presence of this higher current the antiparallel trapped flux will be reduced even more. The effect of these cycles on the wire is shown on a magnetization curve in Fig. 10.15. In this way the transport current reduces the amount of antiparallel flux at each quench and the current will finally reach its maximum value.

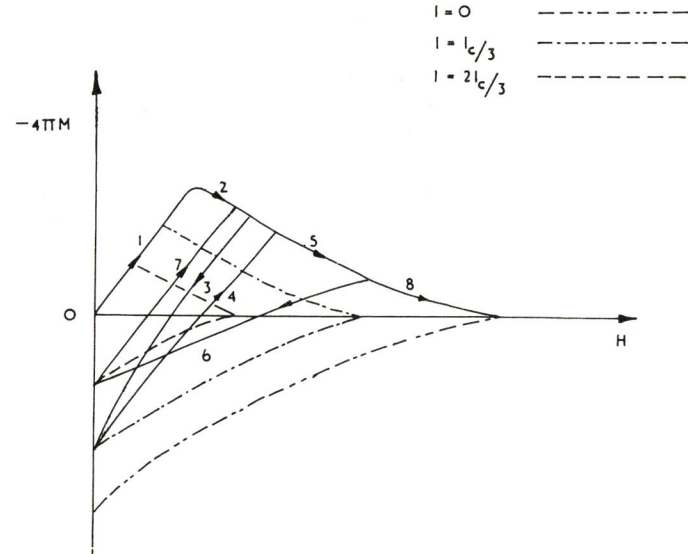

Fig. 10.15. A possible magnetization path followed by the wire in a coil during training. The maximum current is reached in this case with only two flux jumps

When the wire is coated with copper, training is very much reduced and may even be non-existent. There is no indication that the surface sheath plays a part in training and in this case the copper probably acts as a heat sink in good thermal contact with the helium bath and so removes the heat dissipated by the motion of the flux lines before the temperature of the wire is raised above its critical value.

10.3. Degradation

When a type II superconducting wire is wound into a coil, it is frequently found that the critical current is lower than that of a short length of the same material, as shown in Fig. 10.16: this effect is known as *degradation* (Donadieu and Rose, 1962). It is greater in some materials than others: Nb_3Sn shows little effect whereas the degradation is particularly marked in Nb–Zr

Critical Current Characteristics

alloy wires where the short sample characteristic can be reduced by a factor of up to fifty when the wire is wound into a coil.

The usual methods of measuring the critical current–field characteristic of a short sample are either to switch on a set magnetic field and increase the transport current until a voltage appears along the sample or to pass a fixed current and measure the magnetic field at which a voltage appears. This is known as *sequential* testing. In a coil the current and field

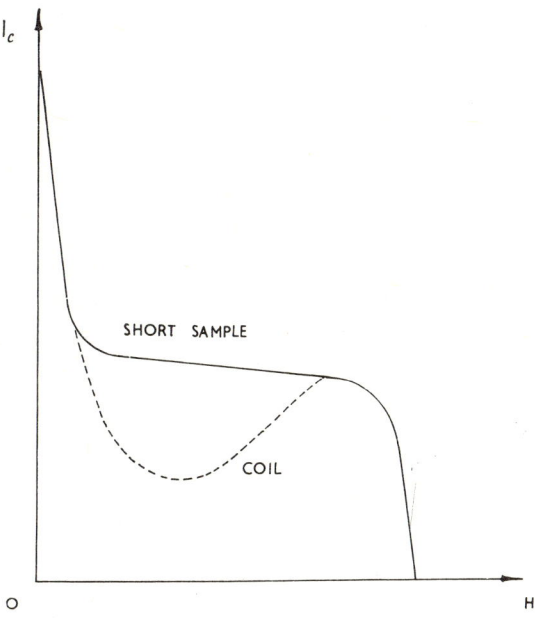

Fig. 10.16. The critical current–transverse field characteristic of a short sample of a wire and when it is wound into a coil showing degradation

rise together and it is found that, when a short sample experiences this current–field effect (known as *coil simulation*), the short wire also has a degraded characteristic (Rosner and Schadler, 1963). This is shown in Fig. 10.17 where the circles refer to short wire testing using the coil simulation technique (the sample having been taken along the lines marked A, B, C, D) and the full line represents the best coil performance of the material. The agreement is good and is seen to be quite different from the characteristic produced by sequential testing. During the coil simulation, it is also found that training occurs. With materials which show much less degradation (e.g. Nb_3Sn), training is not seen and it is apparent that training changes the critical current value by steps from the degraded to the short sample characteristic.

The effect has been investigated by monitoring the voltage across a coil carrying a fixed transport current when the external field is varied (Lubell

Irreversible Properties

et al., 1963). For small coils of Nb–25% Zr wire carrying a low transport current (in this case, less than 2 amps) if the external field is swept up and down, a series of narrow, equispaced voltage pulses is produced across the ends of the coil corresponding to flux jumps and initiated by the Lorentz force. The coil does not go normal, however, up to the maximum field applied (35 kOe). When a transport current greater than 2 amps is flowing in the coil the solenoid always goes normal when the total field (applied field plus self-field) exceeds about 6 kOe. These results show that the premature quenching of the superconductivity of the wire when it is wound into

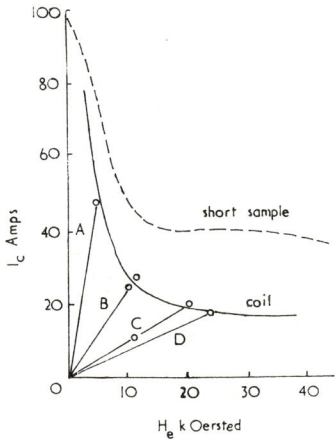

Fig. 10.17. The critical current – transverse field behaviour of short lengths of Nb–25% Zr wire at 4·2°K showing the short sample characteristic measured by sequential testing and the correspondence between coil simulation testing and coil performance (from Rosner and Schadler, 1963)

a coil (which leads to the degraded characteristic) is caused by flux jumping in the presence of a transport current. The flux jumps cause heat to be dissipated in the wire and, if the amounts of heat are large enough, normal regions are formed. Whether the normal region grows or disappears again depends upon the balance between the heat liberated in the wire and the rate at which it is transported away to the surrounding helium bath. The formation of a catastrophic flux jump is much more likely in a coil, where flux jumping from one wire will have an effect on a number of others and in which cooling is hindered by the presence of the other windings. It is apparent, therefore, that degradation will become more severe as the wall thickness of the coil increases.

The reason why degradation occurs only in the intermediate field region follows from the magnetisation curve of the material shown in Fig. 10.18. No degradation in seen up to H_{c1} since flux lines do not enter until this field value is reached and consequently the Lorentz force does not play a part. Above this field value, the flux lines enter but still no degradation occurs since the energy dissipated by moving flux lines is proportional to the square of the line velocity and, near H_{c1} the Lorentz force is small so the resulting final velocity is low. This is compensated, however, by a large number of flux lines (indicated by the large value of M in Fig. 10.18) which take part in the movement and fairly close to H_{c1} sufficient heat is liberated to cause degradation. At first the effect will increase as the field increases. At higher field values, however, although the line velocity will continue to increase, the decrease in M (and, correspondingly, the smaller number of flux lines which enter the sample) will lead to a reduction in the total energy

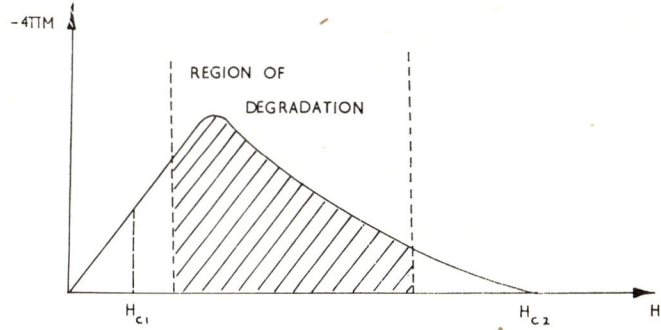

FIG. 10.18. A magnetization curve showing the large values of magnetization in the region in which degradation occurs

dissipated and the critical current in the coil will finally approach the short sample characteristic again in the high field region.

A reduction in the energy of the pinning points in the wire decreases the tendency of the material to degrade since, although the flux jumps will occur more frequently, they will be smaller and less likely to be catastrophic. The reduction in pinning point energy will decrease the short sample characteristic but in many cases an overall gain in critical current is expected.

This effect has been observed in small coils of Nb–65% Zr wire (Meyerhof and Heise, 1965) by increasing the coil temperature. That this lowers the pinning point energy has been shown by direct measurements of the pinning energy at different temperatures using the torque method described in section 8.4. On raising the coil temperature, it is found that, above a critical value, the coil no longer shows degradation but quenches at the short sample critical current level which corresponds to that temperature.

Two methods may be used to prevent degradation limiting the maximum current of a coil. Two concentric superconducting magnets wound from different materials may be used: the material for the outer coil is chosen to show no degradation (although this will probably lead to a lower short sample critical current); that of the inner coil is of high critical current though degraded) material. The outer coil is designed so that when it is

Irreversible Properties

energized it will raise the material of the inner coil above its degradation region so that the flux jumps can take place in the inner coil when no transport current is flowing. The inner coil can then be energized to produce the high field. Another method of avoiding degradation is to warm the coil above its "critical temperature of degradation" and to energize it to a field level above that of the degraded region. If it is then cooled, its ultimate critical current is raised and it may be energized to this higher critical current level.

Bibliography to Chapter 10

AUTLER, S. H., ROSENBLUM, E. S. and GOOEN, K. H. (1962), *Phys. Rev. Letters*, **9**, 489.
BERGERON, C. J. (1963a), *Appl. Phys. Letters*, **3**, 63; (1963b), *Appl. Phys. Lett.* **3**, 171.
BERLINCOURT, T. G., HAKE, R. R. and LESLIE, D. H. (1961), *Phys. Rev. Letters*, **6**, 671.
BLAISSE, B. S. and DE JONG, L. N. J. (1965), *Physica*, **31**, 326.
CALVERLEY, A., DAVIS, M. and LEVER, R. F. (1957), *J. Sci. Instr.* **34**, 142.
CODY, G. D., CULLEN, G. W. and MCEVOY, J. P. (1964), *Rev. Mod. Phys.* **36**, 95.
DE SORBO, W. (1964), *Rev. Mod. Phys.* **36**, 90.
DONADIEU, L. J. and ROSE, D. J. (1961), *High Magnetic Fields*, Wiley.
DRUYVESTEYN, W. F., VAN OOIJEN, D. J. and BERBEN, T. J. (1964), *Rev. Mod. Phys.* **36**, 58.
DRUYVESTEYN, W. F. (1965a), *Phys. Letters*, **14**, 275.
DRUYVESTEYN, W. F. (1965b), Thesis.
EL BINDARI, A. and LITVAK, M. M. (1964), *Rev. Mod. Phys.* **36**, 98.
GRASSMAN, P. and RINDERER, L. (1954), *Helv. Phys. Acta*, **27**, 309.
HAKE, R. R. and LESLIE, D. H. (1963), *J. Appl. Phys.* **34**, 270.
HAUSER, J. J. and TREUTING, R. G. (1963), *J. Phys. Chem. Solids*, **24**, 371.
HEATON, J. W. and ROSE-INNES, A. C. (1964), *Cryogenics*, **4**, 85.
KAMPER, R. A. (1963), *Phys. Letters*, **5**, 9.
LE BLANC, M. A. R. (1962), *IBM J. Res. Develop.* **6**, 122; (1963), *Phys. Rev. Letters*, **11**, 149.
LE BLANC, M. A. R. and LITTLE, W. A. (1960), *Proceedings of the 7th International Conference on Low Temperature Physics, Toronto*, p. 362. Edited by Graham, G. M. and Hollis-Hallett, A. C., University Press, Toronto.
LESENSKY, L. and NEURATH, P. W. (1963), *J. Appl. Phys.* **34**, 710.
LUBELL, M. S., CHANDRASEKHAR, B. S. and MALLICK, G. T. (1963), *Appl. Phys. Letters*, **3**, 79.
LUBELL, M. S. and MALLICK, G. T. (1964), *Appl. Phys. Letters*, **4**, 206; **5**, 39.
MEYERHOF, R. W. and HEISE, B. H. (1965), *J. Appl. Phys.* **36**, 137.
RIEMERSMA, H. (1964), *J. Appl. Phys.* **35**, 1802.
ROSNER, C. H. and SCHADLER, H. W. (1963), *J. Appl. Phys.* **34**, 2107.
SEKULA, S. T., BOOM, R. W. and BERGERON, C. J. (1963), *Appl. Phys. Letters*, **2**, 102.
SWARTZ, P. S. and HART, H. R. (1965), *Phys. Rev.* **137**, A, 818.
TEDMON, C. S., ROSE, R. M. and WULFF, J. (1965), *J. Appl. Phys.* **36**, 829.
YASUKOCHI, K., OGASAWARA, T., USUI, N. and USHIO, S. (1964), *J. Phys. Soc. Japan*, **19**, 1649.

CHAPTER 11

Applications

11.1. High field magnets

Magnetic fields greater than about 30 kOe are produced conventionally by passing a large electric current through a water-cooled copper or aluminium solenoid. This technique has been highly developed over the last 30 years and is still capable of some improvement (Bitter, 1963). The difficulties in producing magnetic fields using conventional conductors, however, seem to increase proportional to the square of the maximum field required. An idea of the power P (in megawatts) needed to produce a field H (kilo-oersteds) in a water-cooled copper solenoid with a bore diameter D (inches) is given by the equation

$$P \sim 10^{-4} DH^2.$$

It is seen that power of the order of a megawatt is required to produce 100 kOe in a 1 in. diameter bore. Most of this power input is released as joule heating in the coil and elaborate precautions must be made for its removal if the magnet is to be run continuously.

Neither the large power supply nor the means of removing the heat are needed when the resistance of the coil is zero, i.e. if it is wound with superconducting wire. The only power required by a superconducting magnet is to keep the magnet cool (so that the wire remains superconducting), to overcome the small amount of dissipation in the normal leads and to set up the magnetic field. Once the required field has been established the two ends of the wire in the coil can be joined with a superconducting link (the magnet is then in the *persistent mode*) and the field will remain even when the power supply is removed. In this case the only power required to maintain the field is that needed to cool the solenoid and for high fields this is several orders of magnitude less than that required to maintain the same field using a conventional resistive solenoid.

The electrical power requirements of a small superconducting magnet are modest (a supply giving 30 amps into a $0\cdot 1 \, \Omega$ load is one frequently used in the laboratory to produce up to 60 kOe in a 1/2 in. working space) and the magnet current may be swept or modulated with relatively little trouble.

Applications

A suitable battery driven power supply, which includes a protective circuit to interrupt the current if the magnet goes normal, is described by McAvoy (1963). A further advantage of a superconducting magnet is that, in the persistent mode, the coil forms a superconducting ring so that any spurious magnetic field changes outside the ring are compensated by changes in the current flowing in the wire and the magnetic flux within the coil will remain constant. Hence the effect of the external field changes will be greatly reduced and the region inside will be almost free from noise. The absence of large pumps which are required to circulate the cooling fluid in conventional

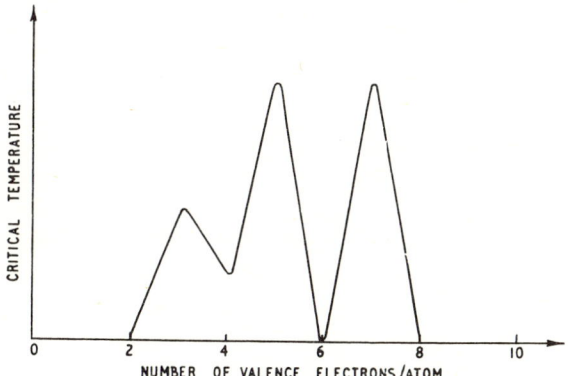

FIG. 11.1. The variation of the critical temperature of a superconductor with its number of valence electrons/atom

high field magnets also means absence of the acoustic noise–a great advantage for the magnet operator.

The low power requirement and simple cooling have lead to a large reduction in the cost of producing high magnetic fields and a high magnetic field facility need no longer be restricted to a few laboratories.

To produce higher fields over larger volumes, magnets are wound using stranded cable containing large amounts of copper (Laverick, 1965) since this makes the magnet more *stable* (less likely to go normal). Then, since current is fed to a larger cross-sectional area cable, the power supply may be required to produce up to 200 amps. This adds appreciably to the cost and also to the difficulty of taking the current into the low temperature region of the magnet coil.

The possibility of a superconducting magnet was first suggested by Kamerlingh Onnes but superconductivity in the materials known at that time was quenched in a field of less than 1 kOe. De Haas and Voogd (1931) found that some Pb-Bi alloys remain superconducting in fields of greater than 20 kOe. Subsequent measurements showed the critical current densities to be low, however, and a prohibitively large amount of wire would have been required to produce high fields. The element helium was scarce and also difficult to liquify and, since this material would have produced only fields already

Irreversible Properties

available using conventional magnets, it was not applied to magnet production. A further 24 years passed before the possibility of high field production using superconductors was demonstrated by Yntema (1955) who produced an electromagnet with niobium wire carrying a current density of almost 10^5 amps/cm² in a field of 5 kOe.

The major advance in the application of superconductivity to producing high fields came with the discovery, by Kunzler *et al.*, (1961), that Nb_3Sn will carry a current density of greater than 10^5 amp/cm² in a field of 88 kOe.

Since then, as mentioned in section 10.1.1, a large number of alloys, and particularly those having a β-tungsten structure similar to Nb_3Sn, have been examined for high field superconductivity. Other materials which may be useful in the future are V_3Ga ($H_{c2} \sim 350$ kOe) and V_3Si ($H_{c2} \sim 150$ kOe). In the absence of a completely detailed theory of type II superconductivity, the search for new materials has been largely empirical, based on the rules put forward by Matthias (1957) that superconductivity is found only in metallic elements with between 2 and 8 valence electrons and that, of these materials, the ones with the highest critical temperatures (and, therefore, probably the highest critical fields) are those with valence electron/atom ratios of 3, 5 or 7 as shown in Fig. 11.1.

Only Nb–Zr and Nb–Ti alloys with critical fields of the order of 100 kOe are normally ductile and can easily be wound into coils. Attempts have been made to overcome the brittle nature of Nb_3Sn by pressing niobium and tin powders into niobium tubing and firing this to form the compound after the coil has been wound, while a ductile form of Nb_3Sn can be made in the form of a thin film by vapour deposition (Heraeus). A number of other methods of producing Nb_3Sn in a ductile form have also been proposed. Thin films of pure materials separated by insulated layers may be grown by evaporation and thin filaments can be produced using Bean's method (Section 8.2) of pressing the material into a porous matrix. Finely divided superconducting/non-superconducting metal structures have also been suggested (Berlincourt, 1963) as possible magnet materials.

The particular design problem of a superconducting coil is to minimize the length of the wire which will produce the maximum field with the required homogeneity, since the cost of the wire is a major factor in the design. The method of designing rectangular cross-section superconducting solenoids has been considered in detail by Boom and Livingston (1962) and a pair of design curves which allow the optimum shape and minimum length of wire required to be found particularly easily are given by Thomas and Bright (1966). Design considerations of Helmholtz pair arrangements of superconducting coils have been considered by Day (1963). A detailed description of solenoid design and construction for those who wish to make their own magnets may be obtained from Supercon,† one of the firms manufacturing superconducting materials suitable for winding magnets.

---

† Norton International Inc., 9 Erie Drive, NATICK, Massachusetts, 01760.

Applications

Although the power required to keep the magnet cool is relatively small it can still be appreciable and attempts must be made to minimize it. To reduce the heat leak from the outside, small diameter leads are required going into the cryostat, while to keep the amount of joule heating in the leads small, low resistance (i.e. large diameter) leads are needed. This dilemma has been overcome, at least on a laboratory scale, by "pumping" the flux into the solenoid (Goedemoed et al., 1965). The principle of a simple flux pump

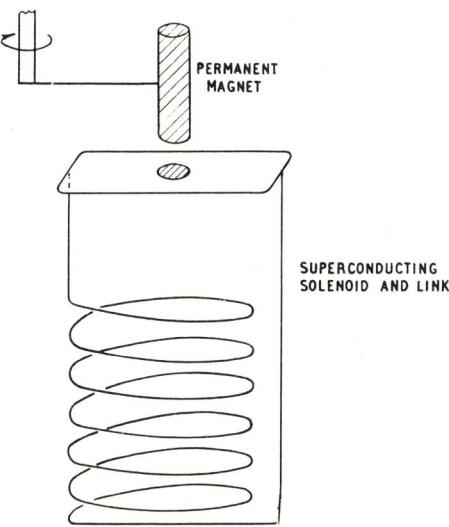

Fig. 11.2. The principle of a simple flux pump. The permanent magnet revolves on an arm. Each time it passes over the superconducting strip the small region under the magnet is driven normal so that the magnetic circuit of the solenoid is broken and another "packet" of flux added

is shown in Fig. 11.2. The two ends of the superconducting coil are joined with a superconducting strip. A magnet strong enough to drive the part of the strip directly beneath it normal, is held on an arm. With each rotation of this arm, the magnetic circuit of the solenoid is broken and flux added, inducing more current in the wire. A refinement of this device has been built (van Beelen et al., 1963) which will induce 175 amps (corresponding to a field of 25 kOe in this particular case) in a Nb–25% Zr wire wound coil.

Moving parts have a practical disadvantage in a cryostat since their movement will generate heat. Laquer (1963) has shown that flux can be introduced into the solenoid using a series of thermally switched circuits. This method is successful but the thermal switches still introduce unnecessary heat into the cryostat. A much smaller heat loss has been achieved (van Houwelingen et al., 1964) by producing the moving pattern of magnetic flux using an a.c. arrangement. By modifying the stator windings of a squirrel

Irreversible Properties

cage three-phase motor, the flux pattern is made to rotate with no mechanical moving parts in the liquid helium.

With all the methods of flux pumping so far used the amount of energy stored in each cycle is small. For this reason they are confined to use with small magnets since the time taken to energize a large magnet would be prohibitive.

As already mentioned, a trip circuit must be built into the power supply of a superconducting magnet to interrupt the current if the wire in the

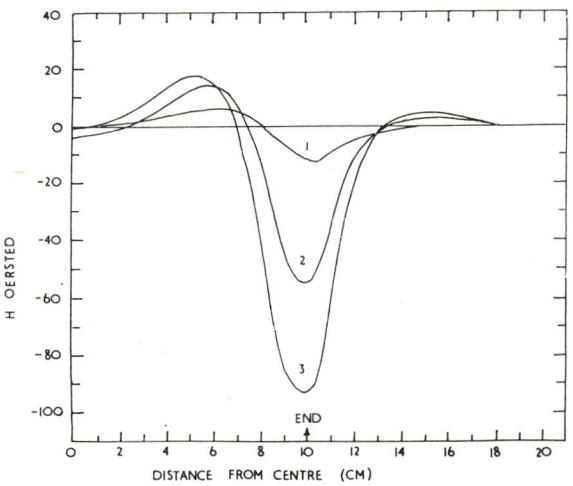

Fig. 11.3. Remanent axial magnetic fields in a 20 cm long superconducting solenoid. The curves are symmetric about the centre and were obtained after producing magnetic fields of (1) 2·5 kOe, (2) 5·6 kOe, and (3) 12·5 kOe (from Anderson and Sarwinski, 1963)

solenoid goes normal. Even so the heat produced by the large current in the windings during the decay may cause considerable damage. There are a number of ways of protecting the windings: a low resistance may be placed in parallel with the coil to carry a large part of the current when the coil goes normal; an inductance may be coupled with the windings; or the coil may be divided into a number of small sections. Analysis of the possible circuits (Smith, 1963) has shown that, for coils with inductance less than about one henry and for a quenching rate of 10^3 Ω/sec, the coil temperature does not increase by more than 500°K and it can be assumed that no damage will be done. Hence small coils are self-protecting and one way of improving the protection of a large superconducting magnet is to make it up from a number of solenoids each with an inductance of one henry or less. For larger coils inductance protection may best be achieved by interleaving large quantities of copper strip between the superconducting windings (Kantarowicz and Stekley, 1965) or by winding stranded cables with several copper strands as well as the superconducting ones. The flux

Applications

jumps are then much smaller and less likely to drive the coil normal. The copper lengthens the decay time and also absorbs some of the energy released. In this case the magnet can be completely stable, i.e. part of the windings can go into the normal state without causing the catastrophic transition of the rest. A combination of all three ways of protecting the coil may be used although the high voltage induced when the current decays can damage an external shunt and this method is not favoured.

Compared with a conventional magnet, the use of a superconducting magnet to produce high magnetic fields is not entirely without disadvantage. Probably the most important is that the superconducting magnet must be cooled to liquid helium temperatures in order that the wire should be superconducting although this is no longer a major problem in a labotatory. A number of other disadvantages of superconducting magnets arise from the presence of pinning points which will, of necessity, be in the wire. The pinning points impede flux motion through the wire and cause the flux movement to be irregular so that the field in the coil will not rise smoothly even if the current through the wire is increasing at a steady rate. But these variations are very small in coils containing a large amount of copper (i.e. highly stabilized). (These irregular field variations will also be smoothed out to some extent by currents induced in the metal former on which the magnet is wound.) More important is that the flux jumps may be large enough to cause the magnet to go normal at fields much lower than the critical value. The pinning points will also cause flux to be trapped inside the wire so that the field in the magnet will not return to zero when the current is removed. The remanent flux along the centre of a solenoid will depend upon the maximum field to which the magnet is taken. Profiles which give some idea of the remanent flux in a 20 cm long magnet, taken to three different fields, are shown in Fig. 11.3. It is seen that the remanent field at the centre of the magnet is low in this case even after cycling to the highest field but the central field does increase as the coil becomes shorter. Furthermore, because of the presence of this trapped flux, if the current is reversed in order to reverse the field round the sample, unless the coil is highly stabilized a large amount of flux jumping will take place in the magnet (see section 9.2). These are relatively small difficulties, however, and may be neglected in many applications. At the present state of the technology, superconducting magnets producing 100 kOe or more in a $^1/_2$ in. gap are available commercially (see Appendix II).

11.2. Other magnetic field applications

11.2.1. *Flux trapping and shielding*

When a type II superconductor is cooled in a magnetic field or when a magnetic field is applied to a type II superconducting cylinder and is then decreased again, if the field is large enough to penetrate into the cylinder

Irreversible Properties

then, in each case, persistent currents are induced in the cylinder walls and a magnetic field is trapped in the ring. The variation of the field within the walls of a hard superconducting tube when the external field is changed has already been described in detail in section 9.1. In the bore of a small cylinder (with about 0·5 cm wall thickness) of suitably treated Nb_3Sn, fields of greater than 50 kOe have been trapped. The field inside the cylinder will remain effectively constant as long as it is superconducting and it may, therefore, be used as a very high field permanent magnet.

The field in the bore of the cylinder at the moment that it becomes superconducting will remain constant until a very high external magnetic field value is reached. Hence, if a suitable high field superconducting cylinder is cooled in zero field it can maintain a field free region in the bore until the external field reaches at least 50 kOe.

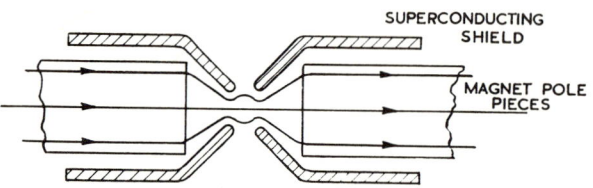

FIG. 11.4. The magnetic flux concentrator. The flux cannot pass into the superconductor so it is compressed into a space governed by the size of the hole in the shield. A narrow slot (not shown) is necessary to allow the flux into the cylinder initially

11.2.2. *Flux concentration*

The field which is trapped in a superconducting cylinder when the external field is increased around it and then reduced again depends upon the maximum value to which the external field can be taken. A particular cylinder may be capable of trapping a much higher field than this. Greater fields can be produced in a cylinder which already contains trapped flux by inserting a superconducting plunger to fill part of the bore (Swartz and Rosner, 1962). The flux is then compressed into the volume of the bore remaining unfilled and the field increases. As the internal field rises, more and more flux is lost into the walls of the cylinder, however, so that the ratio of the final to the initial field is reduced (Goldsmid and Corsan, 1964), and this sets a practical limit to the amount of compression which may be achieved by this method.

The field of an iron-cored magnet can also be concentrated by fitting superconducting shields over the pole pieces. The principle is shown in Fig. 11.4.

11.2.3. *Magnetic bearings*

The possibility of flux trapping and flux exclusion suggests that a superconducting cylinder and disc may be used as a magnetic bearing (Buchhold,

1960). The principle of the device is shown in Fig. 11.5. The bearing is restricted in its lateral movement but can spin on a cushion of flux. If the material of the moving part of the bearing is chosen so that no flux will penetrate into it, the only friction in the bearing will come from the viscosity of the surrounding medium and this may be made very small.

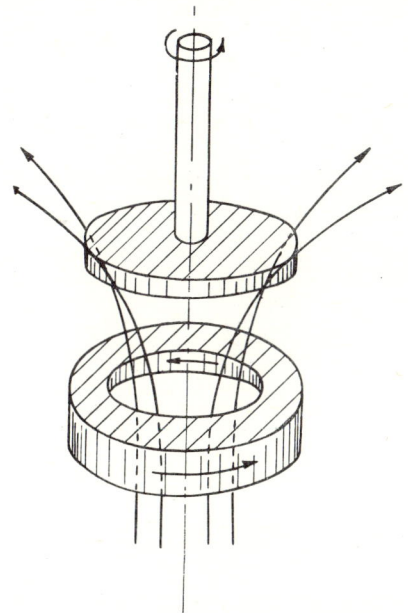

Fig. 11.5. The superconducting bearing. Regions shaded are superconducting

11.2.4. *Energy storage*

Another possible use of a superconducting coil is as an energy store (Weiderhold and Ameen, 1963). A trapped field of H oersteds stores an energy of $H^2/8\pi$ ergs/cm^3 so that a trapped field of 100 kOe corresponds to a stored energy of about 10^8 ergs/cm^3. A conventional capacitor bank stores about 10^6 ergs/cm^3. There is, however, difficulty in regaining the electrical energy quickly from the superconducting energy store since the inductance of the coil may be high and this will lead to a long time constant in the withdrawal circuit. An analysis of the problem and a review of the factors which may affect the time constant have been given by Ameen and Weiderhold (1964).

11.2.5. *D.C. transformer*

The arrangement for a d.c. transformer using thin films of a type II superconductor is shown in Fig. 11.6. In the presence of a transport current, lines of magnetic flux are moved as if under the action of a Lorentz force (see section 9.1). If a magnetic field is established around the device and the input voltage applied to the wider strip, then the flux lines in the primary

Irreversible Properties

circuit will move as shown in the figure, cutting the secondary and, therefore, inducing a voltage at the output (Giaever, 1965). The efficiency of the transformer is only about 10% but the voltage output may be made larger by increasing the number of thin film units to which the primary voltage is applied.

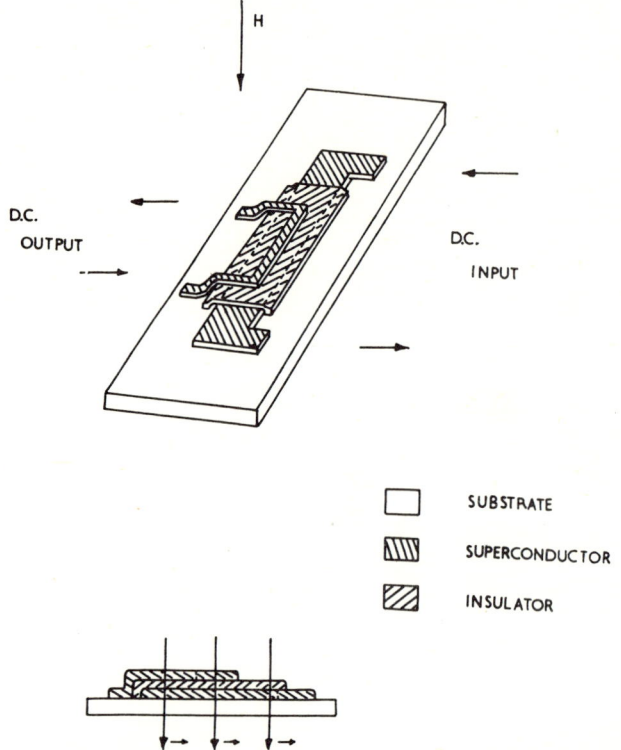

FIG. 11.6. The thin film arrangement of a d. c. transformer using type II superconductors

This device will work using not only type II but also type I superconductors in the same thin film arrangement since the magnetic flux which penetrates into the latter when they are in the intermediate state is acted upon by the current in the same way as the flux in a type II superconductor in the mixed state.

11.3. D.C. power transmission

Since there is no joule heat loss in a superconducting cable it would, at first sight, seem preferable to use superconducting cables for d.c. power transmission. To use such cables, however, large amounts of power would be lost in rectifying the alternating current coming from the generators and also in keeping the cable cool. The power loss in conventional cables is

Applications

fairly small ($\sim 10\%$) so that, with the present materials and cost of rectification and cooling, d.c. power transmission using superconductors appears to be uneconomic at the time of writing. However, the cost of cooling is reducing and in applications where direct current is already used, e.g. in parts of the chemical industry, the time may soon come when the costs of using superconducting and conventional cables will break even.

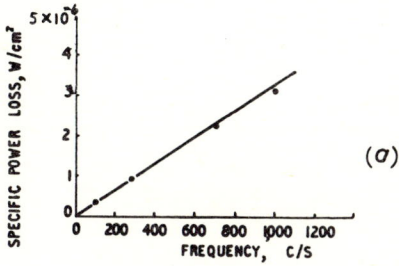

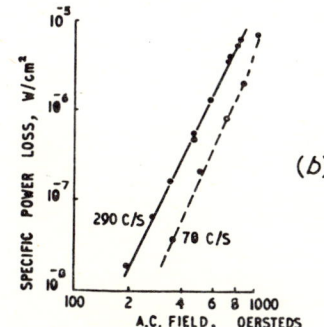

Fig. 11.7. The variation of power loss niobium (a) in a steady field of 300 Oe as a function of frequency and (b) in an a. c. field at 70 c/s and 290 c/s (from Buchhold and Molenda, 1962)

11.4. A.C. applications

11.4.1. Current carrying characteristics

No power loss is expected in type I superconductors until the frequency (ν) of the alternating current is high enough to give the electrons energy ($h\nu$) to jump the superconducting energy gap. This frequency is of the order of 10^{13} c/s for most superconductors. Experiments on type II superconductors show that losses appear at much lower frequencies and may be quite large. The variation of power loss with frequency at constant field and with field at constant frequency for a niobium block are shown in Fig. 11.7 (Buchhold and Molenda, 1962). The calorimetric technique used to obtain these results

Irreversible Properties

is very sensitive and losses can be observed even in well annealed high purity lead with a current density below the critical value.

The variation of the critical alternating current with specimen size for toroidal Pb–Bi eutectic has been investigated by Kamper (1962) who found that the loss factor was independent of the frequency up to 100 c/s. Similar measurements have been made with different sizes of Nb–25% Zr wires each with the same amount of cold work (Young and Schenk, 1964). These results show that there is a logarithmic decrease in the r.m.s. critical current density with increasing wire thickness at all frequencies and also that the critical current decreases logarithmically with frequency when the wires are in coil form. The experiments indicate that the critical alternating current density occurs at the same power level for different frequencies.

Two physical models have been put forward to account for the origin of the a.c. losses. Buchhold (1963) suggests that the losses are due to imperfections on, or just beneath, the surface. These trap flux which has penetrated into the superconductor and therefore cause hysteresis when an alternating field or current is applied. The model has been tested by calculating the voltage variation in a search coil wrapped around a superconducting block. Smoothed out curves of the shapes predicted by the theory have been found experimentally. An analysis of the power loss in a type II superconductor has also been made when a field is applied parallel to a flat slab (London, 1963).

Energy is dissipated in the form of heat when a flux line moves through a type II superconductor (see section 9.1). Hence, in all type II superconductors, once they are in the mixed state, heat is generated when an alternating current is applied.

11.4.2 *Power transmission*

The high cost of rectification suggests that power might be transmitted by alternating current in a superconducting cable. The results surveyed in section 11.4.1 show that such cables have an appreciable power loss, however, so there may be no advantage gained in transmitting alternating current. Once again the cost of insulation and cooling determine the cost of using the cable. Although at the present time, the use of superconducting cables for a.c. power transmission is not economic the time may soon come when they will offer a substantial advantage in size and cost over conventional cables.

11.4.3. *Power transformers and machines*

The reduction in the copper loss which would result from the use of superconductors in heavy electrical engineering machinery and transformers could lead to an appreciable decrease in operating cost as well as allowing the weight and size of the equipment to be reduced. The power of a current generator is limited by the dimensions of the machine. This must be trans-

Applications

ported by road from the factory to the electrical plant and the size is, therefore, limited by the width of the road. A design study of the use of superconducting windings in power transformers (Wilkinson, 1963) indicates that, if a thin film superconductor capable of carrying large currents in high fields were available, then the coil and core weight and, therefore, size could be reduced by almost 40%.

The advantages that superconductors have in heavy engineering applications have to be set against one great disadvantage: they must be maintained at a very low temperature. For power transmission and for heavy machinery and transformers, the cost of continuous cooling more than outweighs the advantages, at present. However, the reduction in the cost of producing high magnetic fields using superconducting solenoids indicates that other major benefits may result from using superconductors. Benefits may also be expected from the use of the high fields produced by superconducting magnets in other devices, e.g. in the production of power using magnetohydrodynamics and possibly by controlled fusion. More efficient insulation, lower cooling costs and, possibly, the discovery of superconductors with much higher critical temperatures will all add to the possibility of widening the applications of type II superconductivity.

Appendix to Chapter 11

Superconducting magnets and the necessary ancillary equipment (power supply, sweep unit and cryostat) may be obtained from:

Centre de Recherches de la Compagnie Générale d'Electricité, Route de Nozay, 91, MARCOUSSIS, France.

Magnion, 144, Middlesex Turnpike, BURLINGTON, 01804, U.S.A.

Oxford Instrument Company, Osney Mead, OXFORD, England.

Siemens-Schuckertwerke Aktiengesellschaft, Technische Stammabteilung Ausland, 8520 ERLANGEN, 2, Germany.

Sodern, 10, Rue de la Passerelle, SURESNES (Seine), France.

Varian associates, 611, Hansen Way, PALO ALTO, California, 326−4000, U.S.A.

Bibliography for Chapter 11

AMEEN, D. L. and WIEDERHOLD, P. R. (1964), *Rev. Sci. Inst.* **35,** 733.
ANDERSON, A. C. and SARWINSKI, R. J. (1963), *Rev. Sci. Inst.* **34,** 298.
BERLINCOURT, T. G. (1963), *Brit. J. Appl. Phys.* **14,** 749.
BITTER, F. (1963), *Brit. J. Appl. Phys.* **14,** 759.
BOOM, R. W. and LIVINGSTON, R. S. (1962), *Proc. IEEE,* **50,** 274.
BUCHHOLD, T. A. (1960), *Sci. Amer.* **202**(3), 74; (1963), *Cryogenics,* **3,** 141.
BUCHHOLD, T. A. and MOLENDA, P. J. (1962), *Cryogenics,* **2,** 344.
DAY, J. D. A. (1963), *J. Sci. Instrum.* **40,** 583.
DE HAAS, W. J. and VOOGD, J. (1931), *Leiden Comm.* 212C.
GIAEVER, I. (1965), *Phys. Rev. Letters,* **15,** 825.
GOEDEMOED, S. H., VAN KOLMESCHATE, C., METSELAAR, J. W. and DE KLERK, D (1965), *Physica,* **31,** 573.
GOLDSMID, H. J. and CORSAN, J. M. (1964), *Phys. Letters,* **10,** 39.
HERAEUS, W. C., British Patent, 1.008.132.
KAMPER, R. A. (1962), *Phys. Letters,* **2,** 290.
KANTAROWICZ, A. R. and STEKLEY, J. Z. Z. (1965), *Appl. Phys. Letters,* **6,** 56.
KUNZLER, J. E., BUEHLER, E., HSU, F. S. L. and WERNICK, J. H. (1961), *Phys. Rev. Letters,* **6,** 89.
LAQUER, H. L. (1963), *Cryogenics,* **3,** 27.
LAVERICK, C. (1965), *Cryogenics,* **5,** 152.
LONDON, H. (1963), *Phys. Letters,* **6,** 162.
MATTHIAS, B. T. (1957), *Progress in Low Temperature Physics,* Vol. II, p. 138, C. J. Gorter, ed., Interscience.
MCAVOY, B. R. (1963), *Rev. Sci. Inst.* **32,** 200.
SMITH, P. F. (1963), *Rev. Sci. Inst.* **34,** 368.
SWARTZ, P. S. and ROSNER, C. H. (1962), *J. Appl. Phys.* **33,** 2292.
THOMAS, E. J. and BRIGHT, C. D. (1966), *Cryogenics,* **6,** 10.
VAN BEELEN, H. and GORTER, C. J. (1963), *Physica,* **29,** 896.
VAN HOUWELINGEN, D., ADMIRAAL, P. S. and VAN SUCHTELEN, J. (1964), *Phys. Letters,* **8,** 310.
WIEDERHOLD, P. R. and AMEEN, D. L. (1963), High Magnetic Fields Conference, Oxford.
WILKINSON, K. J. R. (1963), *Proc. IEE,* **110,** 2271.
YNTEMA, G. B. (1955), *Phys. Rev.* **98,** 1197.
YOUNG, F. J. and SCHENK, H. L. (1964), *J. Appl. Phys.* **35,** 980.

Author index

Abrikosov, A. A. 40, 41, 43, 50, 52, 54, 60, 65, 80, 156, 160, 178, 191, 192, 205
Admiraal, P. S. 285
Alers, G. A. 220, 226
Alers, P. 209, 226
Ambegaokar, V. 191, 192, 204, 205
Ameen, D. L. 279, 285
Anderson, A. C. 276, 285
Anderson, P. W. 139, 156, 160, 183, 192, 230, 231, 233, 247
Antonini, N. 80
Arp, V. 51, 80
Autler, S. H. 80, 259, 271

Baltenspeger, W. 163, 192
Bardeen J. 10, 21, 125, 156, 192, 199, 205, 233, 236, 247
Bean, C. P. 212, 213, 214, 216, 226, 230, 238, 246, 247, 274
Beasley, M. R. 240, 247
van Beelen, H. 275, 285
Berben, T. J. 89, 120, 271
Bergeron, C. J. 262, 271
Berlincourt, T. G. 174, 192, 258, 271, 274, 285
Bitter, F. 272, 285
Blaisse, B. S. 257, 263, 271
Blatt, J. M. 156
Blaugher, R. D. 79, 88, 121
de Blois, R. W. 217, 226
Bogoliubov, N. N. 156
Bon Mardion, G. 88, 89, 121, 147, 156
Boom, R. W. 271, 274, 285
Boorse, H. A. 205
Bremmer, H 198, 205
Bright, C. D. 274, 285
Brown, A. 193, 205
Buchhold, T. A. 278, 281, 282, 285
Budnick, J. I. 224, 226
Buehler, E. 285

Calverley, A. 253, 271
Cambell, A. M. 222, 223, 226, 247
Cape, J. A. 173, 177, 192

Cardona, M. 88, 89, 121
Caroli, C. 81, 121, 136, 146, 147, 150, 156, 172, 182, 191, 202, 205
Casimir, H. B. G. 5, 21
Chandrasekhar, B. S. 158, 197, 247, 271
Clogston, M. M. 158, 192
Cody, G. D. 199, 205, 261, 264, 271
Cohen, R. W. 199, 205
Cooper, L. N. 10, 21, 25, 122, 123, 125, 156, 192, 205
Corak, W. S. 193, 205
Corsan, J. M. 237, 241, 243, 244, 247, 278, 285
Craig, P. P. 89, 121
Cribier, D. 39, 40, 41, 46, 71, 80, 218, 226
Cullen, G. W. 271
Cyrot, M. 156, 191, 192, 203, 204, 205

Davis, M. 271
Day, J. D. A. 274, 285
Deaver, B. S. 45, 80
Decell, R. T. 247
Delrieu, J. H. 74, 80
Deutscher, G. 120, 121
Devlin G. E. 225, 226
Dew-Hughes, D. 219, 226, 247
Doll, R. 45, 80
Donadieu, L. D. 266, 271
Doulat, J. 224, 226
Doyle, M. V. 238, 247
Druyvesteyn, W. F. 16, 88, 89, 121, 230, 233, 247, 248, 249, 250, 257, 261, 271
Dubeck, L. 200, 205
Dzyaloshinski, I. E. 156

El Bindari, A. 258, 271
Essemann, U. 73, 212, 226
Evetts, J. E. 215, 216, 226, 241, 247

Faber, T. E. 10, 21, 90, 121, 225, 226
Fairbank, W. M. 45, 80
Farnoux, B. 40, 71, 80, 226
Fawcett, E. 247
Ferrell, R. A. 159, 160, 165, 167, 168, 192

287

Author index

Fetter, A. L. 62, 80
Fietz, W. A. 247
Fink, H. J. 112, 115, 116, 117, 121
Finnemore, D. K. 156
Fite, W. 75, 80
Fleischer, R. L. 226
Friedel, J. 63, 80, 220, 226
Frohlich, H. 125, 156
Fulde, P. 159, 165, 167, 168, 192

de Gennes, P. G. 26, 40, 44, 65, 74, 80, 81, 90, 93, 117, 126, 143, 145, 156, 180, 182, 184, 185, 186, 187, 189, 191, 192, 216, 217, 226, 234, 245, 247, 255, 260, 261
Giaever, I. 181, 192, 231, 247, 280, 285
Ginzburg, V. L. 23, 24, 25, 26, 28, 29, 35, 40
Goedemoed, S. H. 240, 243, 244, 247, 275, 285
Goldsmid, H. J. 241, 247, 278, 285
Goldstein, Y. 90, 121
Goodman, B. B. 16, 21, 37, 40, 67, 68, 80, 120, 150, 156, 193, 205, 218, 226
Gooen, K. H. 271
Gorkov, L. P. 25, 28, 29, 40, 122, 145, 149, 151, 152, 153, 160, 178, 191, 192, 205
Gorter, C. J. 5, 20, 21, 228, 247, 285
Gossard, A. C. 74, 80
Grassman, P. 261, 271
Grenier, C. G. 247
Griffin, A. 191, 192, 201, 205
Gruenberg, L. 159, 168, 169, 192
Guyon, E. 88, 89, 121, 156, 186, 188, 190, 192
Gunther, L. 159, 168, 169, 192

de Haas, W. J. 198, 205, 273, 285
Hake, R. R. 174, 192, 193, 205, 252, 253, 271
Harden, J. L. 51
Hart, N. R. 89, 100, 107, 109, 121, 226, 256, 258, 259, 260, 261, 264, 271
Hauser, J. J. 258, 271
Healey, W. A. 211, 226, 227, 247
Heaton, J. W. 251, 261, 271
Hecht, R. 209, 226
Heise, B. H. 218, 221, 226, 269, 271
Helfand, E. 153, 155, 156, 168, 192
Hempstead, C. F. 89, 192, 247
Heraeus, W. C. 274, 285
Hohenberg, P. C. 62, 80, 192
van Houwelingen, D. 275, 285
Hsu, F. S. L. 285
Hurault, J. P. 120, 121

Ittner, W. G. 3, 21

Jaccarino, V. 80
Jacrot, B. 40, 71, 80, 226
Jaklevic, R. C. 211, 226
Joiner, W. C. H. 67, 80, 88, 121
Jones, D. P. 214, 226
Jones, R. G. 231, 247
de Jong, L. N. 257, 263, 271
Joseph, A. S. 88, 89, 100, 121, 217, 226
Josephon, B. D. 231, 247
Jurisson, J. 224, 225, 226

Kamper, R. A. 248, 271, 282, 285
Kammerer, D. F. 89, 121
Kammerling Onnes, H. 3, 21
Kantarowitz, A. R. 276, 285
Keesom, P. H. 155, 156
Kessinger, R. D. 112, 115, 116, 117, 121
Khotkevich, V. I. 21
Kim, Y. B. 89, 174, 177, 192, 228, 229, 233, 234, 238, 247
Kinsel, T. 64, 80, 152, 156
Kleiner, W. M. 60, 80
Klemens, P. G. 197, 205
de Klerk, D. 247, 285
van Kolmeschate, C. 247, 285
Kropschot, R. H. 89
Kunzler, J. E. 205, 274, 285

Lacaze, A. 89, 121, 156
Lamb, J. 226
Landau, L. D. 22, 23, 24, 25, 26, 27, 28, 29, 35, 40, 80
Lange, F. 241, 247
Laquer, H. L. 275, 285
Lasher, G. 58, 80
Laverick, C. 273, 285
Le Blanc, M. A. R. 240, 247, 248, 253, 258, 264, 265, 271
Lesensky, L. 248, 271
Leslie, D. H. 252, 253, 271
Lever, R. F. 271
Lewin, J. D. 247
Lifshitz, E. M. 40
Lindenfeld, P. 198, 204, 205
Little, W. A. 253, 258, 271
Litvak, M. M. 258, 271
Livingston, J. D. 215, 216, 226
Livingston, R. S. 274, 285
London, F. 8, 21, 27, 31, 44, 80
London, H. 7, 21, 27, 282, 285
Lubell, M. J. 242, 247, 264, 267, 271
Lynton, E. A. 11, 64, 80, 156, 181, 192, 205, 226

Madhav, Rao, L. 226
Maita, J. P. 205

288

Author index

Maki, K. 90, 121, 143, 145, 146, 153, 156, 160, 170, 173, 174, 175, 191, 192, 204, 205
Mallick, G. T. 247, 264, 271
Marcus, P. 44, 80, 220, 226
Martinet, A. 121, 192
Matricon, J. 51, 62, 70, 80, 121, 156, 190, 192, 217, 226, 237, 245, 247
Matthias, B. T. 274, 285
Mauro, S 186, 189, 192
McAvoy, B. R. 273, 285
McConville, T. 15, 147, 153, 155, 156, 194, 195, 196, 205
McEvoy, J. P. 243, 247, 271
Mendelssohn K. 198, 205, 212, 226
Meissner, W. 3, 21
Mercereau, J. E. 211, 212, 226
Metselaar, J. W. 285
Meunier, F. 156, 172, 186, 192
Meyerhof, R. W. 269, 271
Mochel, J. M. 201, 202, 205
Molenda, P. J. 281, 285
Moore, J. R. 205
Morin, F. J. 194, 205
Morris, G. 204, 205
Muhlschlegel, B. 133, 156

Nabauer, M. 45, 80
Narlikar, A. V. 221, 226
Neurath, P. W. 248, 271
Neuringer, L. J. 174, 177, 192
Niessen, A. K. 235, 236, 247
Novak, R. L. 247

Oakes, J. A. 224, 225, 226
Ochsenfeld, R. 3, 21
Ogasavara, T. 271
Olsen, J. L. 89, 198, 205
van OOijen, D. J. 89, 121, 271
Onnes, H. K. 273
Orsay Group, 147, 156

Park, J. G. 214, 226
Parks, R. D. 201, 202, 203, 205
Paskin, A. 89, 121
Pincus, P. 62, 80, 192
Pippard, A. B. 10, 21, 223, 224, 226
Pratt, I. P. 219, 226

Quinn, D. J. 3, 21

Radebaugh, R. 155, 156
Rao, G. N. 247
Rao, L. M. 40, 71, 80
Redfield, A. G. 74, 80

Reed, W. A. 235, 247
Reese, W. 197, 205
Reif, F. 191, 192
Renard, M. 226
Eeynolds, J. M. 247
Rhoderick, E. H. 247
Rickaysen, G. 156, 199, 205
Riemersma, M. 254, 255, 271
Rinderer, L. 261, 271
Rhabinin, J. N. 21
Rohrer, H. 205
Rollins, R. W. 223, 226, 247
Rose, D. J. 266, 271
Rose, R. M. 271
Rose-Innes, A. C. 247, 251, 261, 271
Rosenberg, H. M. 196, 205
Rosenblum, E. S. 271
Rosenblum, M. B. 88, 89, 121
Rosner, C. H. 267, 268, 271, 278, 285
Roth, L. M. 80

Sandiford, D. J. 214, 226
Saint-James, D. 81, 93, 107, 121, 159, 166, 171, 192, 217, 256, 260, 261
Sarma, G. 74, 80, 159, 163, 166, 191, 192
Sarwinski, R. J. 276, 285
Satterthwaite, C. B. 205
Schadler, H. W. 267, 268, 271
Schawlow, A. L. 225, 226
Schenk, H. L., 282, 285
Schrieffer, J. R. 10, 21, 125, 156, 192, 205
Schubnikov, L. W. 13, 21
Schweitzer, D. G. 89, 121, 214, 226
Sekula, S. T. 261, 262, 271
Seraphim, D. P. 90, 121, 152, 156, 220 226
Serin, B. 15, 18, 21, 64, 80, 88, 89, 147, 153, 194, 195, 196, 205, 226
Shapira, Y. 174, 177, 192
Shepelev, J. D. 21
Sherwood, R. C. 205
Shirkov, D. V. 156
Silcox, J. 223, 226, 247
Silsbee, F. B. 5, 21
Silver, A. H. 226
Sladek, R. J. 205
Smith, P. F. 241, 276, 285
de Sorbo, W. 211, 217, 226, 227, 247, 258, 271
Spurway, A. H. 247
Staas, F. A. 235, 236, 247
Stekley, J. Z. Z. 276, 285
Stephen, M. J. 233, 247
Steyert, W. A. 197, 205
Sirnad, A. R. 174, 176, 177, 192, 230, 231, 232, 238, 247

Author index

Stromberg, T. F. 156
Strongin, M. 89, 90, 121
van Suchtelen, J. 285
Swartz, P. S. 89, 100, 107, 109, 121, 216, 226, 256, 258, 259, 260, 261, 264, 271, 278, 285
Suhl, H. 238, 239, 247
Swenson, C. A. 156

Tedmon, C. S. 253, 258, 271
Tewordt, L. 155, 156, 199, 205
Thomas, E. J. 238, 247, 274, 285
Thomson, R. S. 156, 186, 192
Tinkham, M. 107, 121, 184, 185, 204, 205, 232, 247
Tolmachev, V. V. 156
Tomash, W. J. 88, 89, 100, 109, 121, 217, 226
Toth, L. E. 219, 226
Trauble, H. 73, 212, 226
Treuting, R. G. 258, 271

Ushio, S. 271
Usui, N. 271

van Vijfiejken, A. G. 247

Venktaram, A. 247
Vernon, F. L. 240, 247
Vivet, B. 80
Volger, J. 230, 232, 247
Voogd, J. 273, 285

Waldorf, D. L. 220, 226
Webb, W. W. 220, 226, 247
Weil, L. 226
Wernick, J. H. 80, 205, 285
Werthamer, R. N. 153, 155, 156, 160, 168, 175, 176, 192
Wexler, A. 205
Wiederhold, P. R. 279, 285
Wilkinson, K. J. R. 283, 285
Williams, H. J. 205
Winter, J. H. 74, 80
Woolf, M. 191, 192
Wulff, J. 271

Yasukochi, K. 248, 271
Yntema, G. B. 274, 285
Young, F. J. 282, 285

Zebouni, N. H. 240, 244, 245, 247
Zemansky, M. W. 205
Zimmerman, J. E. 211, 226

Subject index

Abrikosov structure, see Vortex lines
A.C. critical current 282
A.C. power loss 281
Annealing
   effect on Hall voltage 236
   effect on magnetization 215
B.C.S. free energy 132
B.C.S. theory 125 ff, 130 ff

Cerous nitrate 209
Chemical potential 231
Classification of superconductors
   clean superconductors 139, 148, 154, 167
   dirty superconductors 140, 146, 148, 166
   type I, type II 12, 37
Coherence length
   definition 10
   from microscopic theory 141
   temperature dependent
      in Ginzburg–Landau theory 26
      from microscopic theory 150
Coil simulation 267
Collective oscillations of vortex lines 237, 246
   effect of pinning points 238
Condensation energy 6, 133, 158
Cooper pairs
   formation of 122 ff
Coulomb force between vortex lines 227, 228, 241
Critical current 248 ff
   alternating 282
   anisotropy due to lattice defects 252
   at intermediate field angles 263 ff
   effect of lattice defects 251 ff, 255, 257
   surface sheath 255, 256, 261
   for an ideal type II superconductor 19
   in longitudinal field 261 ff
   in transverse field 251 ff
   measurement 248 ff
   of a hard superconductor 221

   of Meissner state 256
   of mixed state 254
   relation with magnetization 257
Critical fields
   angular dependence of 100 ff
   for superconductors coated by normal metals 119
   H_c 3, 5, 14, 133
   H_{c1} 14, 47, 63, 153, 209, 251
   H_{c2} 14, 41 ff, 61 ff, 146, 155
   H_{c3} 85 ff, 251
   in parallel geometries 91 ff
   in perpendicular geometries, see H_{c2}
   in presence of paramagnetic effects H_{c2}, 169; H_{c3}, 171
   variation with temperature, see under the various critical fields
Critical state 238, 239

D.C. transformer 279
D.C. power transmission 280
Decay of magnetic field in coils 234, 235
Degradation 266 ff
   prevention of 269, 270
Density of states in superconductors
   B.C.S. 181
   close to the critical field 187 ff
   in gapless superconductors 186 ff
   relation with the specific heat 193 ff
   relation with the thermal conductivity 196 ff
Energy dissipation in the mixed state 227, 232, 233, 240, 242, 244
Energy of filaments
   interacting 65
   isolated 50
Energy gap
   B.C.S. 10, 131, 180
   effective energy gap 200, 201
   values of 10, 11
   variation with temperature 132
   see also Gapless superconductivity
Energy storage in superconducting tubes 279

291

Subject index

Faber-Pippard relation   12, 13
Faraday effect   209, 210
Field distribution   230, 254
Filaments, *see* Vortex lines
Filamentary superconductors   212, 238, 274
Flux bundles, existence of   210, 233
Flux concentration   278
Flux distribution   209
Flux jumping   227, 238 *ff*, 241
   avalanches   239, 240
   effect of lattice defects   240
      transport current   243
   in coil   268
   in filamentary material   238
   initiation   238
   in type I material   238
   thermal nature   240
   voltage induced by   238
Flux movement   227 *ff*
   thermally induced motion   233, 234
   thermal nature   240
Flux pumps   275
Flux quantization   44 *ff*, 210
Flux quantum
   definition   45
   number of flux quanta for an isolated line   51
   number of flux quanta per unit cell close to H_{c2} 61, 77
   number of flux quanta per unit cell for any value of the field   62 *ff*
Flux trapping   209
   effect of lattice imperfections   224
   in filamentary systems   212 *ff*
      rings   209
      singly connected systems   214 *ff*
      tubes   277
   variation with temperature   223 *ff*
Free energy, *see* B.C.S., Ginzburg-Landau, Microscopic theories
Fulde and Ferrel depaired state   159, 163 *ff*
   in clean superconductors   167, 169
   in dirty superconductors   166

Gapless superconductivity   177 *ff*
   ergodic   185
   non ergodic   186
   see also Density of states in superconductors
Ginzburg-Landau parameter
   in the presence of paramagnetic effects   169 *ff*, 173
   κ 29, 151
   κ_1 144, 154

κ_2 146, 154
κ_3 153
see also Specific heat
Ginzburg-Landau theory Chapters II, III
   derivation from microscopic theory   134 *ff*
   equations   24, 25, 30
   for clean superconductors   148
   for dirty superconductors   148
   free energy   22 *ff*
   validity of   28 *ff*, 141 *ff*
Guided motion of vortex lines   236

H_{c1} determination of   209
H_{c3} determination of   251
Hall effect   235 *ff*

Induced voltage in the mixed state   230 *ff*, 235
Inertial mass of vortex line   238, 239

Josephson junction   210, 211

Knight shift in superconductors   160

Laminar model   37, 73, 258
Lead–bismuth alloys   215, 223, 241, 261, 273, 282
Lead–indium alloys   215, 233, 257, 261
Lead–thallium alloys   217, 258, 259, 261, 264
Lorentz force on vortex lines   221, 228 *ff*, 234, 242, 248, 254, 256, 269
London equation   7
   generalized   64 *ff*
London penetration depth   8

Magnetic bearing   278
Magnetic hysteresis   214, 241
   calculation   222
   effect of lattice defects   215, 216, 243, 251
      surface barrier   216, 217
      surface sheath   214
      sample radius   256
Magnetization, irreversible
   calculation for sponge structure   212, 213
   causes of irreversibility   214
   effect of flux jumps   238, 241
      surface sheath   214
      transport current   242, 249
      relation with critical current   257
   region of instability   242
   size effect   214

Subject index

Magnetization curve
  type I superconductor 4
  type II superconductor 13 *ff*, 62 *ff*
Magnetothermal effect 244 *ff*
  resonant period temperature variation 244 *ff*
Magnets, *see* Superconducting magnets
Magnus force 236
Meissner effect 3, 209
Mercury in filamentary structure 213, 214
Microscopic theory Chapter V
  B.C.S. equation 130
  Bogoliubov transformation 127
  derivation of the Ginzburg–Landau theory 134 *ff*
  equation for the pair potential 129, 131, 135, 139, 140
  free energy 129
  instability of the Fermi surface 125
  Matsubara frequencies 137
  relation with paramagnetic effect Chapter VI — 1
  self-consistent method 126
  transition temperature 131
Niobium 211, 217, 218, 220, 229, 235, 243, 244, 245, 253, 258, 259, 281
Niobium–tantalum alloys 211, 217, 230, 235, 236, 251, 261, 274
Niobium–tin 240, 241, 243, 244, 261, 264, 266, 274
Niobium–zirconium alloys 219, 221, 228, 240, 245, 258, 262, 266, 268, 269, 274, 275, 282
Nucleation *see* Critical fields, Surface superconductivity
Order parameter
  in Ginzburg–Landau theory 22 *ff*
  in surface superconductivity 98, 109 *ff*
  near surface 217
  relation with gapless superconductivity 178 *ff*
  relation with pair-potential 150
  variation close to H_{c1} 47
  variation close to H_{c2} 52 *ff*
Order of the transition 4, 18, 43, 47, 144, 172
  at the critical temperature 4
  at H_{c1} *see* H_{c1}
  at H_{c2} *see* H_{c2}
  at H_{c3} *see* H_{c3}
  in the presence of paramagnetic effects 172

Pair potential
  definition 127, 129
  integral equation for 139
  in paramagnetic effect 162, 174
  relation with energy gap 131 *ff*; see also Gapless superconductivity
  relation with the order parameter 129, 150
Paramagnetic effect Chapter VI — 1
  Clogston–Chandrasekhar limit 158
  order of the transition, *see* Order of the transition
  spin-orbit effect in 174
  transition fields, *see* Critical fields
  see also Fulde and Ferrel depaired state
Peak effect 258 *ff*
Penetration depth
  derivation from microscopic theory 151
  London 8
  temperature dependent 27
Persistent mode 272
Pinning energy 218, 219
  effect on degradation 269
  variation with temperature 269
  magnetic field 219
Pinning force 222, 223
Pinning points 218 *ff*, 236, 239
  density 221
  effective size 218
  mean energy 218, 219
    for point defects 219, 220
    dislocations 220, 221
Power transformers 282
Power transmission 282

Quanta of flux *see* Flux quanta

Resistance transition 231, 248, 250
Resistivity of superconductors 3
Rhenium–molybdenum alloys 261, 264, 265

Sequential testing 267, 268
Short sample testing 267, 268
Specific heat 193 *ff*
  relation between, and Ginzburg–Landau parameter 195
Sponge structure 212—14
Superconducting magnets 251, 272 *ff*
  coil design 274
  power supply 272, 273, 276
  protection 276, 277
  stable coils 273
  suppliers 284
  trapped flux 276
Surface energy 31 *ff*
Surface flux barrier 216, 217, 251
Surface sheath; *see* Surface superconductivity 214, 250, 255, 260
  below H_{II} 109, 111 *ff*

293

Subject index

Surface superconductivity Chapter IV
   H_{c3}, see H_{c3}
   nucleation in a slab 91 ff
   vortex structure in a slab 95

Tantalum–titanium alloys 261
Thermal conductivity Chapter VII
   in gapless regions 202
   variation with magnetic field 200
   variation with temperature 197
Thermodynamics of superconductors
   general 5
   of type II superconductors 16
Tin 210
Tin–antimony alloys 224
Tin–bismuth alloys 224
Tin–indium alloys 223
Titanium–molybdenum alloys 252, 253
Titanium–niobium alloys 253
Training 264 ff
Transition from the superconducting to the normal state
   in finite fields; see Critical fields
   in zero field 18
   order of the transition, see Order of the transition

Viscous force on flux lines 227, 232, 250
Vortex lines
   Abrikosov's theory Chapter III
   Abrikosov's vortex structure 52 ff
   annihilation 241, 265
   collective oscillations of 237, 246
   forces on 227
      experimental observation
         electron microscopy 73
         neutron diffraction 70
         nuclear magnetic resonance 75
   inertial mass of 238, 239
   in the high κ limit 64, 68
   interaction energy 65
   isolated vortex line 50
   number of flux quanta per line 51, 61
   in surface superconductivity 95
   triangular arrangement 60
Vycor glass 213, 214